# Urban Ports and Harbor Management

## Responding to Change along U.S. Waterfronts

*Edited by Marc J. Hershman*

ALERE FLAMMAM

Taylor & Francis
New York • Philadelphia • Washington D.C. • London

| | | |
|---|---|---|
| **USA** | Publishing Office: | Taylor & Francis • New York<br>3 East 44th St., New York, NY 10017 |
| | Sales Office: | Taylor & Francis • Philadelphia<br>242 Cherry St., Philadelphia, PA 19106-1906 |
| **UK** | | Taylor & Francis Ltd.<br>4 John St., London WC1N 2ET |

**Urban Ports and Harbor Management:**
Responding to Change along U.S. Waterfronts



First published 1988
Printed in the United States of America

**Library of Congress Cataloging in Publication Data**

Urban ports and harbor management : responding to change along U.S.
waterfronts / Marc. J. Hershman, editor.
p. cm.
Includes bibliographies and index.
ISBN 0-8448-1547-0. ISBN 0-8448-1548-9 (pbk.)
1. Harbors—United States. 2. Port districts–United States.
I. Hershman, Marc, 1942–
HE553.U73 1988
387.1′0973—dc19 88-1494
CIP

# Contents

# Preface

This book tells two stories: the story of an important and highly visible public resource, and the tale of an obscure but influential governmental institution. The resource is the urban harbor and waterfront, a recently rediscovered asset of the millions of people in cities and towns that front on rivers, lakes, bays, estuaries, and oceans in the United States. The governmental institution is the public port agency, a semiautonomous landlord and public relations agent that quietly builds docks or moves cargo or moors boats, acting as overseer of the waterfront. Ultimately this book asks how the institution affects the resource, and how the public port contributes to the stewardship of the urban harbor.

The urban harbor and adjacent shorelands have taken on new importance in the latter half of the twentieth century. The quality of the water is watched closely and the remaining natural habitat zealously protected. The shorelands have been rejuvenated and people flock to them for shopping, strolling, boat watching, dining. Marine industries like fishing, cargo handling, shipbuilding, and boating are nurtured to retain jobs and gain some new ones, though the battle is uphill. And festivals, regattas, races, and fleet visits adorn the water's edge of the city, making it a new recreational focal point. The harbor has become a city resource as well as a marine resource and vessel haven.

Because of these wide-ranging demands, the numbers of conflicts over use and policy have grown enormously. Environmental protection laws are applied stringently to any new development and lead to sharp differences of view. Coastal management standards for shore development raise questions about the "water-dependency" of a use, or the provision of public access, or the size and siting of a project to protect upland views. And an emerging major issue is the "gentrification" of waterfronts and whether maritime commerce and industry are being sacrificed.

The harbor is far more than a transportation utility, or a community service like those that provide water, buses, or electricity. It is the place where settlement first occurred and where economic roots are found; where people's memories of welcomes and farewells are relived; where the symbols of the city (skylines, towers, statues) evoke pride, loyalty, honor; where major events celebrate the city and its accomplishments; where trade forms linkages with other cultures; where scientific expeditions depart or return. A harbor is for and about people. It is a transport node, but also it is a maritime cultural resource.

Because of conflicting demands and cultural amenities, the harbor needs special attention and great care from those who manage it. Unfortunately, in the United States today it is getting neither. Decisions about the harbor are considered by a dozen agencies with varying mandates. Their jurisdiction is limited to particular resources—the shore, the water, the submerged land, the vessel. Their responsibilities are narrowly defined and the coordination among them deals more with process than substance. Special interest groups manipulate the agencies to advance their narrow goals; those adept at playing the game, win. This cumbersome governance contributes to conflict in harbor use and policy because it is fragmented and adversarial. A holistic perspective is lacking, one that is sensitive to the diverse demands of work and play and to the maritime culture of the harbor. A new form of harbor management is needed where a collective vision can take shape and be implemented.

The public port agency is given primary attention in this book because it has substantial governmental powers and considerable land and financial resources that can directly affect the harbor and its uses. Public ports operate under a broad mandate, and over the decades they have engaged in a wide range of businesses. Public ports are privileged governmental agencies because of their financial and administrative independence and their freedom from day-to-day supervision by higher authorities. As a consequence of their semiautonomous structure, public ports have a stable corps of professional managers and an ability to respond rapidly to changing economic, political, and social demands. For these reasons public ports can make substantial contributions to improved harbor management. In some cases the public port can be the leader, bringing together many public and private actors to form management plans or development strategies. In other cases public ports can be active participants in designing and implementing portions of broader harbor goals.

At first it may seem like a radical proposition to suggest such a broadened role for the public port. Traditionally, the public port has been a visible but small part of the domestic and international transportation world, and the transportation sector is a small part of the much larger system of trade in goods and services. As measured by investment in facilities or rates charged for services, public ports are small compared to the other components of the transport system (for example, rail and ship). In operations, the public port is not a leader but a party that reacts to this larger world around it—the traders, shippers, and carriers—wooing them, making deals, responding to trends in their business. In this pattern of behavior, the public port is acting much as a private service organization.

However, to the people living or working in the city or town surrounding the harbor, the public port's role is much greater. To them the public port is a community leader, an organization that articulates a vision and promotes a better future. To many the public port means jobs, because public port officials seek out the cargo or vessels that use the harbor. For others the public port is looked to for economic renewal and growth, or it is blamed for decline and decay. Most people now ask the public port to improve their quality of life, to make the harbor cleaner and quieter, to provide recreational amenities, and to support cultural and historic projects. Public ports have become emblems of the city, and people have become "port watchers," cheering when affairs go well and criticizing when they do not.

The essays in this book explore the changes that have occurred in the modern harbor in the 1970s and 1980s and the many roles of the public port in stimulating or responding to these changes. The goal is to understand the modern harbor and public port and the contemporary pressures on them. Chapter 1 expands on the concept of harbor management and describes the characteristics of U.S. public port agencies. Chapters 2 through 5 set forth the historical, organizational, physical, and legal settings in which harbor development takes place, how those settings have changed in recent years, and the role of the public port in them. Chapters 6 through 10 review the major activities that consume the time and resources of the public port: their roles in cargo handling, fisheries, boating, economic development, and finance. Chapters 11 through 14 look ahead to the future and explore selected tools of harbor management and governance, such as planning, environmental protection, revitalization, and accountability. Not all subjects important to the harbor could be presented: the subjects of labor relations and water quality, for example, deserve their own full chapters. Each of the essays in the book has a distinct personality. Styles of writing vary and include an interview, a journalist's report, and descriptive, argumentative, and scholarly essays. The authors' disciplines range among geography, law, business, political science, and marine affairs. Each shares a common interest in ports and harbors, though some emphasize port business dealings and others stress the land use or environmental perspective.

The revival of harbors has created a new urban amenity for America. This book will be successful if it brings useful information to bear upon, and sharpens attention to, this special place and its many values. Policymakers and interested publics need to focus more consistently on how the use of harbors is changing. They need to sharpen planning and administrative tools that address multiple use of harbors and waterfronts. They need to look closely at their public port agency, an institution with the potential to contribute substantially to the diverse marine needs of the community. As urbanites become better port watchers, the harbor and its values will be increasingly recognized as an integral and enriching part of city life.

Marc J. Hershman

# Acknowledgments

Primary financial support for research and writing came from the William H. Donner Foundation of New York City. Special thanks to Janet Maughan, formerly of the Foundation staff, for her encouragement and helpful suggestions. Additional financial support is noted at the outset of particular chapters and includes the Washington Sea Grant Program, the William and Flora Hewlett Foundation, and the College of Ocean and Fishery Sciences of the University of Washington.

Institutional support came from the Institute for Marine Studies, College of Ocean and Fishery Sciences, University of Washington. Professor Ed Miles, director of the Institute was particularly supportive. Other institutional support was provided by the Washington Sea Grant Marine Advisory Services, especially the invaluable assistance of Robert F. Goodwin, author of three chapters, and Thomas J. Dowd.

Many individuals assisted this three-year project. Graduate students, now graduated from the Institute for Marine Studies or University of Washington Law School, were helpful in many ways: thanks to Robin Bittner, Pat Jones, Marianne Kory, Jonathan Levie, Robert Levy, Rick Morse, Lee Rees, Lucinda Tear, and Ann Wessel. Staff at the Institute for Marine Studies provided valuable support throughout, especially Chris McCartan, project coordinator for the research phase, Ora Chapman, and Ruth Gainer. The expert and efficient work of Andrea Montclair, the copy editor and production manager for the preparation of the manuscript, made my tasks far easier and much more enjoyable. Thanks to Ray Lincoln, our agent, for her encouragement and assistance.

Many colleagues in industry and academia helped to stimulate and inform this work, for which I and all the authors are grateful. Particular credits are mentioned, as appropriate, in the chapters.

Finally, I am deeply greateful to Carol, Carla, and Jordan for their continuing love and understanding.

M.J.H.
Decatur Island
Washington

# Part I

# OVERVIEW

## Chapter 1

# HARBOR MANAGEMENT:

## A New Role for the Public Port

*Marc J. Hershman**

---

Along the coasts and Great Lakes of the United States, there are 1,435 navigable waterways, harbors, and river stretches—places where boats or ships can take refuge from open water. Most harbors contain ports, those public or private facilities where vessels can moor and exchange people, cargo, or provisions, and most harbors are adjacent to towns or cities.[1] This book is concerned primarily with the harbors of the United States that contain public port facilities and in which a public port agency† operates.

This chapter introduces the harbor, its multiple uses and conflicts, and the current fragmented management system, followed by a description of harbor management. Harbor management is a new concept in the United States, one that stresses multiple use and conservation of harbor resources, and encourages consensus among public and private interests. Harbor management requires vision and leadership and the formation of a common purpose. It contrasts sharply with the present management system which is contentious, costly, and divisive.

The chapter then focuses on the public port agency as one of the major players in harbor management and details its powers and functions. The public port agency

---

*Thanks to Pat Jones, Lucinda Tear, and Lee Rees for research assistance on port numbers, functions, and powers, respectively. This chapter builds on a paper by M. J. Hershman, S. Euston, and M. Rohan (1983).

†Public pcrt agencies are known by many names: port authorities, dock boards, harbor commissions, transportation departments, navigation or terminal districts, city bureaus or divisions, and others. Although their powers and organizational form differ, they are a public organization, created by law, with primary responsibility over waterfront and maritime activities, and in many cases, airports, economic development, etc. Their powers, structure and operations are discussed later in this chapter, and in greater detail in Chapters 10 and 14. When the word *Port* or *Ports* is used alone in this book and capitalized, it refers to the public port agency. When not capitalized, the word has its usual meaning—that part of the harbor where vessels can moor and exchange people, cargo, or provisions.

figures prominently in many activities that take place in the harbor and is often the catalyst for change and the community promoter. The public port can make substantial contributions to harbor management in the years to come.

## DIVERSITY OF INTERESTS IN THE URBAN HARBOR

Virtually all uses of the sea, lakes, and rivers are dependent on harbors. For example, fisheries, boating, marine research, and shipping activities all require harbor facilities and services. Close to one-half of the U.S. population lives near harbors, and demands for shoreside public amenities have increased dramatically in recent years. Harbors are natural environments supporting important marine biological resources, and they are subject to natural hazards and pollution risks affecting resident populations. In addition, increasing dependence on foreign trade (about 25 percent of the GNP in recent years) makes the harbor important to nearly all consumers and producers in the nation. As Hershman and Bittner point out in Chapter 2, harbors in the United States have been used and reused in servicing people's changing needs throughout the nineteenth and twentieth centuries.

People value the harbor for reasons beyond its "use" potential. It is more to them than a marine highway or a fluid environment. It is a source for renewal and refreshment adjacent to the busy urban grid. It provides visual release from the tension caused by dense urban images. It offers exotic experiences that are esthetic, educational, and dramatic—the function and fittings of ships and boats, the joinder of vessel and skyline, the clatter and bustle of cargo transport, the habitat for marine wildlife, the language and customs of maritime workers. The harbor and its symbols—the bridges, statues and towers—evoke for many the memories of arrival, departure, safety, and home; of great historic events; of linkages with faroff lands; of adventure and risk-taking on the sea; of escape and freedom. As Melville wrote in *Moby Dick*, "Thousands of mortal men" throughout the city are "fixed in ocean reveries" and discontented unless they "get just as nigh the water as they possibly can without falling in." The harbor is one of the great cultural resources of a port city because it affects the spirit, energy, and expressiveness of its people. As Fleming observes in Chapter 3, there is a diverse port community that creates and nurtures this cultural resource.

Those who use and appreciate the harbor will demand more from it in the future. Changes in vessels and cargo-handling technology, especially changes associated with the container revolution, require improved navigational capacity and marine terminals. These changes are discussed by Mayer in Chapter 4 and by Chilcote in Chapter 6. The new water resources legislation of 1986, discussed by Hershman and Kory in Chapter 5, authorizes forty-one additional navigation dredging projects in order to meet these new transportation needs.

Harbors are changing in other ways as well. The conversion of obsolete port and industrial facilities into residences, shops, restaurants, parks, and promenades is a well-established trend which continues to change small and large harbors from work places to urban and tourist amenities. Randall and Goodwin present the rationale and results of this conversion process in Chapters 7 and 13, respectively. The maintenance of fish-related industries and the growth of small boat moorages will continue to demand shore locations, as Goodwin describes in Chapters 8 and 9, respectively. In addition, environmental agencies along with civic and conservation organizations increasingly advocate better protection of wildlife and water quality, and the consideration of these issues is now institutionalized. Hershman and Kory discuss the federal environmental regulatory framework in Chapter 5, and Wessel and Hershman outline experience with wetlands mitigation projects in Chapter 12.

Given the diversity and intensity of harbor use, and the way harbors are valued by city dwellers, it should not be a surprise that conflicts are common. Major issues concern: 1) community and environmental impacts from development projects at the water's edge (see Chapter 12), 2) competition between recreation and industry for scarce shoreline space (see Chapter 13), 3) pollution of harbor waters from land or vessel sources, 4) obsolete, vacant, and decayed port and industrial facilities that are dangerous and unsightly, and 5) debates over the expenditure of scarce public funds and the "return on investment" expected from harbor improvement projects, described by Dowd in Chapter 10.[2] As a result of these and related conflicts and issues, new forms of planning have evolved, discussed later in this chapter and by Chasan and Dowd in Chapter 11. And questions of governmental performance are raised with regard to federal agencies (see Chapter 5) and in respect to public port institutions by Olson in Chapter 14.

Since the late 1970s, each issue noted above has been raised in Elliott Bay, the main harbor for the city and Port of Seattle. At terminal 91, which is adjacent to two affluent neighborhoods on bluffs overlooking the terminal, neighborhood activists fought a Port plan to transform the piers into a container operation because of their concern over noise, glare, traffic, and other impacts. At that same terminal, the Port and the state Department of Ecology strongly disagreed over the use of contaminated dredged materials as fill between two adjacent piers. In particular, they disagreed whether the plastic liner separating harbor waters and fill material would adequately protect against leachate, or seepage that might occur from the fill to the water. At the south end of the harbor, a coalition of city officials and community groups successfully pressed the Port of Seattle to provide public access sites along the Duwamish River in at least eight different locations where the Port had plans for developing terminals. They were seeking nature refuges, bike trails, elevated viewpoints, boat launching ramps, and other amenities in the midst of the maritime industrial zone of the harbor. Finally, along the central waterfront, that part of the harbor adjacent to the downtown central business district, there are finger piers and warehouse buildings that are underused or vacant, in disrepair, dangerous and unsightly, crying for demolition or revitalization. While they sit there, sharp debate continues over the nature of the uses that should occur: housing, hotel, park/open space, international trade center, maritime-related office, water-dependent uses. When the Port of Seattle purchased some of the land and disclosed that office and hotel uses were being considered for the site, they were challenged on whether public funds were appropriate for such uses and whether the private sector should be doing this kind of redevelopment.

These issues are not unique to Seattle, nor are they unique to the harbor setting. They are a reflection of structural changes occurring in contemporary U.S. society requiring adjustments in many sectors, including changes in technology, reduction in industrial capacity, consolidation of industrial facilities, demands for in-city recreation, high regard for a quality environment, and greater scrutiny over public expenditure.[3]

Allocation of harbor resources occurs through a mixture of private and public decision making. Private sector decision making plays a critical part in the use of harbors, because the source of capital for many marine terminals and mixed-use complexes is often private (see Chapter 3). However, it is rare to find shoreline investments that do not involve some form of public subsidy and a substantial amount of public scrutiny through agency review. Public agencies usually own the water bottoms and manage the navigable waters as a commons for all the people. Within the public sector questions of the public interest are raised and significant influence is exercised over private investment. Public agencies are often themselves developers, as in the case of port infrastructure, city shoreline parks, and public marinas.

Harbor decision making by public agencies is split by level of government and by functions. Many agencies at the federal, state, and local levels have jurisdiction in the harbor (see Chapter 5; National Academy of Sciences 1976; Marcus 1976). The jurisdiction of some agencies is limited to the land side; for others it is limited to the water side. Some agencies are concerned only about living resources; others are concerned solely with pollution issues. A few public entities have mandates to develop resources and land; others have mandates to protect and preserve natural resources. Even coastal zone management agencies, intended to be comprehensive in their approach, often are not, because they usually have no capital-improvement capability. Conflicts among these diverse harbor agencies are common and often result in long delays, stalemates, and lawsuits. It is fair to conclude that the harbor decision-making process is greatly fragmented, highly adversarial, and more costly to society than necessary.

Society wastes resources, misses opportunities, and consumes energy unnecessarily because of this situation. Many promising resource development opportunities do not get considered. Often only the large-scale and well-funded project developers, with skilled legal assistance, prevail. Project opponents believe they must stall or kill projects because they see no way to achieve their objectives within the harbor, except through obstructing others' projects. The situation in most harbors is similar to that in Elliott Bay. The urban harbor is an arena of conflict over land use and environmental issues, rather than a forum for strategic planning to meet the objectives of many parties and to form a common vision for the harbor.

## EXPERIMENTS IN HARBOR MANAGEMENT

In some harbors, attempts are underway to establish a comprehensive harbor management scheme in order to achieve multiple objectives and resolve multiple-use problems through coordinated land and water use policies and public and private investment decisions within a limited geographical area. Typically, many diverse parties work together to develop the harbor management scheme, to take implementation actions, and to evaluate and revise the scheme as necessary. The primary harbor issues relate to the timing, location, and scale of port and waterfront growth and how this affects environmental and public use values.

Three approaches to harbor management have emerged in the United States since the late 1970s. These approaches are an elaboration of the coastal management framework that is now firmly in place in the nation (see Chapter 5; Hershman 1978; Godschalk and Cousins 1985; Wolf 1985; Archer and Knecht 1987). The approaches use different forms of organization and implementation and they emphasize different sets of issues. Through experimentation, strengths and weaknesses are being identified and the process refined.

***Special area management planning.*** The first approach to harbor management uses a process that has come to be called Special Area Management Planning, or SAMP (NOAA 1981). A SAMP process is useful where there have been persistent problems in a limited geographical area, and where there is a need for greater predictability in planning and decision making in order to meet environmental protection and economic development goals. A number of public agencies with regulatory or developmental responsibilities in the harbor establish a task force that works collaboratively, by consensus, to develop policies and use designations for the shoreline and water areas of the harbor. The task force attempts to integrate the legal requirements of federal, state, and local laws so that each agency can adopt and use agreed-upon policies and plans for its decision making. Frequently an independent facilitator such as a consultant or mediator will be used to guide the process. Federal Coastal Zone Management (CZM) funds have supported the technical work in many cases, and to varying degrees private interests have been involved in the process. Some of the more visible SAMP efforts in harbors are Grays Harbor, Washington (Evans 1980), Coos Bay, Oregon (Dull 1983), Providence Harbor, Rhode Island (Coastal Resources Management Council 1984) and Port Bienville, Mississippi (Mississippi Coastal Program 1984). In each case the public port agency has been an important player. The strength of this process is its simplicity and flexibility. The task force can be formed by agreement of the parties; no new statute is needed. Members can be limited or expanded depending on the problems to be addressed. The issues can be limited to the persistent ones that created the need for greater planning in the first place. Long-term trade-offs can be negotiated using zoning or mitigation strategies that are already at the disposal of the agencies. Long before the ink is dry on the SAMP, agency officials can start to use it in their decisions.

The weakness of the process relates in great part to problems of implementation. One agency may object to an aspect of the plan and may frustrate its adoption. Inadequate public participation can stall the effort. And there may be aspects of the plan that require public investment or subsidy; for example, the provision of public access, acquisition of a refuge or encouragement of particular users, such as those that are water-dependent. For example, the Providence Harbor plan depended a great deal on anticipated federal funds for recreational facilities and for innovative use of dredged materials. These funds were not forthcoming and the plan has been set back considerably.

***The California approach.*** The second approach to harbor management can be referred to as the California approach, because it is used in the major ports of Southern California and in San Francisco Bay. In this approach, port-related development is a specific component of a coastal management plan and it is developed and implemented by the port sector. The public port agencies are asked to identify their land and water needs, and these are then negotiated with the coastal management agency in light of other objectives for the harbor and ultimately approved and adopted by the coastal agency. There are differences in the approaches used in the four Southern California Ports (San Diego, Long Beach, Los Angeles, Hueneme) and the six Ports in San Francisco Bay (San Francisco, Oakland, Richmond, Redwood City, Alameda, Benicia). The four Southern California Ports are required by Chapter 8 of the California Coastal Act of 1976 to produce Port Master Plans for review and approval by the California Coastal Commission. Once approved, the public port agency administers the master plan, building facilities authorized in the master plan and issuing permits to users within the harbor. In some cases, permit decisions can be appealed to the state coastal commission. Thus, the initiative rests with the Port, but their aspirations are tempered through state-level review of the master plan, amendments to it, and the appeals process. A balance of interests in the harbor is achieved through this collaborative process.

In San Francisco Bay, where a coastal management law applies that is different from the one used in Southern California, the six Ports worked closely with a number of public agencies to produce the San Francisco Bay Seaport Plan (Fay 1983). This plan specified the total amount of cargo-related growth expected in the region to the year 2020 and identified how many new port berths would be needed to accommodate this growth. The plan then identified thirteen near-term development sites that could be used to accommodate the expected first-phase growth. Port development could occur only if it were within the growth projections and at one of the identified sites (SFBCDC 1982). This accommodation allows growth, but within limits. It reduces the pressure to fill more of the Bay, but gives the Ports opportunities to compete among themselves and allow market forces to determine which of the thirteen sites is developed first. The Seaport Plan was formally adopted by the San Francisco Bay Conservation and Development Commission and is now being implemented. If changes in the pace of cargo-related growth occur, the Seaport Plan can be amended.

The California approach incorporates port growth into harbor management in specific terms. Areas for port expansion are identified on maps, as are areas reserved for other nonport uses. The questions of need and site location are resolved in the plans even though site-specific issues such as mitigation for disturbance of wetlands and provision of public access must be negotiated. In spite of these positive features, questions remain. Will unforeseen technological changes improve efficiency and reduce demand for more terminals? How accurate or fastchanging are the cargo growth forecasts? If the Port with the best site has the least money for development, could a subsidy be provided, and if so by whom, or could other port authorities assist financially so that the most suitable site gets developed first? These questions get at the heart of the problem of attempting to produce a truly regional plan when the players are independent entities who compete fiercely with one another for business. As an example of harbor management, the California approach focuses primarily on the issue of port growth—a key issue in harbor management, but not the only one.

***Harbor commissions.*** The third approach to harbor management—one that may best be suited where recreational boating is a major use—is the establishment of municipal-level harbor commissions. Connecticut's 1984 Harbor Management Law is the leading example (Rothenberg and Rocque 1986). This law was passed as a result of boating-related problems. In some Connecticut harbors, private moorage encroached into public channels over the years, and maintenance dredging could not be done. In other harbors, congestion and inadequate public access for boats and people were important issues. The Harbor Management Law authorizes the twenty or more coastal towns on Long Island Sound to establish harbor commissions and adopt harbor management plans. The plans would address water use issues in particular, such as protection of shellfish resources, tidal flats, and marshes, the encouragement of water-dependent uses, the allocation of boating moorage, identification of anchorages and navigation lanes, and the specification of public access points. The commission would ensure that the policies and specifications of the harbor plan are consulted and used in other local, state, and federal decisions. A harbor master would be authorized to police and enforce the harbor plan, collect moorage fees, ensure boating safety, and manage public harbor facilities (Connecticut 1985).

This approach combines the concerns over new development, such as siting, size, impacts, and timing, with day-to-day operations, such as transient moorage, public access facilities, fishing and swimming areas, and pollution control. Harbor management in this context is comprehensive. It involves planning, control of development, and operations. Further, this approach gives a municipality a chance to strengthen an existing harbor commission, or to establish a new commission to advocate its interests among other regulatory bodies. For example, under the law, the harbor commission's wishes must be accepted by state regulatory agencies unless the state assumes the burden of proof and is able to show why another result is warranted. Also, the harbor commission is required by state law to seek a general permit from the U.S. Army Corps of Engineers (COE) for its harbor plan. If the COE issues a general permit, the city links its goals with the federal government's interests in managing navigable waters. As experience is gained with the harbor commission approach, its potential use in small, crowded harbors will become clearer.

In the three approaches to harbor management outlined above—SAMP, the California approach, and harbor commissions—a number of useful tasks are undertaken. The problems of the harbor are given primary attention, separate from the more general concerns of the larger coastal zone. The focus is on the use of water areas *and* shorelines, with the special needs of the maritime sector in the forefront. Many problems are viewed concurrently so that a holistic view of the harbor situation is present. The players are looking ahead to anticipate development and conservation needs and appropriate trade-offs. Ongoing coordination among professionals from diverse perspectives (for example, the wildlife biologist, the urban planner, and the port executive) often leads to mutual respect, better communication, delineation of issues in specific rather than ideological terms, and improved compromises (see Evans 1980).

As promising as harbor management appears to be, it is far from common practice in the United States. Most changes in the harbor are fought out on a project-by-project basis without the benefit of harborwide policies and plans. In some locations where harbor management plans have been developed, years pass without full adoption of a final product. Grays Harbor, Washington, is now in its twelfth year preparing a Grays Harbor Estuary Management Plan.

Assuming agreement can be reached on a harbor management plan, implementation can be a problem. Capital investment is often needed before a plan can work, but a source of funds is often unavailable. Also, harbor plans need to be used, interpreted, and revised if necessary, requiring ongoing care and attention by a management entity.

Finally, harbor management should strive to do more than resolve today's environmental protection and multiple-use planning issues. The harbor is a great maritime cultural resource and city managers should promote its historic, esthetic, symbolic, and dramatic values for the people. Too often this human dimension of the harbor and its use is overlooked as the debate rages between the advocates for jobs and the advocates for marine environmental quality. A much broader understanding and appreciation of the harbor is needed in our planning efforts.

## A DESCRIPTION OF THE PUBLIC PORT AGENCY

The public port agency can make substantial contributions to improved harbor management because it can address some of the problems of breadth, implementation, and continuity noted above. It has a wide-ranging mandate, a capital investment capability, and stability and continuity as an organization.

***Numbers.*** It is difficult to estimate the number of ports in the country because of definitional problems and because statistics on ports vary among the agencies[4] (Jones 1985). Nevertheless, there are indicators. The American Association of Port Authorities has seventy-six corporate members, which is a good indication of the number of "majors" (Sherman 1985), although at least twenty important ports are not members. MARAD (1986) estimates 188 deep-water commercial seaports in the United States—those where water depth at the berth is greater than 25 feet for coastal ocean ports and 18 feet for Great Lakes ports. A recent compilation of published port lists showed 643 United States commercial sea and river ports, but concluded that many were missing because small-craft and special-purpose ports are hard to count without a detailed inventory (Jones 1985). If one were to look at berths and terminals—individual facilities rather than a collection of facilities called a port—the numbers jump to 3,018 deep-water berths and 1,841 terminals (MARAD 1986). Adding small-boat marinas to the list would push the number of ports into the thousands. In comparison, Japan lists 1,095 commercial ports, excluding fisheries ports (Inoue 1986), and Canada shows 434 "important" ports, including fish and recreation ports (Ruppenthal 1983, 121). These countries have done a better job at defining and counting ports than has the United States.

***Evolution.*** In earlier days, development of the port was a collective effort of local business interests, such as farmers, manufacturers, and retailers, working with commercial shippers and carriers, such as steamship and railroad companies. But starting in the twentieth century, three important trends converged that led to public ownership and operation of port lands and facilities: the progressive movement in local politics, dissatisfaction with railroad control of the waterfront, and opportunities to assume ownership of surplus federal government facilities (see Chapter 2). Two studies of deep-draft public ports show that in every decade in the twentieth century, with the exception of the Depression years, public port agencies have been established, and the distribution among decades is fairly even (Sherman 1985; Tear 1986). Once established, public port agencies continue to evolve—their organizational structure can change (Sherman 1985), and they can adopt wholly new functions (Stromberg 1982).

***Sources of Port power.*** Ports own, control, or influence a great deal of the waterfront, and this is a source of much of their power.[5] Total acreage owned by public port agencies is available from their promotional literature, but this information needs interpretation before conclusions are drawn. For example, San Diego reports 5,483 acres, but much of this is reclaimed land that has been converted to urban uses. Lake Charles, Louisiana, claims 2,510 acres, but 1,506 remain undeveloped. Massport shows 2,484 acres, but this is split among a wide range of uses: for example, a fishport, a bridge and its approaches, and unused land adjacent to Boston's Logan Airport. Humboldt Bay in Northern California shows 2,060 acres, but over 2,000 acres are preserved marshlands, intertidal areas, and water bottoms. Philadelphia lists 315 acres, a smaller number, but this represents a large percentage of the piers along the Delaware River where redevelopment can occur (Tear 1986). Port land holdings are not large compared to the size of the cities that surround them, but in many cases their lands are strategically placed. Their decisions to buy, sell, develop, or lease lands can have an enormous influence on the use of the harbor.

Another source of power for Ports is their ability to raise money and build facilities. Their operating revenues come from leases and port-related services for which they charge a fee. Funds for capital development sometime come from taxes or tax-backed bonds, and sometime from tax-exempt revenue bonds (see Chapter 10). Comprehensive figures on Port investments are not available. They must be gleaned from diverse sources that cover only part of the industry. Fifty major deep-water Ports surveyed by AAPA reported spending $3 billion between 1973 and 1982 for capital improvements, and those same Ports indicate plans to spend an additional $3.26 billion between 1983 and 1989 (MARAD 1985). Estimates for container terminal needs in the San Francisco Bay Area alone between the late 1980s and 2020 exceed $1 billion of investment (Fay 1983). State government spent $1.1 billion for port facilities and terminals between 1977 and 1984 (AASHTO 1986). This figure does not include expenditures by local government or special districts, and it does not include costs of dredging.

Ports are investing heavily in noncargo handling facilities as well. The San Diego Unified Port District, for example, plans to spend over $30 million (1975 dollars) on infrastructure (roads, utilities, services) to support retail, commercial, office, and public access facilities along the Embarcadero (San Diego Unified Port District 1976). Private investment in this area will be over ten times the Port's infrastructure investment.

Because of their shoreland ownership and their infrastructure development functions, port authorities are perceived as powerful catalysts for economic development, harbor investment, and employment. MARAD (1978) estimates that for every dollar of sales by the port industry, $1.60 in sales results within the economy as a whole (see Chapter 7). Because of public port investments, many private support industries, such as tugs, steamship companies, shipbuilding and repair, agents and forwarders, custom brokers, and chandlers, locate within a harbor and generate further business opportunities (see Chapter 3). Public ports promote their local region in national and international markets through marketing and worldwide trade development offices.

Port officials share their expertise in economic development with other public and private parties and influence the community's overall development (see Chapter 7).

***Activities and functions.*** Public ports have conducted many businesses over the years. Public transit via bridges, tunnels, ferries, and causeways was an early function. Port terminals for general cargo and fisheries developed early as well and have become a major function today. Airports and industrial parks were functions adopted by Ports in the 1940s and 1950s. In more recent times, marinas, waterfront revitalization projects, world trade centers, and economic development programs have been added to the list of functions performed by Ports. This process continues; today, recreational and environmental demands are being placed on Ports, promoting changes in the way Ports operate and the kinds of projects they undertake (see Table 1.1[6]).

## Ports as Public Enterprises

Public ports are public enterprises, a particular form of government agency that has flourished in this country beginning about the late 1940s.[7] Public enterprises are called a hybrid of government and business, containing characteristics of each (see Chapter 14). They are "public" in that government owns them, statutes set them up and dictate objectives, and public subsidies are often provided. They are "private"

**Table 1.1.** Noncargo Activities in 18 Ports*

| Noncargo Activity | Number of Ports |
|---|---|
| Airport | 5 |
| Fishport | 8 |
| Marina | 8 |
| Cruise ship services | 9 |
| Waterfront redevelopment | 7 |
| Parks and viewpoints | 10 |
| Mitigation | 10 |
| Marine resource development | 6 |
| Nonmarine resource development | 2 |
| Civic functions | 5 |
| Foreign trade zones | 13 |
| Sister city (foreign relations) | 3 |
| Land cargo transportation (trucking, rail) | 6 |
| Computer services | 9 |
| Economic development | 11 |

*After Tear (1986). Port sample includes: Canaveral, FL; Cleveland, OH; Galveston, TX; Georgia Ports Authority; Hueneme, CA; Humboldt Bay, CA; Lake Charles, LA; Massport, Boston, MA; Milwaukee, WI; New York/New Jersey; Oakland, CA; Philadelphia, PA; San Diego, CA; Seattle, WA; South Carolina Ports; Tacoma, WA; Tampa, FL; Virginia Ports Authority. Only Hueneme and Cleveland had no noncargo activities.

in that management often operates independently, money can be raised from private sources, revenues can be retained and reinvested, and often there is no dependence on tax revenues. The private character of public ports derives from the competitive marketplace within which they operate.

Although federal laws may influence the actions of Ports (see Chapter 5), the structure and powers of public ports are established under state law, and thus will vary considerably from state to state (Rees 1986; Sherman 1985). However, some generalization about the public port is possible. They are established either as state agencies (the Port Authority of New York/New Jersey is a bi-state agency established by interstate compact), or as political subdivisions of the state such as special districts, or as part of county or municipal government (see Chapter 14). They are given corporate powers, such as the power to sue and be sued and to enter into contracts. Their territorial limits are specified. For example, they can operate statewide, in all or parts of counties or cities, or within limits described by physical features, such as roads (for example, the Port Authority of NY/NJ can operate within an area limited to approximately a 50-mile-diameter circle with the Statue of Liberty at the center). Extraterritorial powers of ownership or annexation are sometimes prescribed. The governing body is set forth in great detail, including the number of commissioners, how they are elected or appointed—nationally about two-thirds are appointed with one-third elected (Sherman 1985), their remuneration, and their procedures.

Other powers assigned to Ports normally fall into three categories: real estate, land use/environmental, and fiscal. Statutes give Ports varying powers to acquire, lease, mortgage, sell, and encumber lands. Many are given the power of eminent domain, which often is crucial for effective harbor development. The powers to use or change land and adjoining waters are sometimes stated in the law, with great redundant flourish. Sherman (1985) quotes from the enabling act of the Mississippi State Port Authority, which authorizes the board to:

> Acquire, purchase, install, lease, construct, own, hold, maintain, equip, use, control and operate ports, harbors, waterways, channels, wharves, piers, docks, quays, elevators, tipples, compresses, bulk loading and unloading facilities, warehouses, floating dry docks, graving docks, marine railways, tugboats, ships, vessels, shipyards, shipbuilding facilities, machinery and equipment, dredges, and any other facilities, required and incidental to the construction, outfitting, drydocking or repair of ships or vessels, and water, air and rail terminals, and roadways and approaches thereto, and other structures and facilities needful for the convenient use of the same in aid of commerce, including the dredging, deepening, extending, widening, or enlarging of any ports, harbors, rivers, channels, and waterways, the damming of inland waterways, the establishment of water basins, the acquisition and development of industrial sites and the reclaiming of submerged lands.

Land use and environmental powers will often include a requirement that comprehensive port plans be prepared and available for public review. In Washington State, for example, the Port statute requires that a "comprehensive scheme of harbor improvements" be in place in each port district and any changes to that plan must be done in a public meeting and preceded by a hearing. Southern California Ports noted above must have approved master plans, and they control other harbor land uses by issuing permits or by regulating or financing facilities to reduce pollution.

The Ports' fiscal powers are often their most important tool. As noted, they can raise money through fees, leases, and borrowing. In some cases, though not the majority of the major ports (AAPA 1986), taxes can be levied to increase revenue flows (see Chapter 10).

The net effect of this array of powers is to offer public ports opportunities and constraints that are not available to the usual government agency or to the private business (Schnelker 1986). These differences define the nature of the Port as governmental business. For example, a Port's eminent domain powers may allow it to condemn property that might eventually be resold at a profit or leased to a private party. Also, it may have advertising, promotional, and lobbying powers, and it may be able to hire staff without following the normal civil service rules. In public financing, its landholdings may be exempt from real estate taxes, it may be empowered to issue general obligation bonds without a vote of the people, and its municipal bonds have the advantage of exemption from federal income taxes. These powers, some of which are not available to other governmental entities, enable the Port to operate with greater freedom from constraint and to compete more effectively in the marketplace.

On the other hand, as a "governmental" agency, the Port is burdened with constraints that distinctly separate it from private business. "Sunshine laws" apply, which require open meetings, public hearings, disclosure of business information, and avoidance of any conflicts of interest among commissioners. Competitive bidding is required on contracts, and in some cases restrictions apply, such as requiring use of in-state banks. In Washington State, Ports are prohibited from "lending the state's credit" and this has inhibited joint financing arrangements with the private sector. Most important, port commissioners, whether elected or appointed, respond to constituent pressures, and this can directly affect actions on particular projects.

Given the balance that is struck between government and business in the structure and powers of public ports, questions often arise whether the balance is correct. How can public ports be made accountable to general government, citizens, or constituents, to ensure that public objectives are met, yet retain the degree of independence that allows them to operate best in a business environment? Walsh (1978) concluded that public enterprise gets high marks for efficient, businesslike management, but that they overemphasize financial return and are too dependent on the views of banking and business interests. Price (1980) argues that maximum management independence is important for public enterprise, but that they should be supervised more closely on financial matters.

Olson (see Chapter 14) provides a more elaborate framework for Port accountability. He believes that accountability criteria should vary depending upon the structure of the Port organization, the goals public ports advance, the goods they provide, and the market structures in which they operate. His analysis recognizes the nature of the multiple objectives that port authorities serve—transportation, commercial, recreational, social, environmental, and political.

## THE TRADITIONAL ORIENTATION OF PUBLIC PORTS

Most public ports view their business as the handling of cargo and the production of revenue that will sustain their operation and help to finance improvements. These activities are done in a way to appease interest groups and regulatory agencies. Thus, public, nonrevenue-generating activities affecting the harbor, such as public access, environmental enhancement, and redevelopment of obsolete facilities for recreational or educational uses, are often considered secondary and done in response to pressures from other agencies or interest groups. They are considered overhead costs and are dealt with in planning and environmental departments that are two or three levels down within the organization. This traditional interpretation by Ports of their mission limits their willingness to participate in harbor management unless it is serving an "appeasement" function.

The mission statements adopted by public port organizations emphasize increased cargo movement and economic development goals. Tear (1986) grouped mission statements from sixteen of her sample Ports into three categories: 1) seven Ports that emphasized increased cargo movements through better facilities and improved management; 2) six Ports which emphasized economic development; and 3) three Ports that emphasized multipurpose development, including public interest and natural resource values. Thus, most Ports see their business enterprise function as their primary reason for being. In their view, they are serving the public interest if they successfully increase cargo movement or spur the local economy.

Economic realities emphasize this business orientation. Ports were established to be independent and to earn their own way. They were not set up to provide free public services as in the case of government agencies receiving annual governmental appropriations. Since Ports generally receive little direct tax support from the government, the ability to produce revenue is a central reality. Given the increased cost of improvements (see Chapter 6), reduction in availability of subsidies (see Chapter 5), and long lead time before many Port development projects begin to pay for themselves, Ports strive to ensure regular and adequate returns on investment (see Chapter 10). And, as Dowd emphasizes, the highly competitive Port industry forces Ports to be efficient, to reduce overhead costs and to stick to their basic businesses rather than tackling new ventures (see Chapter 10). However, changes are occurring in the transportation industry, in public policy, and in constituent demands. These changes directly affect Ports and their business. Public interests in the harbor are expanding, and the traditional business orientation of the Port may no longer be adequate for many agencies.

## TRENDS AFFECTING THE PORT

Three trends reaching well into the next century will influence Port decision-making substantially: 1) shifts in maritime trade and transportation due to economies of scale and deregulation within a highly competitive environment, 2) the scarcity and higher cost of capital with which to build new facilities, and 3) increased pressure from local environmental organizations and public agencies seeking more benefits and better accountability from public ports. The net result of these trends for many Ports will be the expansion of their agendas to include more harbor management and public service functions.

### *Shifts in Maritime Functions*

Many forces are at work that will substantially affect how public ports do business in the future. In the 1960s and 1970s, technological changes in shipping and cargo handling resulted in Ports building new infrastructure within harbors with funds readily borrowed from the private bond market. These were heady days for the port industry and port authorities operated with few constraints and little public scrutiny.

Over the next twenty years or so, it is likely that Ports will have to proceed more cautiously and in a variety of new directions in carrying out their functions. Economic advantages from larger scale operations, and deregulation of transportation industries, will result in load center ports for containerized and other forms of cargo (see Chapter 6). Load center ports may take many forms, based on the commodity, the carrier, or the port (Marcus 1986; MARAD 1985). Many midsize and small ports will either decrease their maritime activities or become highly specialized. Also, maritime industry in general, such as shipbuilding and fish processing, have moved to other countries or to inland locations, leaving some ports with less business and vacant land. Because of the shifts in cargo handling and the decline of some maritime industries, some public ports are scrambling to get involved in new ventures that will bring in revenues and boost the local and regional economy, such as real estate development[8] (see Chapter 7; NERBC 1981). Olson and Denning (1982) have concluded that the end of the era of providing infrastructure for containerized cargo has arrived, and public ports will likely turn to broader economic development missions, such as export trade assistance, tourism promotion, or the construction of teleports (information/communication centers using satellite services; see Chapter 7).

## Capital Scarcity and Higher Capital Costs

Finding funds for major capital improvement is a key problem facing the industry. The causes of this problem include the higher costs of capital construction, reduction in subsidies from federal, state, and local governments, and greater competition for funds for public works projects, especially at the local level where Port projects may have to compete with social service projects. Further, changes in the tax laws have reduced the benefits available to Ports through tax-exempt borrowing (see Chapter 5). There are several important consequences of this situation. First, Ports will be planning more cautiously in the future—perhaps in conjunction with other Ports to reduce overcapacity and ensure better returns on invested capital. Second, Ports will become more self-sufficient, since the tax base and hidden subsidies are rapidly disappearing. Third, joint ventures with other ports, private investors or government agencies may be necessary. Finally, if an appeal to the voters is necessary for capital improvement, good relations with the public will be essential.

## Increased Local Pressures

There is increasing public concern about Port activities within the harbor, and pressures for Ports to provide more local public services and benefits (National Academy of Sciences 1979). Groups such as PortWatch in Seattle, City Club in Portland, Oregon, and the Boston Harbor Associates are concerned with impacts from Port projects on community, recreation, and environmental values, as well as the soundness of investment decisions (PortWatch 1983). Environmental and land use agencies, especially those concerned with coastal zone management, have urged greater regional cooperation in planning new port facilities (Fay 1983; Wardwell 1983; see Chapter 5). And public and agency concern for provision of parks, promenades, public access, and retail/tourist oriented facilities has put enormous pressure on shoreline resources in harbors. In many U.S. harbors, there is keen competition for the parcels available for redevelopment (see Wrenn 1983; U.S. Department of Commerce 1980; National Academy of Sciences 1980).

The combined effect of these local demands and pressures is fourfold. First, Port industrial projects are scrutinized very carefully on economic and environmental grounds, invariably resulting in delays, additional costs, relocation, or redesign. Second, public uses are seen as an alternative to the industrial use: a park, environmental preserve, or leisure use is almost always preferred by some groups, and on occasion there is outright competition between a port or park use. Third, the proposed port or industrial use is often seen as displacing a public resource or use and is therefore required to compensate the public for that loss at the same or another location. Fourth, each proposed new use raises questions of precedent: Is a trend being set that will be more difficult to reverse in the future? Thus, each new port project raises questions about the entire harbor—its environmental quality, the balance between industrial and recreational use, appropriate siting, and long-term direction.

These new pressures raise the question of the breadth of the role of the public port in the harbor. Shifts in maritime trade and transportation can lead to Port diversification and new ventures. The scarcity and higher cost of capital can lead toward more joint ventures and dependence on other sources of funds, public or private. Increased pressure from local environmental organizations and public agencies will result in questions about appropriate use of the harbor as a whole and demands for more comprehensive plans. For many Ports the traditional orientation set out above—cargo handling, revenue generation, and the appeasement of regulators and interest groups—may be an inadequate strategy for the future.

## EMBRACING HARBOR MANAGEMENT

A diversified port that is dependent on many sources of revenue and subject to a wide range of environmental and land use requirements can no longer afford to proceed on an ad hoc, project-by-project basis. Long-term planning is no longer a luxury but a necessity. Assets must be set aside to pay for nonrevenue-generating requirements such as access ways or mitigation wetlands acquired in advance of particular projects. Sites must be made available for future maritime business so that the great demand for leisure-related facilities in urban areas will not foreclose maritime business in the future. Collaboration with other agencies is essential to gain Port project approval or to look ahead at land and water use options in the future. In other words, the intense competition within the industry for business and dollars, and the intense competition for scarce harbor lands and waters combines to require an active long-range planning function that involves many players in the harbor. Public ports really have no alternative but to embrace some form of harbor management strategy.

In some cases it may be appropriate for public ports to take the lead in harbor management. This has occurred in the Southern California Ports, where the Ports of Hueneme, Long Beach, Los Angeles, and San Diego administer master plans. In other cases, Ports can make substantial contributions to harbor management, even though leadership rests with other agencies. In San Francisco Bay, the Ports worked closely with the Bay Conservation and Development Commission (BCDC) and with the Metropolitan Transportation Commission to develop the Seaport Plan. In most SAMP task forces, Ports work as equal partners with a variety of other agencies and interest groups.

Successful participation in harbor management will require Port initiative in four areas, each of which is discussed in detail in Chapters 11 through 14. The common denominator of these initiatives is improved planning within the Port organization, and with other agencies and organizations with interest in the harbor.

***Strategic Planning.*** The concept of strategic planning—a systematic approach to the determination of future directions for an enterprise, has been used seriously in only a small number of Ports, including New York (McKeever and Selinger 1983), Seattle (Port of Seattle 1982), and Portland, Oregon (Port of Portland 1982), even though it is a common tool in the private sector. Despite this, strategic planning as a management tool has great potential for the public port industry because of the broad mandates, independent financing, and management capability of public enterprise (Dowd 1981). It has the ability to sharpen a Port's objectives and actions after analysis of the business and political environment surrounding the Port. Chasan and Dowd (see Chapter 11) study the Port of Portland's strategic planning effort and find value in it for the industry as a whole. A Port that has clarified its own development objectives can be a better leader or participant in harbor management. The AAPA has led the way for more ports to do strategic planning and is publishing a guidebook in 1988.

***Environmental Mitigation.*** New laws require that port development projects at the water's edge compensate the natural environment for loss of habitat for wildlife. This requires establishing healthy environments in other parts of the harbor where they have been nonexistent or degraded. In some cases mitigation in advance of new projects can be provided by a Port, thus providing a "bank" from which mitigation credits can be withdrawn and applied to a new project. Wessel and Hershman (see Chapter 12) describe mitigation experience in locations throughout the country and the role played by the public port. Mitigation will be a central issue in harbor management. Better environmental planning can identify and acquire mitigation sites in anticipation of new development. Such planning would be done in conjunction with other agencies and would look holistically at harbor environmental resources, including the problems of selecting sites for dredging and disposal of contaminated dredged materials.

***Waterfront Revitalization.*** Although the revitalization of obsolete port facilities for leisure and recreation is a major agenda item for cities and some Ports, careful planning is needed because some deep-water sites are scarce and may be needed for maritime uses in the future. Some small marine businesses in boat repair or fishing are displaced and perhaps put out of business by the restaurants, boutiques, and yacht clubs that can pay higher rents. Goodwin (see Chapter 13) argues that the public port can help to resolve the displacement problem through advocating retention of working waterfront sites and purchasing and banking the most appropriate ones for the future.

***Accountability.*** Ultimately, as the harbor shrinks and each proposed use is viewed in conjunction with other potential uses, the Port will be forced to conduct more of its business in conjunction with others and to account more regularly to the public at large. Olson (see Chapter 14) argues that the performance of a public enterprise can be evaluated, but the criteria for evaluation need to be selected carefully. Evaluating a Port's performance can lead to suggested changes in authority, structure, or finance that enhance its capability to meet public and organizational objectives.

Ultimately, public ports should embrace harbor management and contribute in significant ways to the stewardship of the waterfront. Of all the public bodies with jurisdiction in the harbor, Ports have the most comprehensive interest due to their land ownership, their long history of activity there, and the constituent groups they serve. They focus exclusively on the harbor, whereas other agencies have broader mandates. They are an action agency, a catalyst of change, able to acquire and develop shorelands for a wide range of objectives. By planning together with people of diverse perspectives—urbanists, environmentalists, workers, and developers—the narrow views of the participants can be broadened and the interdependence of the uses of the harbor recognized and promoted. The public port represents all the maritime interests of the city and they have the opportunity, as Fleming notes in Chapter 3, to help forge a true port community that has a common purpose, and in which leisure, industry, and maritime culture can develop in harmony.

## *Notes*

1. Some harbors are tiny (Depoe Bay, Oregon contains only 4 acres of water surface at high tide); others are huge (San Francisco Bay, for example, has 435 square miles of surface area). Some harbors contain major deep-water commercial ports; others contain only floats for fishing or recreational boats, or a pier for a single use such as ferry loading, log export, or crude oil off-loading.

2. See Hershman 1978; Kildow 1981; National Academy of Sciences 1979; Armstrong 1981; National Research Council 1985. Armstrong (1981) concluded, after a national survey and six case studies, that "ports have come to face competition for urban space which is unparalleled in American history."

3. There is very little written that attempts to define and describe the harbor as a system in which diverse activities coexist. Nor is there a body of research that adequately describes harbor problems in comprehensive terms. More common are treatments of single aspects of the harbor, such as harbor hydraulics and dredging, estuarine water quality, terminal development and operations, and urban waterfront revitalization. Some geographers have attempted a more holistic vision of the harbor. James Bird's "Anyport" model shows how port installations shift their location in the harbor in response to changes in transport technology and how abandoned port facilities become available for other uses (see Chapter 2; Bird 1971). Other geographers, mostly from western Europe, have explored port and urban change in a series of case studies (Hoyle and Pinder 1981; Hoyle and Hilling 1984). In the United States, exploration of the harbor as a system involving multiple natural and human forces has been virtually ignored, except in the context of suggesting better ways to allocate resources through improved institutional arrangements and decision-making procedures (see, for example, Evans 1980; Kildow 1981; New England River Basins Commission 1981; Wrenn 1983; National Research Council 1985).

4. As Jones (1985) observes, there is a considerable definitional problem: what size and/or function of marine installation is needed before it can be called a "port"? Should ferry terminals, private oil docks, and marinas be included? Also, sources of information about ports are highly fragmented. U.S. government statistics concentrate on port facilities concerned with interstate commerce and foreign trade. Statistics kept by the COE and the Federal Maritime Commission (FMC) are quite different, the COE concentrating on navigable waterways and harbors, the FMC focusing on particular marine terminals where agreements or tariffs of common carriers are in effect. State and local statistics vary with the jurisdiction and are sometimes nonexistent. State transportation agencies keep statistics about port expenditures (AASHTO 1986), but these include only expenditures by state government, not funds raised by port districts. And it is virtually impossible to gather information on municipal recreational facilities such as marinas and landings without conducting a national survey. Even if we had a good definition of a port and an accurate count of U.S. ports, we would have to separate those administered by the public sector from those that are private.

5. Although there are more private deep-water terminals than public ones in the United States (MARAD 1980), a large number of the private terminals are serving single purposes, such as oil, grain, or other bulk commodity movements. Most general cargo facilities are publicly owned, though they may be operated by private firms. Containerized general cargo facilities, for example, have been built by public agencies almost exclusively, but are leased to steamship lines or stevedoring firms for operations (see Chapter 10).

6. Eighteen U.S. Ports, selected for a noncargo activity study, represent examples of Ports with varying functions and sizes. Tear (1986) found that five were a class of mature, highly diversified Ports with large assets which have reached a fairly steady but somewhat flat cargo growth rate and which are trying to increase revenues through diversification (New York, Massport, San Diego, Oakland, Seattle). Interestingly, Tear notes that these Ports are not necessarily the most productive and that the tendency toward lower operating incomes may indicate that expenses are high when many activities are maintained. The other Ports were more difficult to classify, since some were state-level Ports where noncargo functions were delegated to other agencies, some were specializing and trying to find a niche in the trade world; and some were operating at a deficit and contracting their activities.

7. A study published in 1978 estimated that there are over 7,000 U.S. public enterprises, spending about $25 billion annually on such projects as bridges, tunnels, and parkways; public buildings and parks; water, gas, and power services; training, insurance, and mortgage finance; and seaports and airports (Walsh 1978). The amount spent annually today would be much higher.

8. The redevelopment of port lands for commercial/retail/office and in some cases hotel/resort/residential space is becoming an important source of revenue. In 1986 San Francisco earned $14.7 million from commercial property, or 50 percent of its gross revenues. In 1985 Los Angeles earned $23.6 million, or 25.6 percent of its gross revenues from commercial property, sheds, and warehouses. In 1986 San Diego earned $25.5 million, or 43.6 percent of gross revenues from commercial uses. In a recent restructuring at Massport, in Boston, an independent real estate division has been established to convert obsolete piers to revenue-producing facilities.

## References

American Association of Port Authorities (AAPA). 1986. Port authority taxing powers. *AAPA Advisory* **XX**(46):1–2.

American Association of State Highway Transportation Officials (AASHTO). 1986. Survey of state funding of landside port facilities and cargo terminals 1977–1984. Washington, DC, July.

Archer, Jack H., and Robert W. Knecht. 1987. U.S. national coastal zone management program. *Coastal Management* **15:**103–20.

Armstrong, John et al. 1981. The relationship of port development and urban waterfront revitalization. Technical Report No. 2, The South Carolina Sea Grant Consortium, February, 103 pp.

Bird, James. 1971. *Seaports and seaport terminals*. London: Hutchinson University Library.

Coastal Resources Management Council. 1984. Providence harbor: A special area management plan. Coastal Resources Center, University of Rhode Island.

Connecticut Coastal Management Program. 1985. Model municipal harbor management plan. Department of Environmental Protection, Hartford, Connecticut.

Dowd, Thomas J. 1981. Strategic management: A tool for port planning. Master's thesis, University of Washington.

Dull, Laurie J. 1983. Special area management planning: A comprehensive planning process for the Coos Bay Estuary. Master's thesis, University of Washington.

Evans, Nan et al. 1980. The search for predictability: Planning and conflict resolution in Grays Harbor, Washington. Technical Report, Washington Sea Grant, Seattle, 117 pp.

Fay, Dennis R. 1983. Discussion of the San Francisco Bay Area seaport plan. *Coastal Zone '83*. Proceedings of the Third Symposium on Coastal and Ocean Management, San Diego, California, June, pp. 273–86.

Godschalk, David R., and Kathryn Cousins, eds. 1985. Coastal management: Planning on the edge. *Journal of the American Planning Association* **51:**263–336.

Hershman, Marc J. et al. 1978. *Under new management: Port growth and emerging coastal management programs*. Seattle and London: University of Washington Press, Washington Sea Grant, 212 pp.

Hershman, Marc J., Stan Euston, and Melissa Rohan. 1983. The future of U.S. harbors, Institute for Marine Studies, University of Washington, November 11.

Hoyle, Brian S., and David Hilling, eds. 1984. *Seaport systems and spatial change*. Chichester: John Wiley and Sons.

———, and D. A. Pinder. 1981. *Cityport industrialization & regional development*. New York: Pergamon Press. 350 pp.

Inoue, Satoshi. 1986. Japan's experience in port development. Ministry of Transport, Tokyo.

Jones, Pat. 1985. Report on port data collection. Institute for Marine Studies, University of Washington, February 10. Manuscript.

Kildow, Judith T. 1981. *Boston harbor management study*. Cambridge, MA: MIT Sea Grant College Program, 268 pp.

McKeever, E. G., and Carl S. Selinger. 1983. Implementing strategic planning at a diversified public agency. *Transitions*. ATE Management and Service Company, Cincinnati, Winter, 21–27.

Marcus, Henry S. et al. 1976. *Federal port policy in the United States*. Cambridge, MA and London: MIT Press, 371 pp.

———. 1986. Trends in U.S. liner shipping: Past, present, and future. Paper presented at World Conference on Transport Research, Vancouver, B.C., May.

Maritime Administration (MARAD). 1978. *What U.S. ports mean to the economy*. Washington, DC: Government Printing Office.

———. 1980. National port assessment 1980/1990. Department of Commerce, Washington, DC.

———. 1985. Public port financing in the United States. Vol. II. Department of Transportation, Washington, DC.

———. 1986. A report to the Congress on the status of the public ports of the United States, 1985. Department of Transportation, Washington, DC.

Mississippi Coastal Program. 1984. Special management area plan for Port Bienville. Ralph M. Field Associates, Inc.

National Academy of Sciences, National Research Council. 1976. *Port development in the United States*. Prepared by the Panel on Future Port Requirements on the United States, Maritime Transportation Research Board, Commission on Sociotechnical Systems, Washington, DC, 246 pp.

———. 1979. *Public involvement in maritime facility development*. Committee on the Impact of Maritime Services on Local Populations, Maritime Transportation Research Board, Commission on Sociotechnical Systems, Washington, DC, 270 pp.

———. 1980. *Urban waterfront lands*. Environmental Studies Board, Commission on Natural Resources, Washington, DC.

National Oceanic and Atmospheric Administration (NOAA). 1981. The federal coastal program review. Department of Transportation, Washington, DC.

National Research Council (NRC). 1985. *Dredging coastal ports*. Washington, DC: National Academy Press.

New England River Basins Commission (NERBC). 1981. *Harbor Management Strategies for New England,* Vol. 2. Boston: NERBC, 48 pp.

Olson, David J., and Michael Denning. 1982. Economic development and the role of public enterprise. Paper delivered at 1982 Annual Meeting of the American Political Science Association, Denver, September 1–5.

Port of Portland. 1982. *Marine department strategic plan*. Portland, Oregon, March.

Port of Seattle. 1982. *Strategic plan 1983–87*. Seattle, Washington.

Portwatch. 1983. Issue papers regarding the Port of Seattle, presented at Issues and Call for Candidates Night, Seattle, Washington, July 28.

Price, Willard. 1980. Seaports as public enterprises: Some policy implications. In *Making ocean policy: The politics of government organization and management,* ed. Francis W. Hoole, Robert Friedheim, and Timothy Hennessey. Boulder, CO: Westview Press, pp. 217–38.

Rees, Lee. 1986. Port powers: Comparison parameters. Institute for Marine Studies, University of Washington, January 13. Manuscript.

Rothenberg, Daniel, and Arthur Rocque, Jr. 1986. Harbor management: The impetus for action and Connecticut's response. *The Coastal Society Bulletin*. **9**(4):14–17.

Ruppenthal, Karl M. 1983. *Canada's ports and waterborne trade*. University of British Columbia, Centre for Transportation Studies, Vancouver, B.C.

San Diego Unified Port District. 1976. *The Embarcadero development plan*. May, 34 pp.

San Francisco Bay Conservation and Development Commission and the Metropolitan Transportation Commission (SFBCDC). 1982. *The San Francisco Bay Area seaport plan*. San Francisco, California, 55 pp.

Schnelker, Carrie. 1986. Port as governmental business. Port of Seattle, Washington. Manuscript.

Sherman, Rexford B. 1985. Public port agencies in the United States and Canada. American Association of Port Authorities, Alexandria, Virginia.

Stromberg, Rolf E. 1982. The Port of Seattle as a public enterprise: An analysis of the Shilshole Bay Marina development. Master's thesis, University of Washington.

Tear, Lucinda M. 1986. Diversification of function in port authorities. Master's thesis, University of Washington.

U.S. Department of Commerce. 1980. *Improving your waterfront: A practical guide,* 108 pp.

Walsh, Annmarie Hauk. 1978. *The public's business: The politics and practices of government corporations.* Cambridge, MA: MIT Press.

Wardwell, Robert H. 1983. Regional port plans: Government without regulations. *Coastal Zone '83.* Proceedings of the Third Symposium on Coastal and Ocean Management, San Diego, California, June, pp. 336–48.

Wolf, Michael A., ed. 1985. Symposium on coastal zone management. In *Natural Resources Journal.* **25:**7–242.

Wrenn, Douglas M. 1983. *Urban waterfront development.* Washington, DC: Urban Land Institute, 219 pp.

# Part II

# THE SETTING

# Chapter 2

# PORTS OVER TIME:

## Historical Perspectives on the Public Port

*Marc J. Hershman and Robin Scott Bittner**

---

The modern port in the United States can be described as a community of independent enterprises tied together by a common interest in maritime affairs. Central to this community is an entity known as the port agency or authority, which has served, increasingly over the past few decades, as landowner, operator, developer, plus public relations agent for the port. Port authorities commonly assist their customers and constituents with a variety of services, including data collection, labor relations, facility development, government relations, and others in an effort to facilitate the smooth operation of port businesses. Through development, promotion, and coordination of the harbor area, the port authority is able to improve the port's competitive position, attract additional business, and assure its continued success.

Recently, however, Ports have faced challenges when attempting to enhance their competitive position through new development projects. Within the last two decades, environmental, recreational, and nonmaritime business interests have become more successful at competing for land and resources that port authorities intended to develop for the port (Hershman et al. 1978). This new opposition has slowed down the process of port development while increasing development costs. Some communities are voicing a preference for recreational and commercial developments over improved port facilities, and it is no longer uncommon for state and federal agencies to conclude that the value of fish and wildlife resources outweigh the potential benefits from port expansion (Hershman et al. 1978).

These new challenges are compounded by technological advancements and re-

---

*This chapter is a shortened and revised version of R. S. Bittner's Master's thesis completed under the supervision of Professor Hershman. Portions of the chapter are based on bibliographic and research work by Chris McCartan and Pat Jones. Partial financial support came from Washington Sea Grant #NA84 AA-D-00011.

lated changes in the maritime industry, including the advent of containerization and a trend toward load centering (consolidating the flow of goods so they pass through fewer ports). For some Ports, the net effect of these innovations is to intensify the agency's perceived need for maritime development, because a Port must stay up-to-date or risk losing business to competitors. For other Ports, shifts in the maritime industry require that they find new business ventures that will aid their community.

Ports are caught between the need for development and the limitations imposed by new societal values. Their projects are scrutinized more closely by a variety of interest groups. Nontraditional Port development activities—for example, the development of rail service or the construction of office buildings—often raise doubts in some minds about the "appropriateness" of these activities for a port agency. A common dilemma is whether the private or the public sector should take the lead in these new activities. Supporters of private investment feel that private interests can better assess the need and level of investment, and therefore should assume the risks (and reap the rewards) of new transportation ventures. Supporters of public investment argue that the community is better served when public dollars are at stake; long-term commitment to the area is possible and public objectives can be built into the project.

This argument was recently raised in Seattle, where the Port of Seattle proposed to purchase and operate a fleet of railcars designed to carry double-stack containers. The Port wanted to use these double-stack railcars to improve the transshipment of imported goods from Asian producers to U.S. Midwestern markets. The implementation of double-stack rail service from Seattle to Chicago would essentially double the amount of cargo each train carried, thereby reducing the transit time of goods by increasing the amount of cargo that could be shipped through the Port of Seattle per unit of time. The Port, aware of the threat of similar transportation networks that were developing in competing California ports, sought to use its public funds to gain an edge over its competitors to the south. Predictably, the railroad lines and steamship companies operating in the Seattle area argued against Port involvement, citing the traditional role of the private sector in these services. They noted also that the Port would be competing with its own tenants if they developed these businesses.

The Port of Seattle argued that letting private industry, such as a rail or steamship line, develop this new system was not an acceptable option, since it would take the private sector too long to get the system up and running. The Port also questioned the long-term loyalty of these private corporations to the Puget Sound region. If the fleet of railcars were privately owned by a company that subsequently relocated at a competing port, Seattle would lose the business generated by this system. With control over the railcars, the Port argued that secure long-term service contracts with shippers would be possible, thus benefiting the Puget Sound region.

Further studies, however, revealed some unexpected legal, ethical, and economic problems. One was an obscure piece of legislation that had been enacted to enable another Puget Sound port to operate its own railway. Although designed to solve a specific problem at a single location, it legally applied to all port districts in Washington. This legislation did not allow a port district to operate a railway outside of the state. Although the double-stack cars did not constitute an entire railway, their use as part of an interstate transportation network could conceivably be interpreted as a violation of this law. The questionable legality of the project, its $30-million price tag, and the ethical problem of competing with their own customers for the transportation of goods eventually led the Port of Seattle to reconsider its position. The situation was resolved when two of the Port's customers, American President Lines and Burlington Northern, agreed to develop the double-stack system.

The controversy about the Port of Seattle's involvement in the nontraditional activity of rail service, and public hesitation, in general, about unconventional Port development strategies, point to the broader issue: defining the proper scope of business for public ports. The public port enterprise, charged with operating in the common good of the community, is called upon to justify its ongoing activities and new ventures by showing their net benefit to the public. The justification becomes more difficult when such ventures are not clearly a part of the traditional public port industry.

In order to evaluate the appropriateness of these ventures, it is important to examine a broad range of issues as well as Port responses over the course of their existence. The public port is a complex institution, and an historical perspective can help define its nature and explain why it looks and acts as it does. (See the partial bibliography of port authority historical studies at the end of this chapter.)

Hoyle and Hilling (1984) suggest that three forces—physical and technological, political, and economic—shape a port's history. The development of shoreland or upland facilities for handling people or cargo is basic to port operations, yet the facilities must be constructed within the limitations set by the physical environment and current technology. Often tremendous environmental obstacles must be overcome in the process of constructing facilities or making navigation channels safe for the passage of vessels. A sizeable percentage of the expertise, time, and financial resources of the public agency are spent on resolving physical and engineering problems.

Politics play a significant role in the day-to-day operations of a public port. As a public entity, the business philosophy of the public port necessarily differs from private corporations. The public port authority is charged with operating the port in the public interest, which is a highly variable goal defined by local and state political systems. The public nature of the agency requires responsiveness to a wide range of interest groups, which inevitably shapes its behavior.

Economic conditions also help to determine how port authorities will use their physical and financial resources. As part of an industry that serves the transportation needs of shippers, Ports are dependent on the success of their customers for their own economic well-being. In assisting the economic development of their region, their actions often are designed to support existing businesses or to attract new and suitable enterprises.

Physical and technological, political, and economic forces lie at the heart of the events that have shaped the history of public ports. To provide some insight into their effects, the following three sections examine these forces in action. Each force is discussed generally and in case studies, showing how it has produced problems and solutions for particular ports. The first section discusses some of the physical difficulties confronting a "would-be" port when it is just getting started, examining the initial development of the ports in Houston and Milwaukee. The appearance of the public port entity is the focus of the second section, showing the role of politics in port development with a study of the Port of Seattle. The third section examines the influence of regional economies on a Port, with the story of Port Canaveral's response to economic opportunities during the second half of the twentieth century.

## CREATION OF A PORT THROUGH LAND DEVELOPMENT AND FACILITIES CONSTRUCTION

People have typically congregated in geographically accessible places for the purpose of conducting trade. Some waterfront locations became permanent settlements and gradually developed the facilities and services that are commonly associated with ports.

As port communities prospered, the construction of facilities and improvement of waterways became one of their major preoccupations. Physical development of the shoreline was crucial to these communities because they could not exist if they did not provide ships with safe places to moor and transfer cargo. In fact, most ports had to overcome the limitations imposed by the local physical environment in order to construct port facilities. A few locations, such as Seattle and San Francisco, were blessed with deep natural harbors, and these communities found that their first port facilities fit easily into the natural landscape.

The demands on frontier ports were fairly limited. When deep water was available close to shore, inexpensive piers could be built with local timber and landfills created with rocks, ship ballast, and waste materials. As time passed, however, shipping technology advanced to a point where these crude piers were no longer sufficient, and ports found it necessary to replace obsolete facilities in order to keep up with the changing industry. Many times this meant making extensive modifications to the natural environment so that the port would have sufficient level shoreline and deep water to accommodate newer and larger vessels.

The physical development of a port progresses through a series of stages, as discussed by Bird (1971) in his "Anyport" model. Bird observed that British ports went through at least six eras of port development when evolving from a "primitive" to a "specialized quayage" level. Within each era there occur additions or changes to the physical layout of the port as installations are relocated to different sites within the harbor. Over time these changes come together to form a complex modern port serving a variety of port interests.

The physical developments that occur within an era are not necessarily removed or rebuilt in the next era: as Bird notes, "although an era of port development may be recognized as having ended, installations built during that time may continue to have a long working life." This mix of old and new can be observed in the finger piers that are found alongside container terminals in modern ports. The development of port installations in the harbor are the "bread and butter" of the port, and represent the port community's ongoing struggle to develop competitive facilities. As long as the impetus of competition remains strong, the maintenance and evolution of the shoreline will continue to be the primary concern of port communities.

## *The Port of Houston*

In some instances, ports have been created in rather unlikely locations through great effort and expense for the purpose of making a community attractive to business interests. Often such communities have had to undertake phenomenal development projects just to enable ships to reach the port facilities. Houston, Texas, initially did not appear destined to become a port, because it lacked the one feature common to all ports—easy water access. Over time, however, a series of extensive hydraulic engineering projects has essentially brought the ocean to Houston, enabling the community to develop into a major international port.

The Port of Houston, located some 50 miles inland from the Gulf of Mexico on a man-made canal, owes its existence to the persistent efforts of two land speculators. In late 1836 the Allen brothers—Augustus Chapman and John Kirby—purchased a tract of land on Buffalo Bayou, a long, shallow finger of water stretching inland from Galveston Bay on the Gulf of Mexico. The brothers named the townsite after General Sam Houston, a popular hero who led Texas to independence from Mexico, and only a few months later it was selected as the first capitol of the Republic of Texas. In an effort to establish Houston as a center of trade, the Allens widely advertised that the townsite lay at the head of a navigable waterway and would someday be the "great interior commercial emporium of Texas" (Sibley, 1968).

Skeptics doubted the Allens' claims so loudly that the brothers felt compelled to prove that the townsite was indeed at the head of a navigable waterway. In January 1837 the Allens hired the steamboat *Laura* to sail from the Gulf upstream to Houston, and persuaded a boatload of prominent Texans to make the trip. It may have been that the Allen brothers themselves had a few doubts about the success of the trip, for at 85 feet, the *Laura* was the smallest steamboat in Texas (Sibley 1968).

The initial portion of the journey, from the Gulf of Mexico inland to Harrisburg, was relatively easy. The five miles from Harrisburg to the Houston townsite was a different story, however—no boat had ever been upstream from Harrisburg. This leg of the trip took three days, primarily because the bayou was filled with logs, brush, and all manner of debris; and the passengers were obliged to assist the crew in cutting their way through the choked bayou (Sibley 1968). Those who made the journey marveled at the lush greenery and teeming wildlife, including the alligators sunning themselves along the banks. At night, the *Laura* was tied up and the passengers went ashore to dance and frolic with settlers.

The *Laura* finally reached the Houston townsite on January 22, thereby proving that Buffalo Bayou was navigable, if extremely treacherous. A group of Houston's merchants and boat owners formed the Buffalo Bayou Company to collect funds to clear the bayou. Three years later, in 1840, the Republic granted the town permission to construct wharves and charge wharfage fees. And in June 1841 the Houston City Council created the Port of Houston and appointed a wharfmaster (Port of Houston 1986).

In the early 1840s the collective actions of Houston's citizens began to pay off, and over the years the town's economy steadily improved with each navigational improvement made to Buffalo Bayou. The greatest advance in building a nationally recognized port occurred many years later, on November 10, 1914, when President Woodrow Wilson officially opened the Houston Ship Channel by pushing a button in Washington, D.C., which fired a cannon placed in the turning basin of the channel (Sibley 1968). Where once magnolias had blossomed and alligators roamed, there now lay a deep-water ship channel, capable of handling oceangoing ships from around the world. The Port of Houston today ranks third in the nation in total waterborne commerce, and is number one in the handling of wheat and steel (Zullo, 1983). In 1979 the Port of Houston (Fig. 2.1) surpassed the Port of New York in foreign trade tonnage, an achievement that can be attributed to the nearly 150 years of collective action by Houstonians.

## *The Port of Milwaukee*

Milwaukee, Wisconsin, is a city that, although centrally located and eager to conduct waterborne commerce, had to contend with shallow water, strong winds, and unpredictable river currents. The story of the Port of Milwaukee illustrates the many social, economic, and political hurdles that a community may have to overcome when developing a port.

Present-day Milwaukee is in a unique natural setting, at the confluence of three rivers: the Milwaukee, Menominee, and Kinnickinnic. Schenker (1967) provides the following description of the landscape prior to the arrival of the first white settlers:

> The confluence of the Menominee and the Milwaukee offered a potentially fine harbor, but below that junction the river wound a tortuous path to Lake Michigan. At one point

**Figure 2.1.** The Port of Houston, one of the largest river ports in the United States, was once a meandering bayou filled with logs, brush, and all manner of debris. (Courtesy: Port of Houston)

> (the site of the present harbor entrance), only a sand bar separated it from the lake, but then the river turned inland again to meet the Kinnickinnic before eventually reaching the lake, leaving a peninsula about 3,000 feet long and 300 to 700 feet wide connected to the mainland north of it by that sandbar.

This area had been used for several centuries by native American tribes, and from 1670, French Canadian missionaries and trappers frequently came to this location to trade with them. The three rivers were fairly shallow and slow moving, which was ideal for the light canoes used by Indians and trappers for transporting goods and furs. Unfortunately, the very characteristics that made the rivers perfect for canoe travel would later prove treacherous to the heavy, deep boats of the European traders.

Hamming (1953) states that when European settlers arrived (around 1835), they first settled in groups among the bluffs that surrounded a shallow valley cut by the rivers. The primary settlements were located about a mile upstream from the mouth of the river, which meant that the only way to get to the city was to navigate the dangerous shallows downstream from the townsite. The depth of the river limited passage to vessels with a draft of 8 feet or less, but these small vessels were often forced to operate in inclement, and at times severe, weather conditions, as localized storms and gale-force winds swept in off Lake Michigan.

The citizens of the new community devised several ingenious ways of transporting cargo from large lake vessels up to the town. The most popular method employed a lightering service, whereby goods were unloaded from larger vessels onto small ferry craft capable of navigating the river. This method was not without its dangers, however, and local citizens felt that it was unsafe for passengers, especially in rough weather. In addition, promoters of Milwaukee believed that the fee for this service (25 to 50 cents) was too expensive and deterred settlers from stopping at the town (Still 1948).

Later, in 1835, when the population in the area had swelled to about 200, the settlers sent a petition to the War Department in Washington, D.C., asking for a survey of the river mouth (Schenker 1967). Surveyors concluded that the best way to make the river safer was to cut a channel through the sand spit that separated the river from the lake. Unfortunately, the only improvement actually made was the erection of a lighthouse in 1838, for the year before the country had fallen into a depression that lasted several years (Schenker 1967).

Increased commerce toward the end of the decade once again raised the issue of navigational safety, and this time the popular demand for river improvements reached a new high. In December 1839 the citizens of Milwaukee held a "great harbor meeting" where it was proclaimed that since they had been required to pay the federal government large sums of money to gain title to their land, navigation improvements were "a matter of right, justly our due, and too loudly called for by public necessity and the general good to be no longer disregarded" (Still 1948). Sentiments at the meeting ran so strong that some members of the community even called for such radical measures as "exciting slaves in the South to insurrection" or even "secession from the Union" if Congress continued to ignore their pleas for funds (Still 1948).

But there was very little action until 1842, when two men drowned in the river mouth while attempting to transfer a load of wood to a steamboat. The incident prompted a citizen's indignation meeting where it was collectively decided that the current system of lighterage was simply too dangerous. Once again Milwaukee approached the federal government for financial assistance. This time, however, the citizens decided that if Congress refused to appropriate the needed funds, the money would be raised by voluntary subscription.

On March 3, 1843 a long-awaited federal allocation of $30,000 was sent to Milwaukee for harbor improvements (Schenker 1967). Unfortunately, the money was not spent in the way the citizens had originally anticipated. Businesses located around the mouth of the river persuaded federal engineers to improve that area rather than create a new harbor upstream. In spite of charges of fraud and bribery by the townsfolk, the project engineers opted to ignore the proposed "straight cut" plan in favor of improving the natural outlet of the river (Schenker 1967).

This setback inspired two prominent citizens to take matters into their own hands. In June 1843 Byron Kilbourn, one of Milwaukee's pioneer promoters, sent 100 men to dredge a channel and create a "straight cut" from the lake into the river (Still 1948). Kilbourn had planned that the flow of the river would act as a natural dredge, thereby deepening and widening the initial cut, but the river currents in this area were too weak, and the plan failed.

A few years later Horatio Stevens devised a plan to construct a 1,200-foot pier out into Lake Michigan. Stevens' pier proved to be extremely successful, and it became the model for several other facilities. The lakefront around the piers experienced a tremendous surge in growth, and by 1854 the commercial success of the lakefront area convinced the federal government to abandon its plans for improving the mouth of the river (Schenker 1967). The city was then free to take the matter of harbor development into its own hands. In 1857 Milwaukee appropriated $50,000 of city funds to begin construction on the much desired "straight cut" (Schenker 1967). A channel was dug across the sandbar that had separated the river from the lake in order to divert the flow of the river; and the swampland and sandbar between the old river mouth and the new harbor entrance was filled to create an area that later became known as Jones Island. The project was completed within the year and produced a rapid migration of businesses away from the lakefront as shipping activity shifted to the much calmer and safer harbor waters inside the shoreline.

During the second half of the nineteenth century, Milwaukee experienced a boom in waterborne commerce, and responded by expanding its harbor facilities. Around the turn of the century, technological improvements within the maritime industry brought about the need for major expansion once again. The Great Lakes bulk carriers had evolved into ships so large that they could not navigate the narrow river channels inside the harbor. This caused the local harbor community to shift its attention back to the lakefront, where there was enough room to construct the huge facilities needed to service these ships.

The key to the development of the large bulk terminals rested on appropriation of the Jones Island area, which was actually a peninsula of land south of the inner harbor entrance. Over the years Jones Island had developed into a shantytown; and at the turn of the century it consisted of eleven saloons, seven stores, its own post office, and a population of around 2,000. (see Fig. 2.2.) A public eyesore harboring criminals, drunks, and a small fishing fleet, Jones Island had a reputation so tough that the local police refused even to go near the area. As one visitor described the local scene: "All they do in the summer is fish, drink, and fight—and in the winter, they don't fish" (Still 1948). The city initiated condemnation proceedings in 1914, but due to legal complications, the acquisition and development of the land were very gradual. Although the city controlled all the land by 1938, it was not until 1961 that the last of the shanties were demolished (Still 1948).

The development of the Jones Island area from the mid-1920s was the beginning of a new era for Milwaukee's port. The present port authority, the Board of Harbor Commissioners, was established in 1920, and spent the better part of their first decade preparing to open a public port. The Port of Milwaukee has concentrated on

**Figure 2.2.** Between 1914 and 1961, the Port of Milwaukee faced the difficulties of acquiring and clearing the shantytown that had developed on Jones Island. (Courtesy: Port of Milwaukee)

the development of specialized facilities as large-scale industrial and bulk cargoes have become the predominant goods moving through the port.

The evolution of the Ports of Milwaukee and Houston are similar in many respects. Although they experienced different economic and social pressures, these cities can attribute the establishment of their ports to community action. In Houston the major development project consisted of turning Buffalo Bayou into a ship channel capable of handling oceangoing vessels. Milwaukee's long-term development project was the creation of a sheltered harbor to service the lake vessels calling at the port. The efforts of these communities, starting as early as 1841 in Houston and 1842 in Milwaukee, have now become institutionalized in the form of the Harris County Houston Ship Channel Navigation District and the Milwaukee Municipal Port.

## THE DECLINE OF RAILROAD CONTROL AND THE RISE OF THE PUBLIC REFORM MOVEMENT

At one time a "port" was nothing more than a group of businesspeople seeking to improve their waterfront interests. Over time, however, these independent businesspeople were replaced by loose-knit groups of community-minded activists that, in turn, metamorphosed into independent organizations such as the Harris County Houston Ship Channel Navigation District and the Milwaukee Municipal Port.

The collective port community passed through a number of phases of incorporating formalism into its structure. One of the most significant phases in this process was the emergence of the formal public port authority in many cities during the early twentieth century. The appearance of public ports reflected a variety of economic and political changes in the United States around the turn of the century. Many of these changes were the products of movements that had been gaining momentum for several years. One such movement was the collective groundswell of opposition to the dominance of the railroads.

Many historians consider the nineteenth century to have been a period of laissez-faire capitalism in the United States, resulting in the unrestrained growth of those sectors of industry, such as the railroads, that were unregulated in the economic marketplace. Those corporations that were successful rapidly expanded in size as they engulfed their smaller competitors, leading to the appearance and rise of the monopoly.

The development and takeover of port facilities by railroads was so rapid and extensive that, by the early twentieth century, most harbor areas were wholly owned by railroad companies and other private business interests (Olson 1980b). Ports are essentially cargo transfer points within transportation systems, and therefore it was only logical for railroad companies to develop their own port terminal facilities to gain exclusive control of cargo from the time it left a ship until it was delivered to its destination. (Some railroad companies even owned the steamships.) The rail companies incorporated port service charges into their rail freight rates rather than quoting these fees separately (Cuzan 1983; Schwimmer and Amundsen 1973), which signified the complete assimilation of port terminals into the rail system.

During this period, port areas became a jumble of warehouses and piers, as railroads, dock and trading companies, and shipping lines all competed with each other for the much-too-scarce land along the harbor.

In many of the larger cities served by a number of rail lines, such as New York and Seattle, the problems associated with the congestion along the waterfront assumed almost crisis proportions. In the case of New York, port facilities located on the New York side of the Hudson River were not only in competition with themselves, but also with the facilities on the New Jersey side. Because most facilities were independently owned and no common transportation and delivery system existed, the entire harbor area was highly inefficient and extremely exasperating for shippers. This distribution problem was compounded by the lack of a common railroad belt line. Each terminal and warehouse was serviced by a different rail line, and the streets around the harbor area were a maze of mainlines, access routes, and delivery sidings. Often, trains would travel down main lines laid in the middle of streets, backing up delivery trucks for blocks as they conducted the laborious task of placing a car or two on warehouse and pier sidings along the street (Bard 1942). As long as the private interests controlled the waterfront, congestion and high rates would prevail, much to the dissatisfaction of shippers.

During the latter part of the 1890s and into the early 1900s, a gradual shift in the nation's attitude toward the railroads became noticeable. When railroads first started their expansion across the country, they were the only means of long-distance transportation. As a result, the fate of many towns depended on a rail line coming to their community. Those towns that were able to attract rail service were nearly guaranteed economic survival, for once a railroad company invested in a town, business flourished and real estate values increased substantially. As time passed, the self-interest of the railroads diverged dramatically from the community interests of the town.

Resentment toward the railroad companies was very similar in both coastal and inland cities. Over a period of years, waterfront areas had developed into dirty and congested sites that imposed intense social and economic costs on communities. Citizens were afraid to go near these areas. They were typically viewed as rough, unhealthy places, infested with criminals, and unsafe. When civic leaders realized that the public interest, as they conceived it, could not be served with continued private control of the port, many communities, at varying times and rates, began a process of shifting harbor ownership and control from the private sector to the public sector (Olson 1980b). This signified the beginning of the public port entity, created to oversee harbor development and effectively manage port operations.[1]

The road to the present system of public control was long and rocky for most communities as illustrated in the Port of Seattle story that follows. Private interests did not sacrifice their holdings without a battle. Although it took several years before the process of conversion from private to public control developed momentum, an estimated 49 percent of the 2,939 deep-draft marine berthing facilities in the United States are now publicly owned (MARAD 1980), and nearly all general cargo facilities, the primary port facilities within city harbors, are under public ownership.

## *The Port of Seattle*

Since its founding in 1851, Seattle has relied upon maritime commerce for its survival. The first permanent settlers arrived in the area with aspirations that were quite different from other pioneers of their time—they had come to trade, not to farm. This initial attitude set the stage for the future of the new settlement, and in 1852 Seattle's first export, a load of timber, was placed on the brig *Leonesa* bound for San Francisco. During that same year, a bankrupt physician from Ohio, Dr. Maynard, arrived in the new community and realized that another natural resource, salmon, abounded in the waters of Puget Sound. He set about to take advantage of this bounty, and by the end of 1852 had packed a thousand barrels of salted salmon to be sent to San Francisco. Unfortunately, Dr. Maynard's skills as a physician did not apply to fish processing, and all the fish spoiled before they reached their destination. Despite this initial setback, the fishing industry, like the timber industry, was soon to develop into a multimillion-dollar business, and together these two industries would turn Seattle into a major maritime city.

During the 1860s Seattle began building a local shipping network with several small farming and logging communities to help satisfy the growing city's ever-increasing appetite for produce, timber, and other raw materials. Because no roads or rail service had yet been developed, a ragtag collection of independently owned and operated steamships congregated in Seattle to service the intersound trade routes. The size of this fleet slowly grew over the years, encouraging trade throughout Puget Sound, and their habit of scrounging up business wherever it could be found earned them the name of Mosquito Fleet. The Mosquito Fleet enabled Seattle to grow and prosper during the late 1800s, soon becoming enshrined in local folklore when it was adopted as the symbol of the city.

By the late 1880s and early 1890s, continued growth in maritime commerce and a tremendous population expansion transformed Seattle into the major city of the Pacific Northwest. Seattle nurtured its maritime trade and eventually developed into the premier port on Puget Sound; but marine connections alone could not guarantee the economic survival of a city without a way to transport goods over land. This was the era of the railroads, and Seattle needed rail service to ensure that its hard-earned commercial success would continue (see Fig. 2.3).

Unfortunately, the first transcontinental railroad to arrive in the Northwest, the Northern Pacific Railroad, chose not to service Seattle, and extended its line only as far north as Tacoma, a competing community some 30 miles to the south of Seattle. In those days a railroad was so important that towns would offer incentives such as free property and buildings to attract railroad companies. City leaders in Seattle had offered the Northern Pacific an incentive package of 7,500 town lots,

**Figure 2.3.** Seattle's harbor in the late 1880s showing local railroad line trestles. (Courtesy: University of Washington)

3,000 acres of land, $50,000 in cash, and $200,000 in bonds (Burke 1976). Seattle was confident that no other community on Puget Sound could come close to matching this bid, but it was soon revealed that Tacoma had offered the entire western half of its townsite to the Northern Pacific. Much to the dismay of Seattle's citizens, the railroad chose Tacoma as the western terminus of the line.

For nearly twenty years, Seattle suffered without transcontinental rail service while the Northern Pacific ignored the city, apparently in the hope that it would gradually wither away. But Seattle's loyal businesses and citizens were too stubborn to give in, and they began to dominate shipping on Puget Sound in an effort to keep the city economically viable. As Burke (1976) points out, "the railroads frequently did determine the destiny of a town, but by being denied a railroad in 1873, Seattle was forced to shape her own destiny." Maritime commerce kept the city alive until 1893 when James J. Hill selected Seattle as the western terminus for his transcontinental Great Northern Railroad.

The arrival of the Great Northern signaled the beginning of a new era for the city. No longer could competing railroads ignore Seattle, and within two years the vital waterfront area became a bustling center of activity served by four transcontinental railroads. By the turn of the century, Seattle's railroads had established the city as one of the premier ports in the United States. This remarkable economic growth did not come without costs, however, and the price that Seattle paid was the congestion and deterioration of its waterfront. Each of the four railroads ran its own lines down to the harbor, producing a tangle of tracks along the piers. The private interests developing the waterfront had their hands on the throat of Seattle's maritime commerce. It soon became all too clear to local businesses that the railroads that had brought the city enormous prosperity were beginning to bar its way toward future growth.

One of the first major community efforts to support the public port movement was the battle over ownership of the harbor tidelands. Washington had not yet become a state, so the federal government held in public trust all the territory's aquatic lands below the line of ordinary high tide. Theoretically, these lands could not be built on or occupied in any way under penalties of trespass. Ironically, Seattle had supposedly deeded the railroads ownership of tidelands to build railroad trestles and maritime facilities, but the city did not own or even have authority over the lands offshore. The owners and operators of waterfront terminals were therefore particularly interested in this issue, for if the tidelands were declared to be under public ownership, then anyone owning a structure on tidelands could occupy it only at the discretion of the state. At the Washington State Constitutional Convention of 1889, the new state was given control of dispersing these lands.

The tidelands controversy developed into a very heated issue, and the delegates attempted to formulate a compromise solution to the opposing perspectives of public and private ownership. Unfortunately, the products of their labors only confused the issue further. The constitutional articles implied that the tidelands would be designated as public property, yet at the same time alleged that those who claimed title to tidelands could contest their claims in state court (Burke 1976). The lawmakers also called for the creation of a Harbor Lines Commission to address the tidelands ownership issue and establish designated harbor areas for incorporated cities and towns. The Harbor Lines Commission worked diligently to arrive at a fair decision about where to draw the line of public ownership. In late October of 1890 the commission handed down *The First Report of The Harbor Lines Commission,* which contained maps of the harbor lines for Seattle and endorsed the concept of municipal ownership. This incited outrage among the private interests, and one man in particular, Judge Thomas Burke, set out to make sure that the commission's plans were derailed. Judge Burke was serving as the personal attorney for the Great Northern Railroad. His strategy was quite simple: he obtained an injunction to prevent the Harbor Lines Commission from formally filing the harbor lines plan for Seattle. Burke continued his legal maneuvers for over two years, and time eventually ran out for the commission. It passed ineffectively out of existence.

The state legislature then created a new commission; but this time private interests filled it with commissioners who agreed that *any* form of privately owned property was better than public ownership. Predictably, the new Harbor Lines Commission set out to redraw the maps produced by the first commission, and in the revised plan the harbor lines were placed just outside of the longest dock in Seattle. Drawn in this manner, the submerged lands inside the lines were classified as "first class tidelands," which could be sold by the state to private parties. This flagrant disregard of the public interest by the commission set off a new wave of unification among the public against the monopolistic private interests controlling Seattle's waterfront. During this same period, the political environment of the state was undergoing dramatic changes as the people sought to rid government of its inefficiency, bribery, and subordination of worker's welfare. In a popular tidal wave of revolt, the People's Party came to power with the 1896 election of a Populist governor and Populist majority in both houses of the state legislature. Although the Populists remained in power for only one term, they laid the groundwork for the formation of a public port authority by strengthening the state's constitutional provisions supporting the right of public access on the waterfront.

Two city engineers were instrumental in publicizing the congestion on the waterfront and the dilapidation of slowly rotting maritime facilities. R. H. Thompson and George Cotterill appeared repeatedly before civic groups to present the findings of their research on European harbor development and administration. Based on their reports, the citizens of Seattle came to a very progressive conclusion for their time:

> . . . the only feasible solution was the creation of a new "political division" entirely independent of state, county, or city organization, devoted to the single purpose of port development and administration. Once created, this new political entity would then need the power to tax people within its jurisdiction, issue bonds, and be given the right to acquire land by eminent domain (Port of Seattle 1981).

The Port District Act of 1911 was exactly the legislation that supporters of a public port had been seeking. The act granted permission to any locality in the state to form a port district and outlined the powers and rights of the district once it was created. In September 1911 the voters of King County approved by a 3-to-1 margin the county proposition to organize a port district. At the same time, three commissioners were elected for one-, two-, and three-year terms on the newly established port commission. As *Port in a Storm* states:

> The machinery of self help had been placed at the people's command: the Port District Act had granted the new Port of Seattle the right to exist, giving it the power to carry out its goals and objectives. But Seattle was to find out that there was a big difference between laws on the books and actual publicly owned port facilities (Port of Seattle 1981).

Only the wise selection of the first port commissioners, who would nurse and protect the infant authority from destruction at the hands of the private interests, could ensure that the unified efforts of the public would not go wasted.

Fortunately for the Port of Seattle, it was led through its first crucial years by three dedicated men. General Hiram Chittenden, an engineer, proved to be a very valuable asset to the commission when complexities of harbor development arose, and his unfailing integrity earned him a strong reputation within the community. Robert Bridges, a former coal miner, had become well known for chasing down corporate tax evaders and those who had taken land without paying for it, making him a prime enemy of powerful private interests. His philosophy brought independent thinking to the commission as well as a determination to uphold the public interest. Charles Remsberg, a banker and attorney who was initially assumed to be willing to compromise with private interests in the redevelopment of the waterfront, soon proved beyond a doubt his commitment to the idea of municipal ownership.

The commissioners decided that at first it would be unwise to tackle the private interests head-on. They opted instead to locate the first public facilities away from the central waterfront. Soon construction was underway on a variety of public waterfront projects, including new slips, warehouses, and the Smith Cove Terminal, which, when completed, was the longest pier (one-half mile) on the Pacific Coast.

In 1915 the commission commenced its direct attack on the railroad monopoly by announcing that it was considering claiming all the tidelands in Seattle, a move that would invalidate waterfront landowners' claims and return the land to the public. Around this same time, the commission also unveiled plans for the construction of a public cold storage facility, available to local fishermen for a nominal fee. The only existing cold storage facilities were privately controlled and charged exorbitant fees. In yet another attempt to undermine the private facilities, the Port announced that it was lowering dock and wharf charges at public facilities, thereby making the Port of Seattle the least expensive port on the coast. In a futile attempt to save their businesses, railroad representatives and dock owners responded by taking the Port to court, but the lawsuits were struck down and the commission was on its way to breaking the private stranglehold on the central waterfront.

The commissioners' hard work began to pay off in 1916 when the Port's busines skyrocketed. By 1918 Seattle had catapulted into the position of second largest port in the United States. This success was the best possible tonic for the criticism that had hounded the commission since its inception. By 1918 even the newspapers and the most skeptical businesspeople had joined the community in praising the commissioners for their staunch advocacy of the necessity of a well-equipped, diversified port.

Since that time, the Port of Seattle has gone through a series of changes. Following the boom period of World War I, the previously hostile private interests rallied behind the Port. During the 1920s this unity proved to be extremely profitable to the merchants and Port alike, for the Pacific Asian market was just beginning to come into its own and the Port was able to capture 40 percent of the Japanese trade alone. Unfortunately, this prosperity and the maritime commerce of the entire country came to a screeching halt during the years of the Great Depression. The Port of Seattle managed to survive these lean years, however, and when World War II broke out, the Port once again burst into activity. When the war business died down, the Port of Seattle found itself with a drop in operations and a harbor full of deteriorating facilities. The 1950s became a period of rebuilding, but it took citizen interest to pull the Port of Seattle out of the doldrums and inaugurate a new phase in its history.

Voters passed a set of measures that revamped the commission and provided it with a $10-million bond issue to finance the rebuilding of port facilities. The new commission embarked on speculative investment in container facilities in the early 1960s, giving the Port of Seattle a head start on its competitors and initiating a progressive management style still followed today.

## THE PURSUIT OF ECONOMIC OPPORTUNITIES

For the most part, the success of a public port rests on the port authority's ability to recognize and respond to economic opportunities. Intense competition forces port authorities to become better port managers by adopting a deterministic business philosophy and increasing the responsiveness of the management organization. This is accomplished through aggressive marketing campaigns, pricing strategies, special financing arrangements, and custom-designed facilities—each intended to anticipate or create new economic opportunities.

The final case study, Port Canaveral, looks at how a port authority can use its management "tools" to take advantage of new or anticipated economic opportunities. Port Canaveral is a relatively new entity created in the 1950s, well after the public reform movement. Its activities reflect a managerial flexibility that was absent in the old regime of privately owned and operated ports. Whereas much of Port Canaveral's success is based on being "in the right place at the right time," the responsiveness of the port is the result of careful planning and preparation by the port authority. Interestingly, Port Canaveral is not a container port: its growth, a 640-percent increase in operating revenues from 1977 to 1985, can instead be attributed to the development of facilities designed to meet the specialized needs of its customers.

## *Port Canaveral*

Port Canaveral officially opened for business in 1953, but its history dates back to 1878 when the Navy Department went before Congress to recommend that a deep-water harbor be constructed on Cape Canaveral. The Navy felt that it would be important to have a deep-water port centrally located on the eastern coast of Florida. However, Congress did not share this viewpoint and rejected the proposal on the grounds that it lacked "economic importance" (Cocoa Tribune 1953). Six subsequent proposals over the next 61 years were also rejected by Congress. Finally, in 1939 a U.S. district engineer in Jacksonville was convinced at a local hearing that there was reasonable "justification" for a port, and by 1941 the U.S. Army Corps of Engineers (COE) had recommended to Congress that an appropriation of $1,661,000 be made to construct the port (Cocoa Tribune 1953). Congress partially approved the recommendation in 1945 as part of its rivers and harbors bill for postwar construction, inserting stipulations that local interests construct certain facilities at the port and also match a federal appropriation of $830,000 for dredging. In 1947 the voters of the Brevard County Canaveral Port District enthusiastically approved a bond issue to raise the funds necessary to meet the government's demands, and by October of 1949 the Canaveral Port Authority had sold almost $1.4 million in development bonds (Cocoa Tribune 1953).

The site chosen for the new port was a parcel of land on the barren peninsula separating the long sandy beaches of the Atlantic Ocean from the marshy banks of the Banana River. The nearest town, Cocoa Beach, was located several miles down the coast. Unfortunately, the area was devoid of a natural harbor or bay, and so before terminals could be built, the COE had to complete the enormous task of cutting away a large section of land to create a harbor area. This basin was constructed by dredging a channel from the Banana River eastward to the Atlantic, giving Port Canaveral one of the only harbors ever dug from the inside out.

The dredging of the harbor was completed in 1951, and soon thereafter local fishing companies from Cocoa Beach erected processing facilities and an 800-foot temporary wharf for the use of fishing and charter boats. The port was officially dedicated on November 4, 1953, consisting of the fishing companies, a 400-foot general purpose marginal wharf, and a crash boat station for the U.S. Air Force Missile Test Center located just north of the port (Canaveral Port Authority 1984).

When Congress initially approved the construction of the harbor, Port Canaveral was to be a transfer center for the export of fresh citrus fruit grown in central Florida. By the time the port opened in 1953, however, the waterborne export of fresh citrus fruit had practically disappeared as a direct result of improvements in the trucking industry and the advent of citrus concentrate (Cocoa Tribune 1953). Without the fresh fruit export business, the new port had to rely on the local fishing industry for support. It was not until 1955 that the port got its first commercial cargo, a shipload of cement from Europe (Canaveral Port Authority 1984). The first few years were financially difficult, and the members of the port authority sometimes used their own financial resources to keep up interest payments on port bonds.

This dedication to the continued operation of the port gave Port Canaveral an edge over its competitors. During the 1950s Brevard County, where Port Canaveral is located, experienced a 370-percent increase in population. In the late 1950s Port Canaveral began to expand: four areas in particular—housing, fishing, the military, and tourism—have created a succession of growth opportunities for the Port Canaveral Authority.

Florida (in the late 1980s) has the fastest growing population of any state in the nation, with the central part of the state forecasted to grow at a much higher-than-average pace for several years to come. Port Canaveral's central location on the coastline makes it a natural point through which building materials for the burgeoning construction industry can flow. The port authority has aggressively advertised the benefits of the port to the construction industry, and as a result has been able to tap into the flow of dollars from Florida's housing boom. Once customers are attracted, the port authority continues to work with them to meet their specialized needs. For example, the Port erected silos for storing imported concrete and expanded general cargo terminals to service ships offloading gypsum board and lumber.

The region's fishing industry has also experienced tremendous growth, primarily in response to an increased demand for seafood products by the growing population, technological innovation in the industry, and the discovery of new fishing grounds. Before the creation of Port Canaveral, the main fishing industry in the area was headquartered in Cocoa Beach, where locally caught jumbo shrimp were processed (Cocoa Tribune 1953). The dredging of Port Canaveral's harbor created a protected homeport for the shrimping fleet and gave the seafood processors an opportunity to locate their facilities near the shrimping grounds. As soon as the harbor was opened, the port authority leased several parcels of land to seafood companies for the construction of processing facilities and boat moorage. This quick action to secure tenants soon provided revenues from leases that enabled the Port to survive the first few years of development.

Within the last decade, a new high-value item, the calico scallop, has become the focus of attention for Port Canaveral's seafood industry. Large stocks of these scallops, found in the deep, cool waters off Cape Canaveral, have combined with recent developments in shellfish processing techniques to make the large-scale processing of these small scallops economically worthwhile. The scallop industry located at the port has boomed in the last few years, experiencing an impressive tenfold increase in production since 1979 to reach a 1985 wholesale value of over $70 million (Canaveral Port Authority 1986).

The Port itself has been able to capitalize on the success of the scallop fishery by providing homeporting facilities for over fifty scallop boats and leasing waterfront sites to three major scallop processors. Attracting processors that would build local plants has been the key to the Port's success in the seafood industry; yet concentrating the plants within an enclosed harbor poses the problem of water pollution caused by processing discharges. The port authority was concerned with the disposal of unused scallop parts after the shellfish were processed (see Fig. 2.4), but fortunately a clever and unique solution to this problem was developed. The Port struck a deal with the scallop processors to buy scallop shells at a rate that is about one quarter of the cost that the processors would have to pay to dispose of the shells at a county landfill several miles away (Jones, 1985). By using the crushed shells for roadbeds, building foundations, and parking lots, thereby transforming discarded scallop shells into a valuable commodity, the port authority annually saves thousands of dollars in landfill costs (Canaveral Port Authority 1985).

Another relationship valuable to Port Canaveral is its longstanding tie to the military, dating back to the late 1800s when the U.S. Navy first called for the construction of the port. The first official military operations began at the port in 1954 with the arrival of the Military Sealift Command (MSC), now located on U.S. Air Force property on the north side of the harbor (Canaveral Port Authority 1984). Several different military services have since located some portion of their operations at the port, primarily because of its strategic location and proximity to the Caribbean. The port also serves as a support center for ballistic missile weapons systems testing and is the homeport for all Trident and NATO submarines during their shakedown cruises. Located just south of the Kennedy Space Center, Port Canaveral also provides homeport facilities for the missile tracking ships used by NASA as part of the space program.

Encouraging the military to locate within the harbor has brought the Port additional revenue when excess military cargo is loaded over Port docks. More importantly, the federal government maintains the depth of the harbor entrance channel at 40 feet for national security purposes. This ensures that the Port will always have deep-water access to its facilities. As in any good relationship, the benefits have been reciprocal. The Port has been able to return these favors to the military, such as the time when the port authority gave over 3,000 cubic yards of scallop shells to the Cape Canaveral Army Outport. This donation provided a Sea Bees reserve unit with valuable training experience while giving the Outport over $11,000 of free landfill to stabilize a cargo staging area (Port Canaveral Capsules 1985). Port leadership has also benefited from close association with the military: two MSC commanding officers later became chief executives of the Canaveral Port Authority after retiring from the military (Jones 1985).

**Figure 2.4.** An unusual enterprise of the Port of Canaveral is the purchase of scallop shells from local processors for use as construction material by the Port. This subsidizes the fishing industry, which would ordinarily have a high waste disposal cost. (Courtesy: Port of Canaveral)

Many of the Port's recent successes are attributable to tourism, specifically the cruise ship industry. Port Canaveral's entry into the tourism market began shortly after the opening of Walt Disney World in 1971. With this attraction 60 miles to the west and the Kennedy Space Center immediately to the north, the Port has been able to market itself as a stopover on cruises sailing out of the traditional cruise ports of Miami and Port Everglades. These stopovers soon became so popular that Port

Canaveral began aggressively pursuing the cruise industry, and these efforts paid off in 1982 with the opening of a new passenger terminal and the arrival of the first cruise ship based in Port Canaveral.

Within the past few years, Port Canaveral has turned into a tremendously successful cruise port (see Fig. 2.5) and is currently ranked fourth in the United States in terms of total revenue from cruise passengers (Canaveral Port Authority 1984). The Port plans to construct some twelve additional cruiseship terminals, and the lucrative cruise business now provides the majority of Port Canaveral's operations revenues.

One, area of tourism that usually does not occupy a place on the income statement of ports is public access. Port Canaveral is somewhat unusual in this area, for throughout the port people can fish or sun along the bulkheads, sometimes even coming within 100 feet of the cargo operations (Jones 1985). Three boat launching ramps, marinas, parks, camping areas, and fishing areas combine to give Port Canaveral's visitors extensive recreational opportunities.

Undoubtedly, much of the success of Port Canaveral can be attributed to good timing; but, of course, this is not a guarantee of success. It has been the creativity and responsiveness of the port authority that have enabled Port Canaveral to capitalize on economic opportunities and continue to increase its revenues.[2]

**Figure 2.5.** The Port of Canaveral's location just 60 miles from Disney World spawned a new cruise ship business, making it the fourth largest cruise ship port in the nation, measured by revenue. (Courtesy: Port of Canaveral)

## THE PORT'S ABILITY TO RESPOND AND ADAPT

The three cases studies—Houston/Milwaukee, Seattle, Canaveral—illustrate the three major preoccupations of the public port entity: physical and technological, political, economic. These forces are highly integrated and help explain why port development occurs. Physical facilities are built and rebuilt in response to economic, political, and technological changes. The political environment determines the timing, location, and type of facilities built and the economic interests they serve. The economic forces of the region provide the stimulus for political action and facility development at the port.

The case studies also show that over time the public port deals with these same sets of forces regardless of the stage of technological development, the type of economic stimuli, or the size and diversity of the political forces to which it must respond. Even before the evolution of the public port as an independent public enterprise, the port community responded in a collective fashion to advance common interests in the harbor.

As mentioned at the outset of this chapter, today's public ports face a variety of new environmental, social, and economic pressures, and as a result, the public port institution has had to adapt in important ways.

Ports have always been confronted with environmental challenges, but in the past they were caused by physical, not social, obstacles. Most of these problems dealt with making waterways safe, such as in Houston, where the waterway was clogged with snags and overgrowth, and Milwaukee, where the river was too shallow to allow the safe transit of deep-draft lake vessels. In Port Canaveral the environmental challenge consisted of creating and maintaining a deep-water port in a site that was entirely above ground. Now, environmental laws and social and political pressures to protect water and land resources require Ports to revise their goals. Demands for economic development must now be balanced with public concern for efficient use of existing facilities and the preservation of unexploited natural areas. Engineering and planning functions become more costly and time-consuming because there may be fewer options for development and new responsibilities for resource protection. Some Ports are acquiring and banking land for the purpose of meeting wildlife and environmental needs.

Along with environmental pressures, Ports today are being pressed by a new group of leisure industries demanding cruise ship terminals, tourism facilities, marinas, parks, fishing piers, boat launches, and other recreational opportunities. In some harbors, this new user group is quickly outpacing the traditional cargo-related and manufacturing sector. In Port Canaveral, the shrimping industry was the primary user of the port during the initial years of its operation. Over time, military and general shipping activity increased, and fishing was displaced as the primary user. Within the last five years, the cruise ship industry arrived and soon became the most prominent port customer. This change of users and their shifting importance have caused the port authority to repeatedly modify its development plans. In addition to a plan for many new cruise ship terminals, the Port has incorporated parks, marinas, and fishing areas into the port layout to answer the public demands for access to the water. Port Canaveral is not alone; virtually every Port in the country is faced with demands for public access to the shore, even in industrial zones. They are responding to political pressure for open space and access, and to economic opportunities from leisure and tourism.

Ports in urbanized areas are diversifying at a rapid pace, planning "nontraditional" ventures such as office buildings, trade centers, retail/commercial activities, and so on. Many public ports got their start by meshing with nearby shore-based manufacturers who relied on marine transportation for receiving raw materials or shipping products. In an earlier day, the port was the break-of-bulk point, the assembly area for cargoes, and a cargo-processing center. The advent of containerization and the pressures from urbanization have radically altered the activities in many ports. Ports are now capital and information intensive. Their technical and financial resources are being called upon to help solve a wide range of transportation and urban problems. Seattle, for example, provides a sophisticated transportation services program, including truck consolidation and computerized cargo tracking, and in 1987 Port officials were exploring marine-related office buildings, inventory control services through satellite communications, and cruise ship terminal development and promotion. Urban pressures have pushed ports into "nontraditional" ventures as well. The Port Authority of New York/New Jersey has perhaps the longest list of nontraditional ventures, including a teleport, world trade center and training institute, industrial park for hydroponics, mass transit, tunnels and bridges, export trading company, garbage-to-energy facilities, and more (see Chapter 7).

In sum, environmental concern, public recreation, and nontraditional ventures are new issues for public ports. Yet they are strongly related to the past because, as we have seen, Ports have always been managing the physical environment and responding to the political and economic pressures that surround them. As these pressures change, the successful Ports respond and adapt as necessary.

A public port is charged with operating in the public interest. It must be attentive to new ideas and opportunities, and to changes in the community's values. Furthermore, a public port's organizational structure and operating procedures should encourage flexibility, for the organization must have the power, financial resources, and administrative mechanisms to act on new opportunities. At times the organization's structure will have to be changed to meet new challenges. For example, the Port of Seattle increased the number of commissioners from three to five in 1960 to revive the declining port, and the Southern California Ports added coastal management authorization to their list of organizational powers in 1976 in order to respond to environmental concerns. Ultimately, port commissioners and administrators must be community leaders who are willing to take initiatives that meet the new challenges in the harbor. The responsive Port serves its constituents well when it is aware of trends and needs, when it can operate flexibly to meet those needs, and when its leaders are motivated to take action to serve the public interest.

## *Notes*

1. Olson (1980a) notes that four key values provided the driving force behind the reform movement: maximizing economic efficiency, attaining popular participation and control, decentralism and localism, and increased competition. By introducing public control, it was anticipated that uniform access and standardized rate schedules, public entity credit worthiness, and coordinated development of Port facilities would transform waterfronts into safe and efficient districts of the city. There appear to be six particular circumstances that induced communities across the country to create public port entities: physical congestion of waterfront areas which blocked efficient operation of harbor commerce; private control of terminal facilities, which resulted in lost economic opportunities for local businesses; the inability of private corporations to replace insufficient or dilapidated facilities; eligibility of public entities for federal funds to undertake extensive capital development projects; ability of the community to develop comprehensive plans to maximize economic opportunities with public control of the harbor; inability of privately operated ports to compete effectively with their publicly owned and operated competitors.

2. Another factor contributing to Port Canaveral's success is its longshore force. Florida is a right-to-work state, and the Port has capitalized on this situation by having a nonunion stevedore handle all the cargo movement operations in the port. Under this system the cost savings to port customers have been substantial. This cost saving has been one of the major reasons that the port continues to attract new customers.

## *References*

Bard, Erwin W. 1942. *The Port of New York Authority*. New York: Columbia University Press, 352 pp.

Bird, James. 1971. *Seaports and seaport terminals*. London: Hutchinson University Library, pp. 66–74.

Bittner, Robin S. 1987. The public port institution in the United States: Using past experiences to understand modern issues. Master's thesis, Institute for Marine Studies, University of Washington, 95 pp.

Burke, Padraic. 1976. *A history of the Port of Seattle*. Seattle: Port of Seattle, 134 pp.

Canaveral Port Authority. 1984. Fischer's: Five decades and still growing. *Directions*. November/December, pp. 4, 13.

———. 1984. Military cargo on the move. *Local Scallop Industry No. 1*. September/October, 5–6, 11.

———. 1984. *Port Canaveral News*. February 7.

———. 1984. We've come a long way: Port Canaveral then and now. *Directions*. May/June, pp. 6–7.

———. 1985. Directions. In *Port Canaveral Capsules,* ed. Smith and Pennie. September/October, p. 4.

———. 1986. Directions. In *Port Canaveral Capsules,* ed. Smith and Pennie. May/June, pp. 4, 13.

———. 1986. *Port Canaveral News*. January 29.

*Cocoa Tribune*. 1953. Talk of a harbor at Canaveral began way back in 1878. Cocoa Beach, Florida. November 3, p. 2.

Cuzan, A. G. 1983. The political economy of ports in the United States and Great Britain. In *Coastal Zone Management Journal* **11**(3):149–67.

Hamming, Edward. 1953. *The Port of Milwaukee,* Rock Island, IL: Augustana College Press, 162 pp.

Hershman, Marc J. et al. 1981. *Seattle's waterfront: The walker's guide to the history of Elliott Bay*. Seattle: Waterfront Awareness, p. 50.

Hershman, Marc J., Robert F. Goodwin et al. 1978. *Under new management: Port growth and emerging coastal management programs*. Seattle: Washington Sea Grant, pp. 1–22.

Hoyle, B. S., and D. Hilling. 1984. Seaports and development strategies. In *Seaport systems and spatial change*. New York: John Wiley & Sons. pp. 461–69.

Jones, Pat. 1985. Report on trip to Port Canaveral. For Institute for Marine Studies, University of Washington, January 8. Unpublished.

Maritime Administration. 1980. National port assessment 1980/1990: An analysis of future U.S. port requirements. Office of Port and Intramodal Development, Department of Commerce.

Olson, David J. 1980a. California ports: Observations on their historical origins and evolution; relations with parent governments; internal structural characteristics; profitability and capital construction; and planning for facilities development. Unpublished.

———. 1980b. Regionalism in port development: The Pacific Northwest experience. Paper delivered at Regional Port Institutions Seminar, Boston, August 4–5. Unpublished.

Port of Houston Authority. 1986. 1837–1850: Despite growing pains, Houston becomes center of trade. *Port of Houston Magazine,* February 20–21.

———. 1986. The birth of a new port. January, 28–29.

Port of Seattle. 1981. *Port in a storm: A historical review of the founding of the Port of Seattle.*

Schenker, Eric. 1967. *The Port of Milwaukee*. Madison: University of Wisconsin Press, 212 pp.

Schwimmer, M. J., and Paul A. Amundsen. 1973. Management of a seaport. Report #MA-RD-900-75025, Kings Point, NY: National Maritime Research Center, p. 162.

Sibley, Marilyn M. 1968. *The Port of Houston: A history*. Austin: University of Texas Press, pp. 12–59.

Still, Bayrd. 1948. *The story of Milwaukee*. Madison: State Historical Society of Wisconsin, 638 pp.

Zullo, Don N. ed. 1983. *1983/84 Houston port handbook and industrial guide,* Gulf Commerce International, Inc., pp. 9–13.

## *APPENDIX: SELECTED HISTORICAL BIBLIOGRAPHY OF U.S. PORT AUTHORITIES*

Albion, Robert G. 1963. *Naval & maritime history: An annotated bibliography*. Mystic, CT: Marine Historical Association.

Amundsen, P. A., J. S. Smith, and H. Bentley, eds. 1961. *Ports of the Americas: History and development*. Washington, DC: American Association of Port Authorities.

Babcock, Robert H. 1982. The rise and fall of Portland's waterfront: 1850–1920. In *Maine Historical Society Quarterly* **2**(22):63–98.

Bard, Erwin W. 1942. *The Port of New York Authority*. New York: Columbia University Press, 352 pp.

Barker, Thomas T. Jr. 1979. Partners in progress: The Galveston Wharf Company and the City of Galveston, 1900–1930. Ph.D. diss. Texas A & M University, 206 pp.

Beck, Bill. 1984. Seaway Port Authority of Duluth: The early years. In *Minnesota's World Port* **19**:2. *See also* The early years—Part 2, *Minnesota's World Port,* **19**:3.

Binns, Archie. 1941. *Northwest gateway: Story of the Port of Seattle*. Portland, OR: Binford & Mort.

Bird, Frederick L. 1949. *A study of the Port of New York*. New York: Dunn & Bradstreet.

Buker, George E. 1983. Tampa's municipal wharves. *Tampa Bay History,* **2**(5):37–46.

Burke, Padraic. 1976. *A history of the Port of Seattle*. Seattle: Port of Seattle.

———. 1977. Struggle for public ownership: The early history of the Port of Seattle. In *Pacific Northwest Quarterly*.

Case, George Baxter. 1983. The history of the Port of Coos Bay, 1852–1952. Master's thesis, Pan American University.

Cass, Vera Cobb. 1933. The Port of Stockton, California: A study in geonomics. Master's thesis, Department of Geography, University of Washington.

Cellineri, Louis E. 1976. *Seaport dynamics: A regional perspective*. Lexington, MA: D.C. Heath & Co.

Georgia Ports Authority. 1984. Savannah—the city. In *1984 Georgia Ports Handbook*.

Hardeman, Nicholas P. 1981. Overland in cargo ships: The inland seaport of Stockton, California. In *Journal of the West,* **3**(20):75–85.

———. 1983. The Depression-born Port of Stockton, California. In *Pacific History,* **4**(27): 50–59.

Hickman, James H. 1972. The Port of Bellingham: 1920–1970. Center for Pacific Northwest Studies, Western Washington University (formerly Western State College).

Kilgore, Dan E. 1972. Corpus Christi: A quarter century of development, 1900–1925. In *Southwestern History Quarterly,* **4**(75):434–60.

Laurent, Jerome K. 1976. Sources of capital and expenditures for internal improvements in a developing region: The case of Wisconsin lake ports, 1836–1910. In *Explorations in Economic History,* **13**(2):179–201.

Little Rock Port Authority. 1983. History of the Little Rock Port Authority: 1959 to the Present. Unpublished report.

Magden, R., and A. Martinson. 1982. *The working waterfront: The story of Tacoma's ships and men*. The International Longshoremen's & Warehousemen's Union (I.L.W.U.) and The Washington Commission for the Humanities.

Morris, James. 1969. *The great port: A passage through New York*. New York: Oxford University Press.

Niagara Frontier Transportation Authority. (undated) The Port of Buffalo's History. In *Port of Buffalo: Handbook of the 80's*.

Nickell, Lehman Patton. 1928. History and development of the Port of Hampton Roads. Ph.D. diss. University of Virginia.

Pickens, William H. 1978. A marvel of nature: The harbor of harbors: Public policy and the development of the San Francisco Bay, 1846–1926. Ph.D. diss. University of California, Davis.

Platt, Harold L. 1979. Houston at the crossroads: The emergence of the urban center of the Southwest. In *Journal of the West,* **18**(3):51–61.

Port of Baltimore. 1983. History of the Port of Baltimore. In *Port of Baltimore Handbook, 1983–84.*

Port of Baltimore. 1981. 25 years of port progress: 1956–1981. In *Anniversary Issue, 1981–1982 Handbook.*

Port of Lewiston. 1980. History of the Port of Lewiston. In *Development Plan.*

Port of Los Angeles. 1983. From wilderness to world port. 75th Anniversary commemorative publication.

Port of New York Authority. 1967. A chronology of the Port of New York Authority: 1921–67. Public Affairs Dept., Port of New York, 21 pp.

———. 1967. The story of the Port of New York Authority.

Port of Portland. (undated). A brief history of the Port of Portland.

Port of Miami. 1985. Port of Miami: 1841–1985. In *Port of Miami 1985 Port Handbook.*

Port of Sacramento. 1984. Port plan: Port of Sacramento.

Port of San Diego. 1976. Port of San Diego: History and development.

Schenker, Eric. 1957. A port authority for the state of Florida. Ph.D. diss. University of Florida.

———. 1967. *The Port of Milwaukee.* Madison: University of Wisconsin Press, 212 pp.

Schweitzer, James P. 1984. Waterborne: Louisiana's ports & waterways. Louisiana Sea Grant, Center for Wetland Resources. Baton Rouge: Louisiana State University.

Sibley, Marilyn McAdams. 1968. *The Port of Houston: A history.* Austin, TX: University of Texas Press.

Stein, Abraham. 1980. The Port Authority of New York and New Jersey and the 1962 Path-World Trade Center Project. Ph.D. diss. New York University.

Stevenson, Shanna. 1982. Superior shipping service: A history of the Port of Olympia. Commemorative Publication on the Port of Olympia, Washington's 60th Anniversary.

Ventura County Historical Society. 1972. City of Port Hueneme Centennial—1872–1972.

Woodman, Lyman L. 1976. Nome Harbor. In *Alaska Journal,* **6**(4):199–209.

# Chapter 3

# THE PORT COMMUNITY:

## Seaport Character and Public/Private Tensions

*Douglas K. Fleming*

In every large American seaport, thousands of people work for hundreds of companies or agencies engaged in scores of different port-related services. These people belong to the *port community* and, most often, quietly mind their own businesses. The seaport is their bread and butter, their common interest.

Customs inspectors, immigration officials, port directors, and other public servants have a role in port affairs, but the main rhythm of U.S. seaport activity seems to have been set by private commercial enterprise. Profit seekers in a variety of middleman services figure prominently in our port histories. Thus the historian Albion (1939) remarked that the New York port community "grew rich from the profits, commissions, freights and other excuses for levying toll," whereas the novelist Katherine Anne Porter (1945) suggested that extraction of these tolls makes the typical port town "cynical by nature, shameless by experience, hardened to showing its seamiest side to strangers." There is no way of capturing the essential moods of the American seaport without reference to the ups and downs of the community's competing capitalists.

## A MICROGEOGRAPHY OF THE COMMUNITY

In the late 1960s British writer James Morris began his research for a book about the Port of New York with a helicopter ride to inspect the physical layout.[1] Neither the aerial picture nor the map can fully reveal the microgeography of a port community. One needs to go beneath the surface and inside the buildings to discover who locates where and who owns what.

Although there is a considerable body of literature on the internal geography of cities, only a few geographers have thoroughly dissected the seaport community to analyze the distribution of port-related services. Explanations of seaport morphologies have focused mainly on those highly visible trappings of the port, the waterfront industries and dock facilities.[2] Kenyon (1968) ranged a few city blocks inland to examine "land use admixture in the built-up urban waterfront." Little attention, however, has been given to the locational clustering of the various port services and their microgeographical shifts throughout the port city region.[3]

The obvious starting point for a port community survey *is* the waterfront, however. In the 1970s the U.S. Maritime Administration reported that more than 60 percent of the 2,401 major marine nonmilitary waterfront terminals in the United States were privately owned and operated.[4] Recently, though, the domain of the public port authority has expanded impressively, especially in the large container ports where each containership berth requires backup space ranging from 20 to 40 acres (depending on the method of container storage). The pattern of public and private land ownership per se tells only half the story. One must also consider the intended use and operational control of the waterfront property.

## Public Land in the Seaport

Constrained by Article 1, Section 9 of the Constitution of the United States, the federal government has been reluctant to meddle in commercial port affairs except for national security reasons.[5] However, federal government ownership of military waterfront installations and, significantly, operational control of many other port facilities have had much impact on port communities. The Military Sealift Command (MSC) and its predecessor, the Military Sea Transportation Service (MSTS), have had long-term terminal lease arrangements whereby they could control the logistics of great volumes of government-generated cargoes moving overseas. During the Korea and Vietnam conflicts, for example, Seattle's terminal 91 was almost exclusively dedicated to noncommercial shipments, albeit often accommodated by space-chartered or time-chartered privately owned vessels. During certain stretches of the 1950s and 1960s, MSTS cargoes constituted more than half of the Pacific Coast breakbulk cargo loadings for some U.S. steamship lines.[6] Considering the various attendant serices that all ships and cargoes need, this "unrecorded" port traffic had a very large impact on West Coast port communities. The government's intentions here, though essentially noncommercial, had strong commercial repercussions among the port service enterprises. The recent trend has been toward the divestiture of federal government interests in port terminals and warehouses, concomitant with declining military cargo shipments and, of course, diminishing impact on commercial port services.

As Olson discusses in Chapter 14 and in previous studies (Olson 1980), there are public port authorities at various government levels in the United States. In most large seaports the public port has been an aggressive waterfront land acquisitor. The port authority is usually empowered to "acquire land, property, property rights, leases or easements necessary for its purposes through purchase or by condemnation."[7] Evidence of the use of these powers can be found in the vast container terminals equipped with giant cranes and in the massive harbor alterations of modern U.S. seaports. Once again, however, the ownership factor can be deceptive. Just as the military has rented private terminals, so have private operators long-term leased public port terminals. Rather ironically, then, the enlargement of public port property holdings has sometimes been accompanied by an expansion of the private sector's operational control. This "landlord port" phenomenon, as Dowd (1981) pointed out, has significant implications.

Public sector interests in waterfront and near-waterfront properties are not restricted to military facilities and the public terminal operations. Many federal government agencies, such as the Coast Guard, Army Corps of Engineers, Customs and Immigration have traditionally sought locations close to the waterfront in seaport cities. The old New York Custom House near Battery Park, once heavily used by shipping enterprises, has been a prominent landmark near the southern tip of Manhattan.

Parks, esplanades, public fishing piers, historic maritime monuments, and the like reflect another important category of waterfront land use for which public ownership in U.S. seaports has been deemed appropriate.

## Private Land in the Seaport

The private sector also has various real estate interests at the waterfront and nearby. The individual homeowner may covet the marine vista, the waterfront shop proprietor has an eye on tourist spendings, and the large corporation calculates economic advantages in a coastal site. Admittedly, not all these attractions are available at all U.S. seaports.

Traditionally, the two major categories of private sector seaport users have been the transportation carriers and the commodity shippers. Little wonder then that steamship lines such as Sea-Land or grain shippers such as Cargill have controlled (usually via lease) significant stretches of the waterfront for their exclusive corporate purposes. What is more, these companies sometimes provide investment capital for essential seaport installations. Notably, bulk commodity terminals are often privately financed and owned, whereas expensive new container facilities are generally financed by bond issues and belong to public port authorities.

## Port Services

Among the vessel, crew, and cargo-related service enterprises essential to seaport operations, one finds patterns of location that range geographically from the waterfront inland, and from bottom floors upward in port city skyscrapers. Some of the patterns have an obvious spatial logic. Visiting ships have a set of specific needs such as berthing space, fuel, provisions, crew replacements, repairs, immigration formalities, and so forth. The steamship agent orders such requirements from the seaport's long list of service enterprises, some of which are government agencies but more are private companies. Traditionally, these vessel-oriented services have located at or close to the waterfront. Repair yards and fueling facilities, for example, *have* to be at the waterfront; but steamship agency offices can often afford to drift inland for proximity to important shippers, or to banks, or perhaps to join their own competitors in a favored office building.

Various cargo-related services must also be ordered for the visiting commercial vessel. Warehousing, stevedoring, freight forwarding, customs work, and cargo documentation are examples. Again, both private and public sector enterprises are involved and there are explanations for their office location patterns. The shore labor union halls are generally close to the water as are the stevedoring company offices. A waterfront location for large warehouses is customary in port cities. The linkage in function and shipping documentation between steamship agents, freight forwarders, customhouse brokers, the Customs House, and customs inspectors is often reflected in the patterns of office location. Proximity of one function to another is convenient: there are economies of agglomeration.

The clustering of service enterprises providing a like function, for example, freight forwarders, along one or two favored streets or in one or two office buildings is a curious example of camaraderie among competitors. In Houston in the late 1940s, the favored spot for steamship offices, freight forwarders, and customhouse brokers was the Cotton Exchange Building, miles from the Houston ship channel. Why? Houston's greatest export commodity, by value, was cotton. The cotton factors and shippers were on top of the spot cotton exchange and the steamship lines and freight forwarders wanted to be next to their favorite customers.

In the long course of seaport development, certain locational shifts are logical, if not inevitable. As the heavy focus of Seattle's port activity shifted south to Harbor Island, so too did a number of the ship chandlering and warehousing services. As the cargo ships abandoned Manhattan piers for Long Island and New Jersey, so too did *some* of the service agencies and enterprises. However, two important factors tend to perpetuate locational clusters of shipping services. One is the inertia created by sunk costs and habit. "The City" in London has been "the" location for a variety of shipping services for centuries, even though large ships no longer venture that far up the Thames. The other factor is the characteristic daily mobility of many of the service suppliers. Those dealing with documents, monies, and various ship's supplies can easily operate quite far afield from a centrally located office. Furthermore, modern communications and transportation expand their operational range. This means, sometimes, that they can afford to stay put in a cluster with their competitors. All the better for camaraderie, storytelling, and commercial espionage in the port community!

## INTERNAL FUNCTIONS OF A SEAPORT

The seaport is treated as a *place* in the opening section of this chapter and the internal geographic patterns of this place are reviewed. What then about the functions of the various seaport enterprises? What sorts of activities are wanted, needed, tolerated?

### *Commercial Shipping Enterprises*

It has been mentioned that the two main categories of commercial seaport customers are the transport carriers and the commodity shippers. It is not surprising, therefore, that most of the seaport service enterprises are designed to support these two sets of customers. That is to say, most of the services are ship-related, crew-related, or cargo-related. To demonstrate the involvement of various members of the port community, let us trace two ships, a liner and a tramp, through a Seattle port call.[8] Both vessels required services that are the steamship agent's business to order, coordinate, and supervise. What follows are modified port logs of visits of the *Tokyo Maru* (Fig. 3.1) and *Ming Autumn* (Fig. 3.2) to Seattle.

**M.V. Tokyo Maru—Voyage #22AB**

| | |
|---|---|
| Registry: | Japanese |
| Vessel type: | Container ship in liner service |
| Owner: | Mitsui O.S.K. Lines |
| Seattle agent: | Williams, Dimond & Co.[9] |
| Arrived from: | Tokyo, Japan |
| Departed for: | Vancouver, B.C. |
| Cargo: | Containers |

| | |
|---|---|
| 6/17/85 | Vessel departs Tokyo.<br>Mitsui O.S.K. telexes Williams, Dimond (W.D.) departure information and Seattle fueling arrangements.<br>W.D. sets up voyage file.<br>Cargo documents mailed from Japan. |
| 6/21 | Vessel telexes W.D. advising ETA Port Angeles and listing routine needs: light lists, ship supplies, cash advances, etc. for which W.D. makes appropriate arrangements. (W.D. has open account with owners to cover port expenses.) |
| 6/24 | W.D. notifies pilots of vessel's ETA Port Angeles.<br>W.D. notifies U.S. Coast Guard (USCG) of vessel's ETA, details of registry, tonnage measurements, berth to be occupied, etc.<br>W.D. notifies terminal operator, T-37, Seattle of ETA. (Terminal operator will provide utilities hookup.)<br>W.D. notifies stevedoring company (in this case the T-37 terminal lessee and operator) of cargo to be worked, gangs needed, shifts required.<br>W.D. orders tugs and linehandlers for vessel's arrival 6/26. |

**Figure 3.1.** *Tokyo Maru* at Pier 37, Seattle. (Courtesy: Mitsui O.S.K. Lines)

W.D. orders customhouse brokers to join boarding party, handle documentation and formalities for vessel entry and clearance through U.S. Customs, and to inform Customs and Immigration of vessel's ETA Seattle and berth assignment.

W.D. telexes owners the ETA, loading arrangements, and itinerary of vessel.

6/25 W.D. confirms bunker delivery arrangement with oil company for 6/26 fueling.

W.D. notifies U.S. Department of Agriculture (USDA) inspectors of ETA.

Stevedores order two longshore gangs for 6/26 day shift.

Stevedores order checkers and supercargo (for supervision of cargo handling in conjunction with owner's agent, chief officer of the vessel and stevedores).

6/26 0300: Vessel off pilot station, Port Angeles.

0710: Vessel docks T-37, Seattle.

0720: Boarding party: USDA, Customs, Immigration, customhouse broker, W.D. agent.

0800-0300 (on 6/27): Two longshore gangs working containers day and night shifts: total 32 gang hours.

0805-1400: Vessel bunkered from barge alongside: 1,101 metric tons.

**Figure 3.2.** *Ming Autumn* (Courtesy: Yangming Marine Transport Corp.)

1700: W.D. notifies tugs, linehandlers, USCG, U.S. and Canadian pilots of vessel's 6/27 ETD. (Both pilots necessary for Seattle-Vancouver run.)

1700: W.D. telexes Mitsui O.S.K. vessel arrival details. Customhouse broker enters and clears (for Vancouver, B.C.) vessel simultaneously at U.S. Customs.

6/27

0300-0800: No work ordered for "hoot-owl" shift.

0800-1115: Two longshore gangs work part of day shift (16 gang hours paid for) to complete loading. Total containers handled on 6/26–27: 586 discharged, 250 loaded, 2 rehandled.

1300: Vessel sails for Vancouver. (U.S. pilot returns by taxi!)

6/17 onward

Williams, Dimond and their appointed customhouse broker have provided ongoing *cargo-related* services for the owner. Import cargo documents (mailed from Japan) and import declarations are filed with Customs on vessel's entry. W.D. not only performs cargo booking, tracing, and expediting services for owner, but also prepares export bills of lading, manifests, stowage plan, etc. Export documents including export declarations must be filed with Customs within a week of vessel's departure.

**M.V. Ming Autumn—Voyage #37W**

| | |
|---|---|
| Registry: | Taiwanese |
| Vessel type: | Bulk carrier in tramp service |
| Owner: | Yangming Marine Transport Corp. |
| Seattle agent: | Transmarine Navigation Corp.[10] |
| Arrived from: | Los Angeles |
| Departed for: | Kaohsiung, Taiwan |
| Cargo: | Soybeans in bulk |

| | | |
|---|---|---|
| 9/29/84 | Solar International (owner's general U.S. agent) telexes Transmarine Navigation (T.N.) appointing T.N. owner's agent, Seattle, for Voyage 37W. Solar telexes T.N. main terms of grain voyage charter, Tradax (Geneva) charterer, Cargill supplier, expected loading Pier 86, Seattle; T.N. also appointed charterer's agent. | |
| 10/1 | Cargill confirms vessel will load Pier 86, Seattle. | |
| 10/2 | T.N. telexes Solar estimated port charges requesting advance funds concurrent with vessel arrival from California. | |
| 10/3 | Solar advises vessel ETA Port Angeles PM 10/10. | |
| 10/3–10 | Periodic information exchange between vessel, Solar, T.N., and Cargill regarding vessel's ETA. | |
| 10/4 | T.N. submits six-day notice and grain berth application to Cargill. | |
| 10/5 | Based on soybean stowage factor, Master of vessel advises T.N. estimated load 26,600 metric tons. T.N. advises Cargill and Solar. | |
| 10/7 | Solar advises T.N. time and details vessel departure Los Angeles. | |
| 10/8 | T.N. telexes vessel giving entry instructions and orders proceed Seattle anchorage pending berth availability. T.N. notifies pilots, USCG, USDA, National Cargo Bureau (NCB), stevedores, supercargo, customhouse broker, and launch service (to meet vessel at Seattle anchorage) of vessel's ETA. | |
| 10/9 | T.N. advises all parties of adjusted ETA. | |
| 10/10 | 2220: | Vessel off pilot station, Port Angeles. |
| 10/11 | 0310: | Vessel anchors Elliott Bay, Seattle. |
| | 0830: | Boarding party: T.N., USDA, NCB, customhouse broker, cargo superintendent. |
| | 0930: | USDA, NCB inspectors clear all holds except #4 which needs fumigation. T.N. orders fumigation and tug to bring equipment alongside. Based on draft limitations at expected discharge port, master declares tonnage allocation each hold. Stevedores and Cargill advised. |
| 10/12 | 0910: | Fumigation completed. |
| | 1500: | Hold #4 reinspected and cleared by USDA/NCB. |

1500: Notice of readiness tendered by T.N. and accepted by Cargill. Principals advised.

10/13–18 Vessel idle awaiting grain berth. No bunkers, stores, repairs, laundry, immigration, or consular services needed. T.N. arranges medical attention for one crew member. T.N. orders and delivers requested new charts, tide tables, light lists, etc.

10/17 Cargill advises berth ready 10/18. T.N. orders pilot, tugs, linehandlers, and notifies USCG of expected shift 10/18.

10/18 0518-0636: Shift to Pier 86 (see Fig. 3.2).

0800-1130: Two longshore gangs pouring soybeans in #2 and #4 holds.

1110: Solar, on behalf of owners, refuse to order/pay for Saturday (10/20) overtime work.

1130-1700: Emergency longshore stop work because of longshore death on waterfront.

1700-0800 (on 10/19): Vessel idle.

10/19 0800-1700: Two longshore gangs alternately working all five holds. Customhouse broker clears vessel for Taiwan. Cargill advises will order Saturday work for own account (to expedite vessel). T.N. orders pilot, tugs, linehandlers, and notifies NCB, USCG of 10/20 ETD.

1700-0800 (on 10/20): Vessel idle.

10/20 0800-1630: Two longshore gangs alternately working #2, 3, 4, 5 holds. Loading completed 1630. Total cargo laden: 26,600 metric tons soybeans.

1545: Pilot on board.

1600: NCB issues certificate of loading.

1711: Vessel sails for Kaohsiung (ETA 11/8).

1800: T.N. telexes departure details to Solar and Tradax.

10/22–23 Telexes between T.N., Solar, Tradax, and Cargill regarding freight payments and release bills of lading.

10/24 Customhouse broker files export declaration and other documents with customs.

10/29 T.N. dispatches shipping documents (bills of lading, manifests, stowage plan, statement of facts, etc.) to Yangming Marine Transport in Taipei via air courier.

There were few unanticipated vessel needs or voyage disruptions in these two port calls of the *Tokyo Maru* and the *Ming Autumn*. Even so, quite a large segment of the port community was actively involved in getting these ships in and out of the port, not to mention attending to the personal needs of officers and crews while ashore. In a large seaport, the carriers and shippers usually find a long list of individual firms that offer *each* type of port-related service—many steamship agencies, many customhouse brokers, many ship chandlers, and so on. In the past several decades there have been important advances in maritime transport technology and in cargo-handling techniques. Yet a vessel port call still requires the same sorts of basic services that still provide thousands of jobs in the large seaport community.[11]

Relatively minor schedule changes, breakdowns and "mistakes" can have reverberative effects among carriers, shippers, and their attendant services. Fleming (1974) noted somewhat cynically that a liner voyage is often marked by "false expectations, broken promises, minor catastrophes and both pleasant and unpleasant surprises," most of this unintentional. If there are major accidents, damages, overbookings, expired letters of credit, crew injuries, and such, many legal, insurance, repair, banking, consular, and other services may need to be called into action. At large seaports the banks, law firms, and insurance companies need employees who are maritime specialists, and, of course, they too belong to the port community.

## *Waterfront Amenities*

The human needs of those who inhabit, work in, or pass through the waterfront district have barely been touched on here. In each big seaport thousands working in maritime services need food, shelter, recreation, meeting places, union halls, and so forth. Crews ashore, as mentioned, have similar requirements, so that seafarers' restaurants, taverns, and assorted "dens of iniquity" become very much part of the historical seaport scene.[12] The barroom, of course, is a favorite mariners' meeting place, in fiction and in fact. In the 1940s, an old friend, then an agent for Irish Shipping, Ltd. in New York, used to collect crew members who had strayed before ship's sailing by sweeping through four or five popular bars not far from the Hudson River terminal. His bait was a bottle of Irish whiskey, which he waved before them and opened only when the stragglers were safely aboard. In Galveston, Texas, in the late 1940s any sailor or dockworker could purchase for a dollar a lifetime membership in one of many "private clubs" where hard liquor was "legal." This was part of Galveston's seaport culture as were the frivolities on Post Office Street.

Most seaport cities have waterfront parks, paths, public piers, esplanades, benches, historic monuments, and maritime exhibits to serve public needs for recreation, relaxation, esthetics, and preservation of the maritime heritage. The users of these lands and facilities, which are usually public property, include citizens and tourists, workers and loiterers, seamen and landlubbers, Americans and foreigners, all contributing to the character of the seaport scene. The private sector, which gets more publicity for its polluting activities, can attend also to the public's cultural needs. Corporations have contributed to historical preservation, to waterfront beautification, and so on. Isbrandtsen Lines used to offer free Tuesday noontime band concerts on Bowling Green opposite the New York Custom House.

Many U.S. seaports now cater heavily to tourists' needs and inclinations which, of course, are served by the public amenities mentioned above. In addition, a whole array of commercial services for passers-by flourishes "on the docks." For example, on Seattle's downtown waterfront where piers are now too short, small, and antiquated for modern shipping, all sorts of sea-flavored restaurants, bars, and shops are found. Such enterprises serve foreign and out-of-town visitors, but are also heavily frequented by local citizens who are "micro-tourists" on lunch breaks or sometimes absent-without-leave from their usual daily occupations.

A recreational feature, which is ever more obvious at U.S. seaports, is the small-boat marina, discussed by Goodwin in Chapter 9. In the ports that harbor fishing fleets, commercial fishing combines with recreational boating to support a variety of small-boat building, repairing, supplying, chartering, and sheltering services. In several Washington and Alaska ports, for instance, this is a very important source of jobs and incomes, and certainly a striking element of the seaport's character.

In most large city ports there is a "lived-in" portion of the waterfront district. Residents, in turn, require nearby shops, laundries, and various household services. In many cases, today's favored residential neighborhood was a blighted urban neighborhood of the recent past, as Kenyon (1968) explained in detail. Some of these properties have been restored to their historic splendor, as in Charleston and Savannah, some modernized with marine vista apartment complexes, as along stretches of the Mississippi River behind the levee in New Orleans. In some cities the perceived need is only for a serviceable dock for houseboat moorage, as on Lake Union in Seattle, or for the Queen Mary Hotel, a former luxury liner permanently moored at Long Beach, California. These waterfront residents, certainly part of the broader port community, usually prefer unobstructed marine vistas, reasonably clear water, and protection from noise pollution and from excessive commercial traffic.

## TENSIONS IN THE PORT COMMUNITY

The residential and other human needs of the city dweller on one side, and the space and access requirements of the commercial seaport users on the other present an eternal problem of compatibility, a challenge to seaport planners and to city governors.

Kenyon (1968) studied the morphology of waterfront districts in several southeastern U.S. ports and called particular attention to the conflicts between commerce and esthetics, between economic advantages and social costs, and among residential, recreational, and commercial land uses. There is also a possible conflict between private and public sector control. The public port authority may be tempted to exercise its land-acquisition powers simply as a precautionary measure to ensure adequate expansion room for commercial site development.

## *The Port Authority in Fact and in Fiction*

The port community in U.S. seaports is always larger, older, and more diverse than the public port authority that is supposed to serve its interests. When port commissioners are elected by the public, when the Port has tax-levying, bond-floating, and land-acquiring powers, the port institution is accountable for its activities, as Olson notes in Chapter 14. This implies that the port authority should mirror the entire spectrum of port community interests. Ideally, then, the multifaceted personality of the seaport would be reflected by its port authority. Whether or not individual public ports conform to this model is an interesting question, in fact, an issue of considerable political significance.

Since the port authority is an organizational entity and is staffed by personnel, directors, and commissioners who are empowered decision makers, this component group of the port community is very important and influential. In the "landlord port" era, lawyers and real estate specialists are in demand for Port staff positions. The "old sea-dog" types who used to manage the terminals seem to have vanished, perhaps retired, perhaps gone to the private sector where their experience handling ships and cargoes is valued. There is danger here of a dichotomy of interests and expertise. The public port authority has no business developing secret or separate goals or developing a personality that diverges radically from that of the larger port community. If the port authority *leads,* as it is often expected to do, it should presumably be along lines and toward goals that have been generally endorsed by tax-paying port district constituents. It would be unfair to point to U.S. seaports that have been unsuccessful in handling these internal tensions between public and private sector interests. All seaports have had their ups and downs. And an "up" commercially can mean a "down" in certain other ways. The innate conflicts of interest remain. These are what the port authority is supposed to understand and mediate.

## *An Information Gap in Publicizing the Port*

The public port's facilities and work projects, as mentioned, are usually very much in the public eye. The port authority's headquarters may be in a huge World Trade Center, as in New York, or a well-marked waterfront terminal, as in Seattle. In addition to these physical features, public port developments make news. The Port's marketing efforts, its sister cities, and its visiting delegations from abroad receive media attention.

The port community usually has several voices, not always harmonized. One of them is the port authority's, or more specifically, its public information (or public relations) department, which composes the brochures, harbor maps, facilities handbooks, annual reports, and periodical magazines used for informational and *promotional* purposes. Many of these materials are useful, informative, and beautifully crafted, but there are times when the advertised image does not accord perfectly with the real thing. There may be no intention to deceive, but little effort is made to offer a clear picture of the private sector's role in port activities, or to offer an objective appraisal of the real state of affairs in the Port's competitive struggle, especially when the edge has been lost. It is difficult, of course, to inform fully and promote forcefully at the same time, unless the Port is clearly superior to all rivals.

The local newspapers, radio, and television sporadically touch on most aspects of the seaport's activities, although generally only the major issues are made into news items. Seldom is there full coverage of the multitudes of enterprises quietly minding their own businesses, although many of the daily newspapers' marine editors are well-informed and influential figures in the port community and know what is going on whether or not they consider it printable.

Finally, there are the more specialized maritime periodicals, the freight market forecasts and analyses, the shipping information services, the government agency maritime statistics, and the daily journals of commerce that circulate in the port community and offer a detailed picture of shipping and seaport activities. This is not the general public's favorite reading material.

One consequence of this high visibility of the public port and comparatively low visibility of the private sector is an information gap. Many citizens who are neither active members of the seaport community nor students of it are apt to think that the Port (a public enterprise) is the port. The former, however, never quite encompasses the scope, variety, and history of the latter. By nature and supposed function, both the public port and the communications media ought to reflect the diverse interests of the multifaceted port community. Sometimes they do not. Of course, it is difficult to write an exciting story about bills of lading or take interesting pictures of customhouse brokerage. The port logs shown above suggest various essentials of the port operations that simply are not considered particularly newsworthy. It is the oil spills, the bitter port rivalries, the statistical breakthroughs (a million TEUs!), and the politics of port governance that make better press.

In examining the needs and preferences of the various port users and those who serve them, many potential conflicts of interest become apparent. The waterfront has many conceivable and, from alternative perspectives, desirable uses. The port community, as we have seen, contains so many component parts, persons, and enterprises that mingle geographically. Compromises are required with respect to interests and preferences. The clever planner discovers ways to design compromises. This is also one of the public port administrator's major challenges. The tensions cannot be eliminated. The need to promote cooperation and choose conflict resolution strategies is addressed in considerable detail in Chapter 1.

## CHARACTER AND SPIRIT OF THE COMMUNITY

The names of many port cities evoke salty images of ships and seafaring life: Albion (1939) noted that the names of Liverpool and Hamburg suggest the sea and little else.[13] Galveston, where the essential urban character is founded in its maritime orientation, is on a narrow island where one can almost always hear, see, smell, or feel the sea. And no one there has forgotten the story of its total inundation during the hurricane in 1900.

In many port cities, however, other urban aspects and functions have diluted the seaport flavor. Such is the case in New York or London, both of which have long and rich seaport histories. But today's visitor to either city seldom sees a ship or tastes salt in the air or smells creosoted piling. There still is a district in every port, however, where these waterfront sensations are experienced, and there is a community in which life revolves about the facts and fictions of ships, cargoes, and the sea. *Fiction* should be underlined, since one of the vital signs of the port community is the irrepressible flow of legend, gossip, and outrageous exaggeration. Early escapades on the San Francisco waterfront are legends, perhaps overblown, today, and there's a good deal of artistic license in the Hollywood image of cancan dancers in boisterous dance halls near the waterfront in the city by the Bay.

## The Determinants of Character

The character of a seaport comes from a complex combination of physical, human, and institutional elements, from the atmosphere these things create, from the stories told about them. Thus the legends of an inundation, an explosion, a bridge ramming, the beginning of a gold rush, and the inauguration of the celebrated Black Ball Line's transatlantic packet ship service are imbedded, respectively, in the remembered history of Galveston, Halifax, Tampa, Seattle, and New York.

The visitor's first and perhaps indelible impression of a seaport may come from the physical environment. The natural beauty of Vancouver's harbor entrance may be unforgettable, or the sight of Chateau Frontenac looming above Quebec, or the experience of the Mississippi Delta pilotage to New Orleans, or the industrial aroma along Houston's ship channel, or the welcome of San Francisco's magnificent Golden Gate, or *everyone's* image of skyscrapers rising out of New York's harbor.

In Chapter 4, Mayer describes various physical characteristics of harbors. Geomorphological classifications of natural harbor types have been used by other writers. Natural features can be fundamental elements that give character to seaports. Persistent atmospheric conditions, for instance, can define facets of the seaport's personality. Remember the heavy languor of Conrad's ports in the tropics and the rather bleak, gray, uncomforting aspects of Monsarrat's North Atlantic ports.[14] Regardless of people's backgrounds and dispositions, they tend to share a common impression of a port. It seems to be characteristic.

The human components of the port scene are more elusive, more susceptible to varied interpretation. Merchant princes, shipping magnates, labor leaders, politicians, philanthropists, gangsters, and even priests have left their mark on Boston's waterfront, and San Francisco's, and New York's. For instance, it was said that the

Catholic clergy had considerable influence on Boston's longshoremen in the 1940s and 1950s.[15] Despite the priests' good intentions, their advice to labor led to poor productivity on the waterfront and adverse commercial effects on the port. In this same era, gangsterism on the New York waterfront was not just a figment of the filmmaker's imagination. In 1947 the foreman who supervised the daily "shape-up" (an arbitrary choice of workers) at a Hudson River terminal was successor to a man, *so the story goes,* who was machine-gunned on his front doorstep. The feats and foibles of colorful local personalities are incorporated in each port's iconography. They survive in stories and pictures of the past, are heard or hung in the taverns, clubs, union halls and other meeting places of the port community. To be sure, there are different perspectives of the same incidents and different versions of the stories!

Typically the seaport acquires flavor from its strongest commercial linkages. Oil ports like Valdez, Alaska, or Baytown, Texas, have the look and smell of one-product towns. The character of several Oregon, Washington, and British Columbia lumber ports is emphatically shaped by their timber products hinterland and by the behavior of the controlling corporations. Large oil, steel, and wood-product corporations have, in fact, financed and molded industrial ports almost entirely to their own specifications. The port image is then part of the corporate image and can change as company policy and personnel change. Thus, Longview, Washington, used to be a Weyerhaeuser seaport; Freeport, Texas, a Dow Chemical port, and so forth. Sometimes a dominant steamship line sets the tone in a maritime community as Waterman once did in Mobile, Lykes in Tampa, and Alaska Steam in several Alaskan ports.

Overseas linkages also contribute to the seaport's personality. Thus, Seattle, a "gateway to the Orient," acquires touches of the Orient from the ships, people, cargoes, companies and communications associated with oriental trade. And Seattle's ties with Alaska, symbolized by the legendary ton of Klondike gold that arrived one day in 1897, are reflected now in many aspects of the seaport scene, not least the fantastic bargeloads of Alaska-bound prefabricated structures moving out during the summer. And, of course, no linkage was more obvious than New York's with Europe in the heyday of the great passenger liners, the splendid vessels that once dominated the New York harbor scene.

A seaport's character is also molded by its many commercial enterprises and institutions. Large U.S. seaports contain an impressive number and variety of businesses, agencies, associations, unions, and governments that belong to the port community. Many of these organizations share the general objective of getting ships in and out of the port, but they range widely in function and in the way they deal with the public. Some serve their customers and do their professional duties silently, caring little about their public image. Others care a lot, as their public relations efforts prove. Some terminals, for instance, are off limits to the public, maybe for good reason; others consider visitors' tours useful publicity.[16] The main point here is that the character of a seaport and the visitors' impressions of it can be shaped by institutional attitudes and behavior.

The public port authority is generally the most important single public institution of the port community. It becomes part of the image and it also shapes the image of the seaport. The role of the port authority in this regard can be a large and active one. An aggressive authority, as in Tacoma, Washington, in the 1980s, actively acquires new properties and customers and proposes further expansion, promising a competitive edge both to shippers and to carriers. The Port of Tacoma in this instance is an actor and a promoter in the port community and of the port community. In the performance of this role, the Port has done much to instill a spirit of progress, pride, and enthusiasm in the local community.

## A Community Spirit

Is it reasonable to depict the port community as part place, with an atmosphere, and part human, with a character, personality, and "spirit"? The seaport *is* usually part of an urban place and has definite geographical dimensions; and it also contains people, their maritime pursuits, and their stories, which have definite human dimensions. Consolidated, the place and people have a character that can sometimes be immediately sensed by the visitor, but that can also be gradually learned and carefully analyzed by those with a deeper interest. A community spirit can reflect more than local boosterism. It usually evolves from shared associations in a shared seaport environment. This spirit is not necessarily a self-generating, consistent phenomenon. It needs nurturing. It is something that port directors and commissioners may want to cultivate and "use." Austin Tobin and his port authority were trying to do this when they commissioned Morris to write "The Great Port, A Passage Through New York." Tobin, the port director, told Morris, "Many of us within the Authority have a great pride in our port and what it stands for. It has great romance and historical meaning, you see. It is very beautiful, too."[17]

## Defining the Common Purpose

At a high level of generalization, objectives usually can be found that will be endorsed in principle by most members of the port community. Thus, in the port authority's statements of purpose there may be reference to strengthening the community's economic base, enhancing the flow of commerce, maximizing the productive use of port facilities—all to be accomplished "within the constraints of good environmental planning."[18] Such declarations may seem innocuously noncommittal, but they constitute a platform from which more specific common principles, goals, and methods can be articulated. Typically, it becomes more difficult to reach consensus in the community as the levels of generalization are lowered. An enthusiastically endorsed general statement is a good starting point. To put it in the simplest terms, virtually everyone wants a healthy, dynamic, good-looking seaport, and most would endorse, in principle at least, good environmental planning.[19] Consensus even in very general terms gives a sense of cohesion to the community and reinforces its community spirit.

The esthetics of the seaport and its environmental quality are important to the seaport community on a personal, if not corporate, basis. There are laws, sensitivity to public relations, and sometimes a corporate "social conscience" that induce many companies to toe the popular line. There is no absolute incompatibility between commercial and esthetic needs. Everyone wants a good-looking port. Virtually everyone also wants to see some activity on the water. Gazing from Manhattan office windows in the 1940s and 1950s and watching the extinction of transatlantic passenger line service was a sad experience regardless of how other aspects of the New York waterfront may have improved. It eliminated a vibrant part of the seaport scene and might have permanently damaged the port community spirit had not the container revolution and (to a lesser extent) the cruise trade injected new vitality into the larger New York/New Jersey port complex. Manhattan, though, will never be quite the same.

The greatest difficulty in defining common purpose comes from the information gap, previously discussed. Often one part of the port community literally does not know what other parts are doing.[20] Private enterprises can carry "minding their own business" to extremes. Some participants don't *care* what the others are doing. Some don't *understand* what others are doing. There can be, then, both a communication failure and an education failure that contribute to the information gap. One subtle way in which the gap is widened is in the tendency to rely on gross statistics to describe the port community. If one could "disaggregate" the statistics in order to identify categories of enterprises and categories of purposes, statistics would tell a more specific story about the port. In this respect it could be argued that Morris' descriptive book about the New York port which identifies many of the enterprises and their aims is more valuable than New York's analytical economic impact study which masks many different types of port enterprises in statistical aggregations. Be that as it may, it is important to know who the actors are so that their needs can be incorporated in the port's "common purpose." It is especially important that the needs of the seaport's primary clientele—shippers and carriers—be recognized, since they have alternative geographic choices. They can use *other* ports.

## *Bringing the Community Together*

The finding of common ground and common purpose is one of the port authority's essential functions. There are many other actors in the port community who can help, in fact, *must* help if the information gap is to be closed.

Aggressive public ports have well-developed advertising and marketing strategies. The trade development offices at home and abroad should be prepared to promote the whole port community, not simply the public port component. Obviously, this again calls for ample information flow and coordination between the Port and the various enterprises within the seaport. "Promotion," if properly done, can bring the port community together. The danger, especially for "landlord ports," is that port authority officials either lose touch with details of ships and cargo operations, or are unable to "promote" since the private sector has not provided them the promotional tools.

Seaport promotion might be considered just another aspect of city boosterism in which all chambers of commerce indulge. It is a very familiar American pastime. Sister cities overseas perpetuate this sort of mini-patriotism and introduce a curious form of international bilateralism with possibly significant commercial and cultural effects. Actually this is a rather pleasant way of involving the public in city and seaport relationships, no matter how contrived it may seem at first blush.

Trade, transportation, and maritime clubs or associations provide a comfortable setting in seaport cities for port community gatherings. These organizations, sometimes simply a lunch club, sometimes a nationwide association like the Propeller Club, are forums for information exchange, which includes sea stories and commercial gossip.[21] Strong professional friendships and a spirit of port community cohesion are cultivated here. This is where public and private enterprise can sometimes find common ground and easy rapport.

The preservation of seaport heritage can also bring together those from different segments of the community who share a strong feeling for the past glories or charms of the seaport. The seaport needs historians and those who preserve its memorabilia. Quite often the latter will be tangible, visible symbols of the past. An old wharf in Boston, an old ship in San Francisco, a restored waterfront in Savannah, a cluster of totem poles in Sitka, or a spectacularly revamped statue in New York's harbor evoke rich historical images and enrich the seaport's spirit.

Maritime events provide occasions for speeches, conferences, celebrations, parades, and large assemblages from every part of the port community. National Maritime Week, Navy Day, and national conventions of various ship and port-related industries are serious occasions for finding common directions and purposes. And those occasions are likely to be well reported by the media. Along more frivolous lines, the pirates from Seattle's "Seafair" or the buccaneers from Tampa's celebrations behave in the same outrageous way every year. They have become part of those seaports' cultures. The origins of these antics are rather fuzzy, but no one really cares about that. It is a tradition that people share and enjoy.

A progressive public port can sometimes kindle a flame of pride and enthusiasm in the community. There *is* an element of boosterism in most seaport development programs. The massive container facility projects of the past few decades were accompanied by much publicity from the public port and the local press. Port rivalries were rekindled with the inevitable statistical illustrations. The air of progress was inescapable; and some of the esthetic needs of the port community were relegated to the back burner.

Growth per se is *not,* however, the sine qua non of port community spirit. Nor is the promotional advertising of the public port likely to convince or change the attitudes of professionals in the community who know more about their own port activities than the port authority does. The kind of spirit that holds a port community together comes from common experiences, from joint associations, and from stories told and retold. It touches the mind and the heart, and therefore can be a very personal thing. It is something that a person transferred to another location will not find elsewhere in duplicate form. It is something shared, although your experiences and mine are never quite the same.

## *Notes*

1. Morris (1969) was persuaded by Austin Tobin, then New York port director to visit, survey and write a book about "the universal seaport," New York.

2. See, for instance, Weigend (1956), Mayer (1957), Bird (1957), Verlaque (1975), Vigarie (1979), and especially the essays and illustrations in Hoyle and Pinder (1981).

3. Albion (1939) and Bird (1957) do, however, discuss the historical patterns of commodity exchanges and various commodity trading enterprises in New York and London, respectively.

4. These figures, mentioned in the Marcus (1976) book, came from a Department of Commerce report.

5. Marcus et al. (1976) emphasize the impact of this constitutional constraint on U.S. policymakers.

6. Waterman Steamship Corporation and States Marine Lines were examples.

7. This accords, for example, with legislation for public ports in the state of Washington as quoted in Port of Seattle's "purposes and objectives" listed in their 1980 Annual Report.

8. Actually, these illustrations understate the variety of arrangements that often need to be made. Various types of ship's supplies are needed from ship chandlers. Banking, insurance, freight forwarding and other services are needed by the commodity shippers, and so forth.

9. I am indebted to Mr. Roderick Wagoner of Williams, Dimond & Company, Seattle, and to the vessel owners, Mitsui O.S.K. Lines, for this account of a containership's port call at Seattle.

10. I am indebted to Mr. James Papp of Transmarine Navigation, Seattle, and to the vessel owners, Yangming Marine Transport, Taipei, for this account of a bulk carrier's port call for loading soybeans at Seattle.

11. There was an astounding similarity between Wagoner's and Papp's (notes 9 and 10) list of necessary contacts in the 1980s and my own (as a steamship agent) in the 1940s in Houston and 1950s in Seattle.

12. A number of the pubs along or near the Thames in London are steeped in historic lore connected with smuggling and other nefarious maritime activities.

13. See Albion (1939), p. 1. This, of course, was before the era of the Beatles or the bombs.

14. Most of the ports, real and hypothetical, described by Conrad (1915) were within a radius of about "800 miles drawn around a point in North Borneo," a "magic circle" for the "enchanted" Heyst in "Victory," where the seas were generally "tepid and shallow" beneath

"the inanimate, brooding sunshine of the tropics." Monsarrat (1951) gave a vivid impression of Liverpool, Glasgow, St. John's (Newfoundland) in wartime as perceived by naval personnel.

15. At this time the author worked for States Marine Lines, whose lumber ships deserted Boston because of labor unrest.

16. As the domain of the public port authorities expands, the public may expect to get easier access to the waterfront; however, the terminal lessees often impose restrictions.

17. See Morris (1969), p. 5.

18. See, for example, Port of Seattle 1980 Annual Report, p. 19.

19. There can be a difference, of course, between an individual's personal endorsement (and vote) and the "company line" of the person's employer.

20. The "confidentiality" of certain types of shipments (e.g., military) is a logical reason for part but not all of this.

21. At some of the great port cities of the world, in the commodity exchanges and chartering exchanges—for instance, the Baltic Exchange in London—the exchange of trading and shipping information is, after negotiation, the most important service provided. Brokers, incidentally, are paid for their negotiating expertise, but also for their information evaluations.

## References

Albion, R. G. 1939. *The rise of New York Port, 1815–1860*. New York: Charles Scribner's.

Bird, J. 1957. *The geography of the Port of London*. London: Hutchinson University Library.

Conrad, J. 1915. *Victory*. New York: Doubleday.

Dowd, T. J. 1981. *Strategic management: A tool for port planning*. Master's thesis, University of Washington.

Fleming, D. K. 1974. Reflections on conventional cargo liner service. *Maritime Studies and Management* **1:**178–90.

Hoyle, B. S., and D. A. Pinder. 1981. *Cityport industrialization and regional development*. Oxford: Pergamon.

Kenyon, J. B. 1968. Land use admixture in the built-up urban waterfront: Extent and implications. *Economic Geography* **44:**152–77.

Marcus, H. S., J. E. Short, J. C. Kuypers, and P. O. Roberts. 1976. *Federal port policy in the United States*. Cambridge, MA: MIT Press.

Mayer, H. M. 1957. *The Port of Chicago and the St. Lawrence Seaway*, Research paper No. 49. Chicago: University of Chicago Press.

Monsarrat, N. 1951. *The cruel sea*. New York: Alfred A. Knopf.

Morris, J. 1969. *The great port, a passage through New York*. New York: Harcourt, Brace.

Olson, D. J. 1980. Regionalism in port development: The Pacific Northwest experience. Paper given at Regional Port Institutions Seminar, Boston, August 4–5.

Port of Seattle 1980. *Annual report*.

Porter, K. A. 1945. *Ship of Fools*. Boston: Little, Brown.

Verlaque, Ch. 1975. *Geographie des Transports Maritimes*. Paris: Doin.

Vigarie, A. 1979. *Ports de Commerce et Vie Littorale*. Paris: Hachette.

Weigend, G. G. 1956. The functional development of the Port of Hamburg. *Tijdschrift voor Econ. en Sociale Geogr.* **47:**113–20.

# Chapter 4

# THE PHYSICAL HARBOR:

## New Demands on a Scarce Resource

*Harold M. Mayer*

If one were to compare a drawing of Boston from the 1630s with a contemporary map, or examine the first chart of Seattle's Elliott Bay made in 1841 with a current aerial photo, one fact leaps out: the physical growth of these major cities has been accomplished by encroaching upon and filling shallow water areas. The area encompassing the city of Boston today has four times as much land as it did when the city was first settled. The Charles "River" was, at that time, an expansive bay. Similarly, Seattle's 2,000-acre port and industrial zone were tideflats, oyster beds, and waterfowl areas only a century ago.

Over the past two or three decades, an abrupt halt to massive reclamation in harbors has occurred in the United States. Yet physical change takes place continuously as new uses affect old along the shorelines and in the waters. Even though major landfill projects are rarely proposed today, there are increasing pressures for competing use of harbors from many sectors of society. In addition to traditional port activities and port-related industries, virtually every U.S. harbor is actively used for urban recreation and tourist services. We also see enormous public pressure to protect the harbor's natural environment and to preserve the remaining water areas. As the physical attributes of the harbor become "scarcer" (in the sense that they are sought after for more and perhaps incompatible purposes), physical planning that allocates harbor resources among sectors will be a growing phenomenon. Harbor resource planning will require information about social goals, economic interests, financial resources, and natural conditions. This chapter focuses on the natural conditions of the harbor.

The chapter begins by defining key terms, describing physical characteristics, and classifying harbors by physical type. The inquiry then shifts to an analysis of the trends affecting use of harbor space—shorelands and waters—by the transportation, urban, and environmental sectors. This leads to a set of conclusions that shows how public policy dealing with navigation projects, and addressing the protection of environmental and recreational values, need also to stress greater collaborative planning among affected interests.

## DEFINING PORTS AND HARBORS

The terms *port* and *harbor* are not synonymous. Ports are places where there is the interchange of cargo and passenger traffic among vessels, and between vessels and overland carriers or sites alongshore. Harbors are places that provide protection against winds, currents, and waves, and thus allow port activities to take place. The protection in a harbor is provided by land and/or structures designed to interrupt or minimize the actions of wind, waves, and currents. Harbors are natural or artificial, although the configuration of most harbors includes elements of both.

Ports can span more than one harbor, and some harbors are the loci of more than one port. San Francisco Bay, for example, is a harbor that includes several ports: San Francisco, Oakland, Richmond, Alameda, and others. The Port of New York, in contrast, includes several harbors separated by a series of connecting waterways and channels. Within the Port of New York—which is the area of concern of the Port Authority of New York and New Jersey—are the Lower Bay, the Upper Bay, portions of the Hudson (North) River, the East River, the Harlem River, part of Long Island Sound, and other water areas.

The term *port* is also used to designate a unit of organization or operation of a set of facilities associated with the transfer or interchange of waterborne commerce, or of other maritime activities such as naval installations, or the servicing of fishing fleets or pleasure craft. In addition to, or in some instances instead of, facilities for handling cargo, some ports include specialized installations for handling ferry, cruise ship, or other traffic.

The usefulness of harbors, whether improved artificially or not, depends upon their geographic *situation* with relation to the existing or potential traffic, and upon the conditions of their *site*, which refers to the characteristics and configurations of the land and water within the port area proper and its approaches.

The situation of New Orleans, for example, is ideal for a port, and because of its situation it has developed as the second or third most important port in the United States. It is the principal outlet for the Mississippi basin which, at a million square miles, constitutes one-third of the nation's area. The site of New Orleans, however, could hardly be less favorable for the development of a port. The Mississippi River is subject to wide variations in level and current velocity, and there is need to keep it, including the approach channels from the sea, constantly dredged. Under natural conditions, there is a threat of frequent flooding of the adjacent land area, and the river currents often make the handling of vessels in the stream difficult.

Seattle, on the other hand, has a site that, in most respects, is ideal for development of a port. Its approach from the sea is wide and deep, with complete shelter from the winds and waves of the open ocean. Its principal disadvantage is the relative scarcity of level land in the immediate vicinity of the port. But the situation of Seattle is very unfavorable with respect to its *hinterland*, the inland area that originates and terminates its traffic. The nearby population is relatively sparse compared to that near most of the principal world ports, and much of Seattle's hinterland lies two or three thousand miles to the east, involving either a long circuitous voyage through the Panama Canal, or a "landbridge" transit across most or all of the North American continent. Even to reach the nearby hinterland necessitates crossing the Cascade Mountains by rail or highway. A major competing port—Portland, Oregon—although also with a situation peripheral to the cores of activity of North America, has the low-level route of the Columbia River, with parallel low-gradient railroads and highways, giving access to the regional and continental hinterland. But Portland suffers from the disadvantage, in contrast to Seattle, of difficult and sometimes dangerous access from the sea. Many shipwrecks have taken place at the entrance to the Columbia River where a shifting sandbar constitutes a notorious hazard.

Although the terms are not widely used, some geographers describe the overseas equivalent of a port's hinterland as the *foreland* or the *discontinuous hinterland*. The trade that moves through the Port of Seattle is in large part transpacific. Seattle and the other ports of the Pacific Northwest have the advantage, over the ports of California, of being closer to the Great Circle transpacific route. The *situation* of a port, then, refers to its landward hinterland, as well as to its overseas areas.

Site characteristics of harbors are very diverse and may be quite complex. They are the result of natural forces and processes as well as human action in modifying natural site conditions. As with most human works, harbor development and port operation are easier when modifications work with, rather than in opposition to, natural processes—geological, meteorological, hydrological.

Many locations may be sufficiently attractive for port development to justify considerable expense to modify sites. At the national scale, improvements to the Mississippi River and its principal tributaries rendered many inland river ports accessible to modern barges and towboats and, with the use of LASH (lighter-aboard-ship) barges, to international waterborne traffic. A series of canals in the nineteenth century made the ports of the Great Lakes accessible to small oceangoing vessels, whereas in 1959 the opening of the present St. Lawrence Seaway transformed the principal lake ports into ocean ports. At the local scale, dredging to provide wider, deeper, and straighter channels than those provided by nature is common. Where level land for port terminal and port-associated industrial installations is scarce, it may be possible to increase the land area by fill. In the United States there are some locations not originally suited for modern port development that have been made accessible by artificial channels or improvement of inadequate natural channels. Albany, New York, 150 miles up the Hudson River from New York City, accommodates moderate-size ocean ships. Stockton and Sacramento, California, originally accessible only by shallow-draft river boats, were converted into ocean ports by creation of largely artificial channels. More importantly, Houston, Texas, sited originally on a small bayou, or creek, was rendered accessible to the Gulf of Mexico by a 50-mile artificial canal, enabling it to become a formidable competitor of New Orleans as one of the two leading Gulf ports (see the discussion of the Port of Houston in Chapter 2).

## PHYSICAL CHARACTERISTICS OF HARBORS

The measure of the efficiency of a Port is its ability to maximize "throughput" of traffic at minimum cost and, for general cargo, with minimum delay. The physical characteristics of the harbor may contribute to the Port's efficiency, or they may constitute costly impedances.

An ideal harbor would have the following physical characteristics: 1) adequate depth of water within the harbor, the approaches, and the entrance channels to accommodate safely the largest vessels expected to use the harbor, at the same time providing adequate protection from the open sea, 2) minimum interference from tides and currents or, alternatively, structures and facilities designed to compensate for or overcome disadvantageous tides and currents, 3) freedom from fog and ice, or adaptation, such as ice-breaking technology (see Fig. 4.1), 4) bottom characteristics that provide suitable anchorage for vessels within the harbor, and 5) adequate adjacent land with physical characteristics permitting appropriate development of port terminals, port-oriented industries, and associated facilities and operations. (See the appendix at the end of this chapter for a more detailed discussion of physical characteristics.)

Few areas have all these characteristics, but substantial modification of the natural conditions is possible by dredging, protective breakwaters and piers, and by use of modern electronic aids to facilitate safe vessel movements into, within, and out of the harbor areas.

## CLASSIFICATION OF HARBORS

Ports and harbors may be classified in many ways, depending on the purpose of the classification. Ports, typically, are classified according to the function they perform: 1) industrial ports, handling primarily bulk commodities such as coal, petroleum, grain and ores, 2) commercial ports, in which general cargo, whether containerized or breakbulk, predominates, 3) specialized ports, including those characterized mainly by ferry, passenger cruise, pleasure boating, fishing, and naval activity, and 4) diversified ports. Here we are classifying harbors in accordance with their physical configurations.

**Figure 4.1.** Coast Guard ice breaker vessel underway in Antarctic waters. (Courtesy: U.S. Coast Guard)

***Open roadsteads.*** The simplest form of harbor, an open roadstead, is an area of water where the shoreline configuration offers only minimal protection from wind and wave. Where these are fairly constant in direction, an open roadstead located to the lee of a land mass may offer protection during most of the time for transfer of cargo and/or passengers between vessels (see Fig. 4.2). Typically, open roadsteads function where the water depths close to shore are too shallow for the drafts of vessels. Many open roadsteads are important along the coasts of Africa and South America. In lesser developed nations, investment capital for cargo-handling facilities may be in short supply, whereas labor is abundant and inexpensive. In such instances, vessels may anchor offshore, with transfer to and from lighters, either manually or with use of ships' gear. In the harbor of Hong Kong, one of the world's busiest ports, for example, buoys are provided, anchored to the bottom, for use of ships that employ lighterage.

In the contiguous United States, there are few harbors with sufficient depth to accommodate supertankers (generally defined as those of 200,000 deadweight tons or more), with the exception of Puget Sound. However, in that water body, tankers with loads of over 125,000 tons are prohibited by law because of fears of oil pollution in the confined waters. As a consequence, much of the crude oil imported into the United States, especially from the Middle East, arrives in supertankers at nearby foreign ports, Nova Scotia and the Bahamas, there to be transferred to "handy sized" tankers for delivery to the coastal ports. An alternative, which attracted considerable attention in the 1970s when importation of Middle East crude oil was anticipated in increased volume, was offshore transfer in open roadsteads as in the Gulf of Mexico off Texas and Louisiana. Artificial offshore islands were proposed, but the actual transfer, in one instance at least, is from large tankers tied to a "monobuoy" offshore into a submarine pipeline leading to shore-based storage tanks and refineries.

***Estuaries.*** Many ports, including some of the world's largest and busiest, are located on estuaries. An estuary, for our purposes, is a water area or embayment

**Figure 4.2.** Mutton Bay, Quebec, illustrates an open roadstead type of harbor where larger vessels—here the *SS North Voyageur*—are served by smaller ones for loading and off-loading. (Courtesy: Harold Mayer)

resulting from invasion of a low-lying former land area by a rise in sea level relative to the land. Estuaries may be funnel-shaped, narrowing inland, as those of the Thames, Mersey, Seine, and Weser rivers in northwestern Europe. By contrast, other estuaries have narrow entrances, widening out inland, as the Delaware and Chesapeake bays on the U.S. Atlantic Coast and San Francisco Bay on the Pacific (see Fig. 4.3).

Those estuaries that narrow inland tend to have especially high tidal ranges. Accommodation to the tides is, as mentioned earlier, in the form of "wet docks," floating landing stages and wharf boats or, at some smaller ports, stranding at low tide. Noteworthy examples of the latter are at Hantsport and Windsor, Nova Scotia, where tides of over 40 feet permit loading of gypsum aboard medium-size ocean-going vessels while stranded during low tide. In some estuaries, the funneling of water as the tide rises is in the form of a "bore" or wave front of considerable height, which at times may constitute a hazard to vessels tied up or at anchor in the estuary. Some tidal bores, as in the arms of the Bay of Fundy in eastern Canada or in Cook Inlet near Anchorage, Alaska, are tourist attractions.

Ports located on funnel-shaped estuaries are especially vulnerable to high spring tides and to high water associated with hurricanes, typhoons, and cyclones. London on the Thames estuary, Tokyo on the Sumida estuary, and Providence, Rhode Island, on the estuary of Narragansett Bay are examples of ports and cities now protected by seaward barriers—dams with moveable gates.

Estuaries with relatively narrow entrances and wide expanses of water within have lesser tidal ranges. This is because the tides spread widely over the extensive water areas inland from the entrances. Some have entrances with restricted depths, typically because of sandbars deposited by littoral currents across the narrow entrances. The entrance to Chesapeake Bay has a vehicular tunnel 60 feet below the water surface, which prevents dredging of the entrance to depths sufficient to admit very large bulk-carrying vessels to Baltimore and other ports on the estuary.

Some estuary ports, as well as those on rivers and canals with minimal fluctuations in water level, are provided with artificial basins similar to "wet docks" but without locks. In such instances the water levels in the basins are the same at any given time as the water in the adjacent river or canal. Examples of such ports are Rotterdam—the world's busiest, Hamburg, Bremen, and Duisburg-Ruhrort, the largest inland Rhine River port. In many instances major ports associated with estuaries may develop as multiples, with one or more near the entrance to the estuary and another port (or ports) farther inland, commonly at or near the head of the estuary.

The inner port may specialize in bulk traffic of low value per ton and hence not time-sensitive, if the channel is deep enough to admit the bulk vessels. The port nearer the sea may specialize in time-sensitive traffic, such as passengers, mail, and general cargo, including containers. In North America, Baltimore and Hampton Roads, Virginia, are competitive twin ports at the two ends of the Chesapeake Bay estuary. Houston and Galveston became twin ports with the opening of a ship canal to inland Houston, and river dredging and canalization have made Stockton and Sacramento

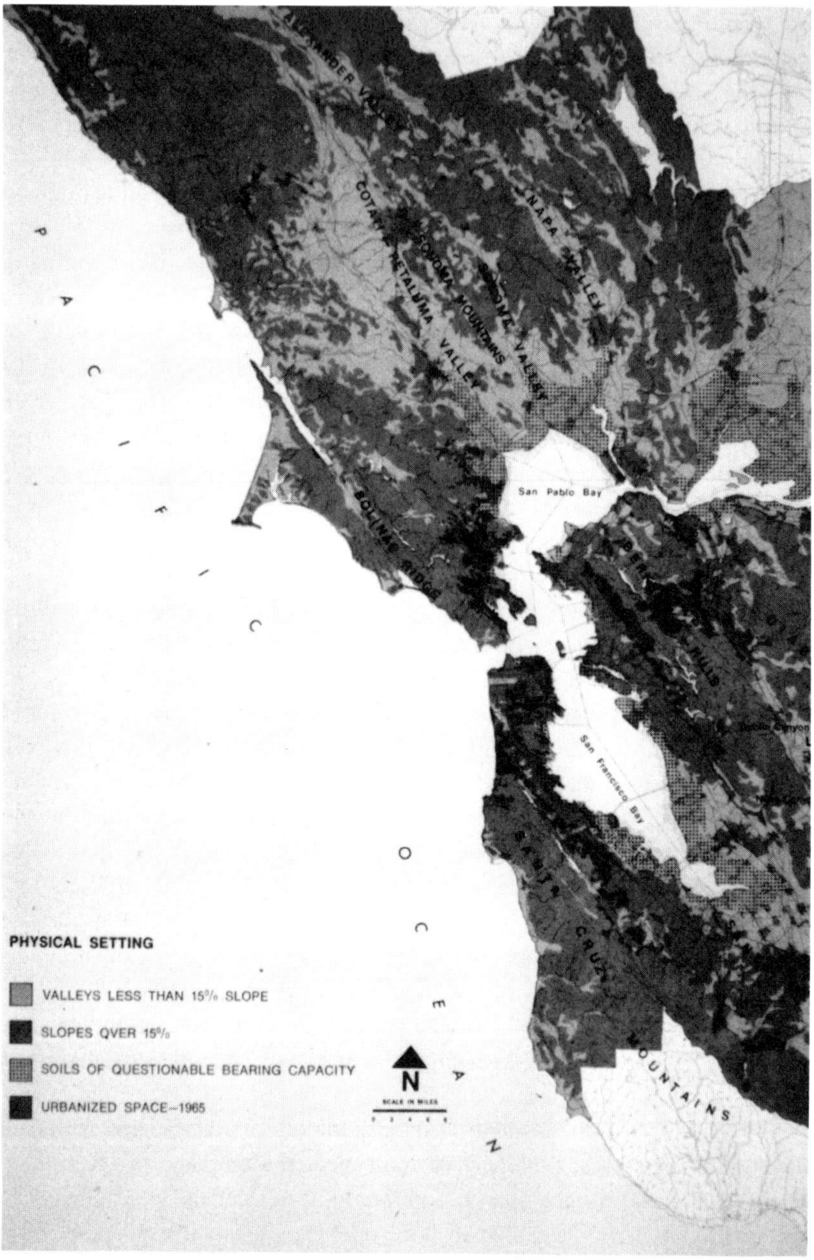


**Figure 4.3.** San Francisco Bay illustrates the estuary-type harbor where river and ocean waters meet in a semienclosed protected area. (Courtesy: Association of Bay Area Governments)

inland ports associated with San Francisco Bay. Unless the estuary is naturally of sufficient depth for large vessels, maintenance and deepening of the channel in the estuary, river, or canal may be so difficult and costly that provision of adequate depth may not be justified. Richmond, Virginia, was once a major port, but the increasing size of vessels greatly reduced its importance.

***River and canal harbors.*** Many ports are located on rivers, canals, and in some instances large lakes, accessible to typical oceangoing ships. Baton Rouge, Louisiana, for example, is 200 miles above the entrance of the Mississippi River and 100 miles above New Orleans, with a dredged channel that enables moderate-size ocean tankers to reach the refineries there (see Fig. 4.4). Major Great Lakes ports, formerly accessible by oceangoing ships carrying 1,600 tons or less, are now reached by "salties" of up to 30,000 tons, whereas "lakers," which are landlocked, can take over 60,000 tons.

Many inland ports are served primarily by vessels specifically designed for operation on the lakes, rivers, and canals along which they are located. Distinctive types of vessels are characteristic of the Great Lakes, the Mississippi River system, the inland waterways of western Europe, the Yangtse in China, and elsewhere. These include vessels such as 9-foot-draft barges and specialized pusher towboats on the Mississippi River system.

River entrances, especially where deltas have formed, are generally protected by parallel piers at their mouths, by offshore breakwaters, or by both. The twin contiguous ports of Los Angeles and Long Beach have harbors that, although largely artificial, are combinations of both. Many Great Lakes ports, including Chicago, Calumet Harbor (South Chicago), Milwaukee, and Cleveland, have harbors with both parallel river-entrance piers and outer breakwaters. Duluth-Superior and Toronto harbors are protected by natural offshore bars, cut through by artificial entrance channels.

Harbors consisting of portions of the lower courses of rivers characteristically occur on deltas, which generally have a number of distributary channels and, because of more-or-less continuous silting, require frequent or continuous dredging. In some instances more than one of the distributary channels may be maintained for navigation. The "passes" at the mouth of the Mississippi constitute several entrances, although the Southwest Pass is the route of most oceangoing ships serving New Orleans and Baton Rouge. Recently a second, entirely artificial channel, independent of the lower Mississippi, has been completed, shorter and straighter than the river and cutting the distance between New Orleans and the Gulf of Mexico by more than 25 miles.

***Artificial harbors.*** Although most harbors serving active ports are partially artificial as the result of dredging, filling, bulkheading and other engineering works, some harbors are almost completely artificial. These may be created by dredging

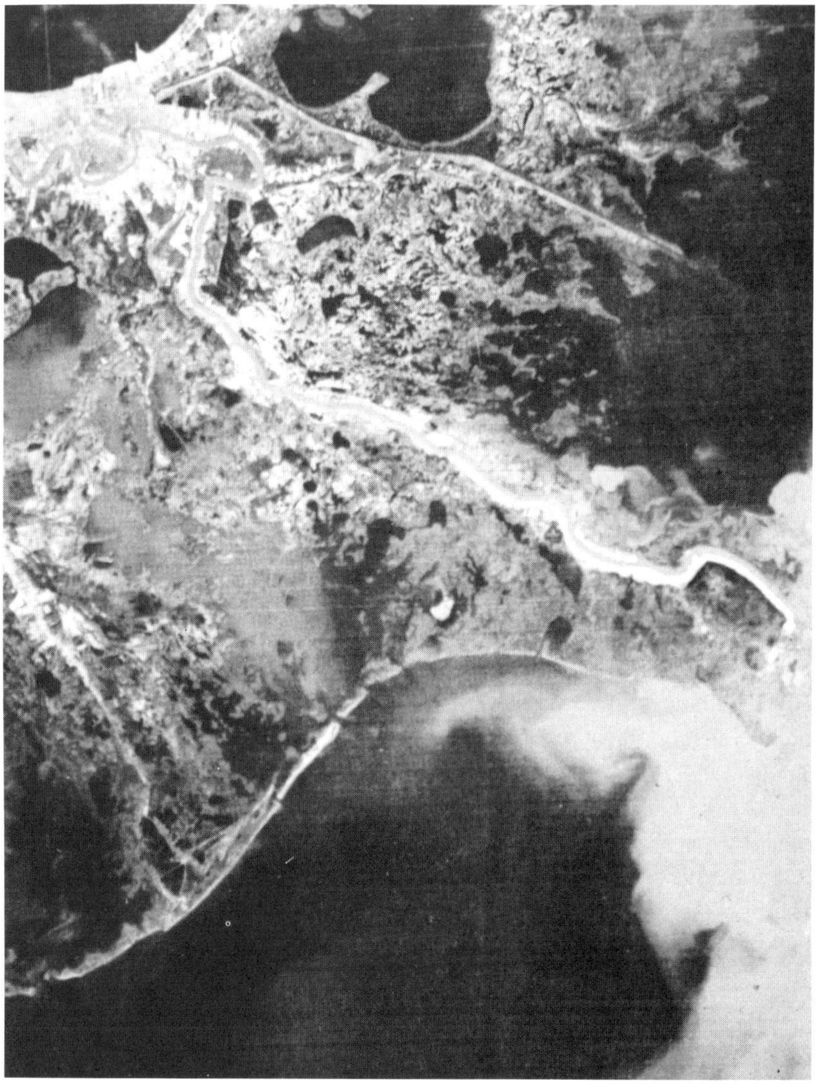

**Figure 4.4.** NASA Landsat view of the Lower Mississippi River and delta. The city and port of New Orleans is at upper left. (Courtesy: NASA)

inland and using the excavated material for adjacent landfill, or they may be the result of construction of offshore breakwaters The Port of Oakland is largely on landfill in San Francisco Bay; Long Beach is mostly on fill (see Fig. 4.5), whereas the adjacent Los Angeles port is partly on landfill behind a breakwater-protected outer harbor and partly inland from the natural shore where dredged channels provided fill for Port facility sites.

**Figure 4.5.** The Port of Long Beach in Southern California illustrates an artifical harbor where both the protected breakwater and inner basins are man-made. (Courtesy: Port of Long Beach)

## TRENDS IN HARBOR LAND DEVELOPMENT

Equal in importance to the areas within harbors are the characteristics and extent of land areas adjacent to the harbors. Some ports are fortunate in having, contiguous to the harbors, abundant and extensive areas of well-drained level land. Ideally, the land's subsurface conditions are adequate to support the loads of buildings, other facilities, and stacked goods, as well as providing sites for railroad and highway access, yards, and staging areas. Other ports are handicapped by inadequate nearby level land, or by existing land uses that are incompatible with port activities. One of the world's largest container ship lines recently moved its Pacific Northwest terminal from Seattle to Tacoma because expansion needs could not be met at the heavily urbanized Seattle harbor. At many ports throughout the world, high-density central developments of office buildings, commercial and industrial establishments, recreation and esthetic uses, and even high-rise residential structures, together with insufficient land for rail and highway access to the port terminals have restricted maritime activities. In response, activity has shifted from waterfronts adjacent to central city areas to outlying locations.

Three trends of the past few decades have converged to alter how harbor land is used and developed: cargo-handling technology is radically different than in former times; urban forces are moving into the harbor and claiming areas for recreation and leisure; and environmentalism puts severe constraints on development at the water's edge.

## *Containerization and Intermodalism*

Eighty percent or more of the general cargo—as distinguished from bulk cargo—on the major ocean routes connecting developed regions is now moved in standardized containers, which are commonly filled and stripped at some distance from the port terminals. Other cargoes are transferred to and from vessels by RORO (roll-on, roll-off) methods across ramps connecting ship and shore. Both methods expedite turnaround of vessels and cargoes by eliminating the labor-intensive and time-consuming older breakbulk operations. The throughput—amount of cargo handled in a given period of time—is vastly increased. Formerly in the older harbor areas, such as in central Boston, New York, Philadelphia, San Francisco, and Seattle, there was a need to maximize the lineal extent of water-land contact; the result was a series of piers and slips designed to provide berthage for the largest possible number of vessels in a given extent of shoreline. With unitized cargo handling—containers and truck trailers—there is increased need for extensive backup areas for fast rail and highway access and for the staging of containers awaiting onward movement (see Chapter 6 for a detailed discussion of containerization).

As noted above, today it is not uncommon to observe twin ports in urban areas: an older one near the city, which is now obsolete because of inadequacy of land areas to accommodate the growing volumes of traffic and the changing technology of port operations, and a newer one on extensive areas of level land farther from the central cities, most commonly seaward from the former locations. For example, the major port terminals in the Port of New York are no longer located along the Manhattan and downtown Brooklyn waterfronts; changes in shipping technology necessitated transfer of the major port activities to Newark Bay and elsewhere. In ports with large volumes of container cargoes, 20 to 40 acres of backup land for every ship berth is common. In a few instances, required sites may become available that were formerly occupied by waterfront industries that declined and closed. In other instances, extensive areas of landfill may be needed.

Although the provision of level land for large-scale port terminals is costly, the technology of creating landfill is well developed. Dutch engineers have been creating "polders" for centuries. Impervious walls or cofferdams are driven into the earth and the fill placed behind them. Commonly, some time must elapse to permit the fill to settle sufficiently to support the anticipated loads, or deep piles must be provided to carry the concentrated weights.

Containerization and intermodalism have resulted in land use changes within entire regions of the country, not just within individual harbors. Only heavy traffic volumes can justify the large investment of public and private funds required for the development of a port's harbor and terminal facilities. It is obvious that every existing harbor cannot be improved to the necessary extent, and that traffic must be concentrated at fewer, but more efficient ports and terminals called "load center" ports. Eventually it is possible that there may be no more than two or three on each coast. The remaining smaller ports may be maintained as "outports" with feeder services by water and land to the load center ports. Such smaller ports need be provided with channels of lesser depth, and with less elaborate and expensive terminal facilities; at the same time the resulting concentration of traffic at major ports would justify extensive improvements to them. The trend is similar to that of the domestic airlines, with the trunk carriers serving primarily the largest hubs and with regional and local feeder services to the smaller and medium-size communities.

## *Reuse of Harborfront Land*

Many communities, large and small, are taking the opportunity to redevelop centrally located former port land for nonport uses. Some of these waterfront areas are used for civic centers and parks, as in Baltimore and Philadelphia. Other cities use such areas for development of valuable commercial and residential projects. The lower Manhattan waterfront, formerly occupied by finger piers and slips, is filled in between the pierhead and bulkhead lines, with the landfill constituting the site of Battery Park City, adjacent to the Wall Street district.

As cities rediscover the waterfront, one can expect urban pressures to compete vigorously, and perhaps successfully, for limited shorelines within harbors. Recreationalists, housing developers, and tourist promotors have considerable political and financial resources to call on. The increased value of waterfront land and the constraints on industrial uses may combine to displace industry and force it to move to other locations. This displacement issue will undoubtedly grow in future years (see Chapter 13 for a detailed discussion of waterfront redevelopment issues).

## *Environmental Issues*

Conflicts between the demands for port-associated industrial developments and conservation of natural conditions are frequent. Wetlands in coastal areas are especially critical natural resources, and their preservation may constrain development. "Docks or ducks" is a frequent issue.

Economies of scale dictate that, where the traffic is available, the largest vessels which are physically able to use the desired habors and channels should be employed. In order to provide economies of scale at the ports, relatively few ports can afford, or could justify, extensive channel deepening. Costs of initial and maintenance dredging tend to increase exponentially with increased depth (see Fig. 4.6).

Dredging to deepen channels or to maintain authorized project depths creates a demand for areas in which to place the excavated material ("spoil"). Deposited behind impervious bulkheads or revetments in water areas, the fill changes the shoreline configuration and affects wildlife habitat, water quality, and currents; thus, even relatively remote areas along coastlines, estuaries, or rivers may be subject to significant environmental or ecological changes. To some extent, hydrological effects can be anticipated by computer or physical models or both, but the complexities of other ecological effects may be so complex as to be in large part unpredictable. Concern for the presence of toxic substances, in particular, is especially critical, and policies requiring compensation "mitigation" for environmental damage provide technical and financial challenges (see Chapter 12 for a detailed discussion of environmental mitigation).

After a lapse of many years, programs of harbor and waterway improvements are accelerating as a result of federal legislation in 1986 (see Chapter 5). Consequently, there is an intensification of concern for the possible impacts of changes in river and harbor configurations upon local and regional ecosystems, and an increased need for research on the extremely complex physical, chemical, geological, hydrological, and biological interrelationships that may be affected by harbor improvement projects. Demands on the several institutions conducting basic research in these fields, and especially those specifically concerned with oceanographic and limnological research, are increasing. At the same time, the financial resources of these institutions are constrained by the current policies of budgetary restraint at all levels of government.

## THE NEED FOR A REGIONAL PERSPECTIVE

In recent years, several trends relative to harbor resources have emerged or intensified. Economies of scale point toward fewer but larger and more efficient ports; hence concentration of the financial resources upon those "load centers" and industrial harbors that can generate larger tonnages of cargo than was typical in the past. Many smaller and medium-size ports have experienced declines in traffic and may have to reconsider the extent to which they maintain their harbors and channels. At the same time, improvements at the fewer ports that can attract concentrations of traffic and larger vessels consist of deeper harbor facilities and expensive channel improvements. Increasingly, physical changes in the configuration and use of harbors and adjacent land areas have been subject to environmental constraints. Impacts of such changes upon ecosystems, both landward and in the waters, are major considerations in determining the feasibility and desirability of harbor and channel projects. New port terminals and waterfront industries that receive and ship bulk fuels and

**Figure 4.6.** U.S. Army Corps of Engineers dredging vessel operating to maintain a navigable channel. (Courtesy: Dept. of the Army)

commodities directly require large land areas free from the landward congestion and high land costs associated with locations adjacent to or near central areas of cities. Many such facilities have shifted to locations on or beyond the urban peripheries, generally seaward. A major consequence is the freeing of centrally located former port areas for redevelopment as sites for nonport functions: civic centers, parks, or intensive commercial and residential use. Waterfronts adjacent to such cities as Boston, New York, Philadelphia, Baltimore, Jacksonville, New Orleans, Portland (Oregon), Detroit, Milwaukee, and many others have recently transformed their downtown waterfronts from port to other uses, or are planning to do so in the near future.

The several trends identified in this chapter call for a reassessment of policies relative to ports at local, regional, and national levels. Trends in U.S. federal policy are discussed in detail in Chapter 5; however, it is important to highlight here some of the key federal policy issues affecting physical change in the harbor.

In the United States, unlike many other leading maritime nations, there has never been, nor is there now, a national policy or program of port development, nor is there any unified organization or management of the nation's ports. Ports are extremely competitive; in that respect they resemble private-sector business, even though in many ways they are essential public utilities. Several comprehensive studies, including notable ones by the Maritime Transportation Research Board of the National Academy of Sciences, have recommended that port planning be on a regional basis rather than national (NRC 1976). Thus, competition would be reduced within a region, but the various coastal regions would continue to compete with each other.

At present federal policy favors local port improvements with virtually no consideration of trends or relationships among ports within a region. The federal government has traditionally been the agency for effectuating waterway and harbor improvements, and it maintains and operates the navigation facilities. When an improvement is suggested, the Congress may authorize a study, under the U.S. Army Corps of Engineers. If as a result of the study a favorable benefit-cost ratio is demonstrated, Congress may authorize an improvement project, subject to subsequent appropriations. Each project is thus subject to individual investigation; rarely are the potential benefits adequately related to the total regional and interregional relationships among the competing waterways and ports. The recent inability of the Tennessee-Tombigbee waterway to generate more than an insignificant fraction of the projected traffic is a conspicuous example of the failure of this approach. Planners and promoters of the project severely overestimated benefits and underestimated costs when they failed to account adequately for a parallel navigable waterway, the Mississippi River. The Mississippi, which serves the same shippers, operates efficiently and has capacity to expand. Thus, regional factors were virtually ignored in planning the project.

More recently, environmental concerns have tended to play a more prominent role in decisions relative to waterway and harbor projects. Detailed environmental impact studies are required for virtually every significant proposal, including the impacts of alternative solutions, one of which may be to not carry out the project. Because of the time consumed in environmental analysis, and the identification of important resources that need to be protected, the pace and scope of change are altered. Some harbors may never experience port expansion because of environmental interests. Although all harbors face constraints on development because of environmental values, the distribution of port services can shift within a region. This underscores again the need for a regional approach to port growth.

Recent changes in the financing of port projects have had, and will continue to have, significant impacts on harbors of the nation. Until recently, improvements and maintenance of waterways and harbors have been entirely financed from general funds of the federal government, if they are to carry interstate or international traffic. These facilities—as distinguished from port terminals—were completely free from user charges. The first modification occurred a few years ago, when a tax was imposed on the diesel fuel used by shallow-draft towboats on the inland waterways. Deeper-draft vessels on the oceans and Great Lakes were—and are—exempted. The revenues are designated for river and harbor improvements. In the mid-1980s, a change in the law now limits the federal government share of these projects to only a portion of the costs of harbor improvements and maintenance, with state, local, and private interests paying for an increasing proportion of the costs. Some of these costs can be met through a user charge on the vessel, but it is doubtful that all costs can be recovered. Because of the many other demands on state and local funds, little money is generally available for port development purposes when the port project competes with, for example, social services or education.

In some instances the availability of redundant former port land, commonly centrally located, may justify local and state financing of site preparation, including landfill, at more peripheral locations for port terminals and associated facilities. If local interests must pay more of the costs, we are likely to see a further concentration of port facilities in the larger harbors where there is a large financial base and a further decline in capacity among small ports.

The effects of federal policy on the authorization, financing, and environmental aspects of navigation projects and port development have somewhat blurred the traditional definitions and categories discussed at the outset of this chapter. Both physically and organizationally, the terms *harbor* and *port* are becoming more intertwined. Intermodalism, environmentalism, and urbanism make the attributes of the harbor far more of a factor in what takes place at the port. It is no longer sufficient to say that a harbor merely accommodates ships and protects them from wind and wave. The economic base, land-use patterns, and environmental concerns of the surrounding urban complex are of increasing importance in understanding how the harbor is used.

Independent and isolated planning for different uses of the harbor is no longer a viable alternative. In a physical sense, each new use of water area may foreclose other options for use of that environment; thus, they must be viewed jointly in a common and neutral forum. Each harbor project competes with other projects for scarce public dollars; public debate about priorities is needed. And each harbor project must reflect the transportation, economic, and social goals of the region; a larger context for planning is needed. The time is at hand to forge institutions that can integrate harbor and port planning, and to relate harbor plans to comprehensive plans at the municipal and regional levels of government.

## References

National Research Council (NRC). 1976. *Port development in the United States*. Report prepared by the Panel on Future Port Requirements of the United States, Maritime Transportation Research Board, Washington, DC, National Academy of Sciences.

## Additional Reading

Hershman, Marc et al. 1978. *Under new management: Port growth and emerging coastal management programs*. Seattle: University of Washington Press.

Karmon, Yehuda. 1980. *Ports around the world*. New York: Crown Publishers.

Kenyon, James B. 1968. Land use admixture in the built-up urban waterfront: Extent and implications. *Economic Geography* **44**(2):152–177.

Ketchum, Bostwick H., ed. 1972. *The water's edge: Critical problems of the coastal zone*. Cambridge, MA: MIT Press.

Klimm, Lester E. 1956. Man's ports and channels. In *Man's role in changing the face of earth*, ed. William L. Thomas. Chicago: University of Chicago Press, pp. 522–41.

Konvitz, Josef. 1982. Spatial perspectives on port city development, c. 1780–1980. *Urbanism Past and Present* **7**(2):23–33.

Marcus, Henry S. et al. 1976. *Federal port policy in the United States*. Cambridge, MA: MIT Press.

———. 1975. The need for a unified governmental approach to port planning. Proceedings, 14th Annual Meeting, Transportation Research Forum, Cleveland, October 15–17, pp. 831–39.

Mayer, Harold M. 1981. The reuse of redundant transportational land: The central Milwaukee waterfront. Proceedings of Applied Geography Conferences **4**:6–22.

National Research Council. 1979. *Public involvement in maritime facility development*. Prepared by the Committee on the Impact of Maritime Services on Local Populations, Maritime Transportation Research Board, National Research Council. Washington, DC: National Academy of Sciences.

———. 1980. *Urban waterfront lands*. Committee on Urban Waterfront Lands, Environmental Studies Board, Commission on Natural Resources. Washington, DC: National Academy of Sciences.

Schenker, E., and H. C. Brockel, eds. 1974. *Port planning and development as related to problems of U.S. ports and the U.S. coastal environment*. Cambridge: Cornell Maritime Press.

Wrenn, Douglas M., J. Eric Smart, and John A. Casazza. 1983. *Urban waterfront development*. Washington, DC: The Urban Land Institute.

## *APPENDIX: A DESCRIPTION OF THE PRIMARY PHYSICAL CHARACTERISTICS OF HARBORS*

***Depth.*** The depth of the harbor determines the draft of the vessels that can use it. Modern general cargo vessels, whether breakbulk or container ships, typically have loaded drafts of 30 to 35 feet. Large oceangoing cruise ships have similar drafts; VLCCs (very large crude carriers: "supertankers") have been built with drafts of up to 90 feet. Inland waterway vessels have, of course, much less draft. In the Great Lakes of North America, for example, the connecting channels and the major harbors have a standard controlling depth of 27 feet, enabling vessels with drafts of up to 25.5 feet to operate, under normal conditions. Ports on rivers, above the head of deep-draft navigation, may be accessible to vessels of considerably less draft. The standard draft for barges and towboats on the Mississippi River and its major navigable tributaries is 9 feet.

Depths in the world's principal straits, canals, and connecting channels have important impact upon the sizes of vessels, including liners that are designed to transit through them regularly, and other vessels that are intended for worldwide tramping. The Suez Canal, for example, has been subjected to constant dredging since its reopening, so that ever-larger ships, notably tankers, can transit with capacity loads. Restrictions have been imposed on the drafts of supertankers passing through the Strait of Malacca, requiring that the largest loaded tankers use Sunda Strait en route between the Middle East and Japan, with consequent additional voyage time. The dimensions of the Panama Canal locks limit the size of ships able to use that waterway, in length and beam as well as draft. Many ships are specifically built to "Panamax" size. The St. Lawrence Seaway has a controlling depth of 27 feet, thus limiting the size of ships able to enter and leave. A declining proportion of the world's merchant ships are able to use the Panama Canal or the St. Lawrence Seaway.

The depths must allow sufficient water under the keels of the vessels. At least 1.5 feet is minimal in still water. Allowance must be made for the vertical motion of the vessel due to waves, and also for squat and scend: the changing draft as the vessel is underway.

***Tides and currents.*** Tides are important physical elements in port operation at most harbors. They vary from virtual absence in the Mediterranean and the Great Lakes to as much as 60 feet in some tidal estuaries as in northwestern Europe and in the Bay of Fundy of eastern Canada. At some ports, vessels can enter and leave only at high stages of the tide; in the interim they may rest on the bottom ("stranded"). On the other hand, the usefulness of the harbor in spite of unfavorable tide conditions may be improved by deeper dredging or by the construction, where adjacent land conditions permit, of "wet docks" or basins. Entrance to and departure from them is through gates, which are opened for passage of vessels when the fluctuating level of the water outside is the same as the constant level of the water inside the basin, either at high tide twice each 24.5 hours or at an intermediate tidal level four times during that period. More usually, the entrance and exit are through a lock chamber with gates at each end, permitting the vessel to be raised or lowered in the lock chamber from one water level to the other. In the latter instance, transit is possible at all times that the depth of the channel leading to the lock is sufficient to float the vessel. In either instance, transfer of cargo to and from the vessel within the basin is possible at all times without the need to compensate for fluctuations in water level within the basin.

None of the major ports of the United States have a sufficient range of water level to justify the provision of "wetdocks," although portions of some are accessible by locks, as, for example, Lakes Union and Washington in Seattle and the Industrial Canal at New Orleans.

Another type of adjustment to fluctuations in the water level on rivers and tidal estuaries is the "landing stage," as at Liverpool and Tilbury in England, and Hamburg in West Germany. On the Mississippi River and its tributaries, the counterpart of the landing stage is called a "wharf boat." These consist of large floating platforms or barges, which rise and fall with changes in water level. They are connected to shore by one or more gangplanks or bridges. At major ports in the United Kingdom and northwestern Europe, vessels may load and discharge in wet docks but handle time-sensitive cargo, passengers, and mail at the landing stages, thus expediting movement of the latter by eliminating the delays of lockage.

Currents, whether riverine or tidal, may interfere with movement of vessels, or may facilitate movement if not excessive. On the other hand, especially in ports with "pier and slip" arrangement such as in the older central portions of the U.S. Atlantic ports of New York and Philadelphia, currents as well as tidal levels may restrict movement of vessels into and out of the slips at times other than at high water slack. Large passenger ships using New York's North River piers with drafts nearly as great as the high-tide depths of water in the slips had to schedule arrivals and departures only at high water slack periods. Because of the short duration of such periods, during peak traffic there was frequent congestion in the river.

Furthermore, riverine and tidal currents commonly result in silting, thus demanding frequent or constant dredging in order to maintain the requisite depths both in the main channels and alongside the terminals.

Littoral, or alongshore, currents in open water may create hazardous navigation conditions and may demand considerable expense for their control. Currents and waves abutting against a shoreline cause erosion of beaches and undercutting of bluffs, and at the same time they deposit the eroded materials causing shoaling elsewhere. Commonly, bars are depositied across river mouths and harbor entrances by such currents. These bars tend to be unstable, and frequent or constant dredging is needed to maintain safe navigable depths through them. Typical examples are the Ambrose Channel, the main entrance to New York Harbor between Rockaway and Sandy Hook, and the bar at the entrance to the Columbia River.

Control of deposition is facilitated by construction of piers, groins, and bulkheads. Parallel piers at harbor entrances may concentrate the tidal and river currents into a narrow area, thus creating a scouring action and reducing the need for dredging. Along shorelines, deposition by waves and littoral currents takes place against such structures, thus forming beaches and at the same time reducing the amount of deposition on the leeward side.

Tidal levels and the times and velocities of tidal currents, unlike those water levels and velocities caused by runoff in rivers or by storms, are predictable with considerable accuracy. Expected tidal levels and the velocities of tidal currents are published by the governments of major maritime nations for all of their principal harbors. They are cyclical and repetitive.

Tides are caused by the gravitative attraction of the moon and sun. There are two diurnal tidal peaks and troughs with a frequency of approximately 12 hours and 25 minutes. But not all high or low tides are equal: when the moon and sun are in a direct line with the earth, their combined force results in higher-than-average tides, called *spring tides*. Conversely, when the moon and sun are 90 degrees apart relative to the earth, the amplitude of the tidal range is less, resulting in a *neap tide*. Erosion and deposition as well as the velocities and volumes of tidal currents are greatest during spring tides. Some harbors, for example, Southampton, England, are subject to double tides.

***Fog.*** In spite of modern electronic devices such as radar, fog continues to be a source of difficulty and a hazard for vessel movements to, from, and within harbors. Fog is caused by evaporation from a water surface that is warmer than the air; by movement of air upslope, as when air moving from offshore is forced upward due to nearshore hills or mountains and consequently cooled; or by movement of air over a cooler surface. Fog may be expected whenever there is a substantial difference in temperature between land and water. Thus, coasts and harbors are especially susceptible.

***Ice.*** In high latitudes many harbors are affected seasonally by ice. Availability of ice-free harbors is an important consideration in the political and commercial strategy of several northern nations. Some ports incur considerable expense in order to extend the navigation season on waterways and harbors subject to ice. In a few instances, such as Montreal, ports that were formerly closed in winter are now kept open year round.

Powerful icebreakers keep the Soviet Union's northern sea route open for longer each year than previously. Churchill, on Hudson Bay, is a major exporting port for

Canadian wheat, which is accessible for only a few weeks during late summer. The St. Lawrence Seaway and the connecting channels between the Great Lakes are normally closed between mid-December and early April. For several successive winters, experiments were conducted to extend the navigation season on the four upper Great Lakes. Although demonstrated to be technically feasible, the economic justification was doubtful and the experiment was discontinued.

Some harbors in middle and northern latitudes, even though open all year, may be affected by ice floes. This is the case at the Port of New York, and ice breaking is sometimes required to keep the Port of Albany, up the Hudson River, open in winter. The upper Mississippi River, north of the junction with the Illinois, regularly is closed in winter, but heating of the effluent water from the electric generating plants in and near Chicago eliminates the need for ice breaking on the Illinois Waterway during some winters.

***Bottom characteristics.*** The characteristics of the harbor bottom may affect use of ports. Obviously, insufficient depth causes loss of traffic. Often shoaling occurs as the result of deposition of sediments from tidal action, erosion of the shorelines, silting from rivers, changes in river courses, and long-term changes in the relative level of land and sea.

In some instances harbors may be too deep. Many of the Norwegian fjords, U-shaped valleys scoured out by glaciers and subsequently subjected to flooding from the sea, are too deep to furnish anchorage.

Ability of harbor bottoms to hold anchors without danger of dragging or fouling is important. If the bottom is too soft or too hard, use of the port may be impaired; most ports require anchorage areas for vessels awaiting berthage or improvement in weather.

***Adjacent land.*** In general, the immediate vicinity of a harbor should be relatively free from topographic barriers that impede or constrain access to the port's hinterland. There should be sufficient level area upon which to develop industrial structures and open stacking areas to handle the anticipated traffic of the port and of port-associated industrial establishments, as well as to facilitate good railroad and highway access.

Port terminals and related industries commonly require firm foundation conditions; otherwise spread footings and deep caissons or pilings may be required at considerable additional expense. Shipbuilding and repair yards, vital to many ports, especially require firm foundations.

# Chapter 5

# FEDERAL PORT POLICY:

## Retrenchment in the 1980s

*Marc J. Hershman and Marianne Kory**

The port is a control point of the modern nation-state, a place where national security, trade, immigration, and customs are handled. Thus, in virtually every nation of the world port development, administration, and finance are important functions of national government. For example, in Canada, a country where ports are spread among Atlantic, Great Lakes, and Western provinces, there has been a substantial national presence in the administration of the port systems for the past fifty years. A National Harbours Board (NHB) was established in 1936 to improve management in many of the existing local port authorities. In 1983 the NHB was replaced by the Canada Ports Corporation which allows more autonomy for local port corporations, but still reviews their three-year corporate plans, major capital expenditures, and annual capital budgets (Canada Ports Corporation 1986).

In Japan the national government plays an even stronger role in port affairs. Port development in Japan is used to achieve regional economic development. It provides the infrastructure that facilitates substantial private investment. Thus, port planning and development, although initially a function of a port management body at the local level, is aided substantially by technical and financial support from the Ministry of Transport and reflects established national goals for industrial growth (Inoue 1986).

*Materials on regionalism are drawn from J. Levie's Master's thesis, cited in the references and supervised by Professor Hershman. Thanks to Stan Payne and Al Hammon for permission to quote from their recent writings.

## THE FEDERAL ROLE: CONFERRING SUBSIDIES AND IMPOSING CONSTRAINTS

In stark contrast to the national approaches in Canada and Japan, the United States has no central agency with any direct administrative or financial review functions over port organizations (as in Canada). Nor does it have a commercial port development strategy that is linked to other national goals and financed centrally[1] (as in Japan). The United States has neither a national port administration law nor a national port development policy. Port administration is primarily a function of U.S. state and local governments. Virtually every coastal or Great Lakes state has a port law that contains provisions on port governance, administration, functions, finance, and so forth (see Chapter 1). However, it would be wrong to conclude that U.S. laws and federal agencies play no role in guiding port development and operations. To the contrary, many federal agencies influence ports. In some cases the influence is a direct subsidy,[2] as when the U.S. Corps of Engineers (COE) dredges and maintains a federal navigable waterway adjacent to port lands, or when the Maritime Administration (MARAD) issues a technical report useful to port officials, or when the U.S. Coast Guard installs navigational aids that ease the flow of marine traffic and minimize hazards. In other cases the influence is a constraint on port action as when rulings of the Federal Maritime Commission (FMC) determine the validity of cargo rates and lease agreements, or when federal environmental agencies establish strict policies to protect natural resources.

Table 5.1 shows the primary federal and state/local agencies that constrain or subsidize the Port.[3] Interaction with government is high because of the many resources of concern to the Port: shoreland, water, vessels, cargo, people, money. The Port is often a developer, an agent of physical and economic change. This change in land use brings with it social conflict and results in increased attention from government regulatory agencies.

This combination of state-level laws establishing port governance and finance and federal-level laws setting up subsidies and constraints results in a complex legal framework in which Ports operate in the United States. First, they have to work with agencies at two levels of government that have similar functions, such as the federal COE and a state transportation, commerce, and/or coastal management agency. Second, the federal agencies are concerned with discrete Port functions, and this requires Port officials to interact with a wide range of agencies. Third, the activities of a federal agency that affect Ports are invariably minor compared to the agency's basic mission, and particular Port problems are sometimes overlooked. Fourth, federal functions can conflict with one another, as when a proposed federal subsidy, for example, COE dredging, conflicts with a federal environmental rule, for example, Environmental Protection Agency (EPA) wetlands protection policy. Fifth, the federal framework encourages the decentralized and highly competitive port system in this country in which there is no assurance of fairness among Ports in the

**Table 5.1.** Primary Federal and State/Local Agencies that Subsidize or Constrain Resource Use in Seaports

| Resource | Resource Activity/Use | Federal Agencies that Subsidize/Constrain Resource Use | State/Local Agencies* that Subsidize/Constrain Resource Use |
|---|---|---|---|
| Water/submerged lands | Navigation aids<br>Anchorage, channels, basins<br>Mitigation<br>Landfill, bulkhead<br>Dredge material disposal<br>Piling: docks, floats, wharves | U.S. Coast Guard<br>U.S. Army Corps of Engineers<br>Environmental Protection Agency<br>U.S. Fish & Wildlife Service<br>Natl. Marine Fisheries Service/NOAA<br>Coastal Resources Mgmt./NOAA<br>Indian nations** | Submerged land ownership<br>Fish and game agencies<br>Water quality<br>Coastal management<br>Transportation & commerce |
| Shorelands | Terminal development<br>Road/rail connection<br>Buildings, structures<br>Public access, recreation | U.S. Corps of Engineers<br>Coastal Resources Mgmt/NOAA<br>U.S. Dept. of Transportation | Coastal management<br>Planning & zoning<br>Building/fire/safety codes |
| Carriers (vessels, rail, truck) | Fair competition/nondiscrimination<br>Access to harbor<br>Fire & hazards safety | Federal Maritime Commission<br>Interstate Commerce Commission<br>U.S. Coast Guard<br>U.S. Dept. of Transportation | Harbor police<br>Marine fire depts.<br>Police<br>State transportation<br>Local engineering |
| Cargo | Import/export laws (trade laws)<br>Sanitation, inspection<br>Customs insp. & duties<br>Hazardous materials | U.S. Dept. of Commerce<br>U.S. Customs Service<br>U.S. Dept. of Agriculture<br>U.S. Public Health Service<br>U.S. Dept. of Transportation | Fire & safety codes |
| People | Longshore labor, other org. labor<br>Civil service<br>Worker benefits | U.S. Dept. of Labor<br>Occupational Safety and Health | State port laws<br>State workers' comp. |
| Money | Tax-exempt borrowing<br>Tax receipts<br>Fees for services<br>Budget & audit | Internal Revenue Service<br>Federal Maritime Commission | State port laws<br>State port laws |

*Listed in generic terms, since names of agencies vary with state and local jurisdiction.

**Listed with federal agencies for convenience. Under the U.S. Constitution, treaties with Indian nations are paramount to federal law.

distribution of subsidies and the imposition of constraints. Ports adept at taking advantage of the system by winning subsidies and minimizing constraints tend to prosper. Even at the state level, where a port law sets up the administrative apparatus, considerable discretion remains at the local level, which fuels the intense competition among ports within a state. Thus, although Ports interact with many governmental players, they retain autonomy because of the highly compartmentalized and individualized nature of that government interaction.

This chapter describes four particularly important shifts in federal government policy that will affect the competitive advantages/disadvantages of certain Ports and change how harbors and Port lands are used in the future. The first two sections show the retrenchment of the federal role in Port affairs accompanying the recent U.S. emphasis on free market principles: the restricted role of the FMC due to deregulation, and the increased local cost-sharing now required on federal harbor improvement projects. The second two sections relate instances where national policy has intervened more heavily in Port activities: the protection of the natural environment within the harbor, and the encouragement of Port cooperation within regions to promote efficient use of land and water resources. The chapter concludes with a plea to revitalize national attention to Port and harbor affairs.

## DEREGULATION: THE RESTRICTED ROLE OF THE FEDERAL MARITIME COMMISSION

The FMC oversees business practices among common carriers by water, and among ports. Decisions about these business practices—cargo shipment, routing, and freight charges—are made by private interests who compete aggressively with one another. Occasionally there are claims of unfair competition or unjust discrimination in these business dealings, and the jurisdiction of the FMC is invoked to ensure that U.S. commerce is not being harmed. Thus, through its policies and decisions, the FMC can influence waterborne commerce and hence the amount of cargo business a Port receives.

The FMC implements the Shipping Act of 1916 as amended and Section 8 of the Merchant Marine Act of 1920 as amended,[4] which gives them both a regulatory and protectionist role with respect to Ports. The Shipping Act provides the regulatory power.[5] It is aimed at common carriers by water and operators of marine docks, wharves, warehouses, and terminals. It attempts to halt unfair competition and unjust discrimination among carriers, shippers, and Ports that adversely affect the commerce of the United States. Agreements among these parties, and their tariffs and rate schedules must be filed with the FMC, which can hold hearings and disapprove or modify the documents filed. Through this public disclosure and review mechanism, shippers can discover their rights and obligations vis-à-vis common carriers and Ports, and receive some protection from unfair business practices.

Section 8 of the Merchant Marine Act of 1920 sets forth the protectionist role.[6] It describes the duty of the FMC as "promoting, encouraging and developing ports and transportation facilities . . . in connection with foreign and coastwise trade and to promote and encourage the use by vessels of ports adequate to care for the freight which would naturally pass through such ports." This latter phrase has given rise to the "naturally tributary" doctrine that the FMC has used to protect port interests (discussed below).

Taken together, these two provisions provide the FMC with broad authority to affect a Port's business. However, the 1984 Shipping Act and recent FMC decisions suggest that there will be less public sector influence over the choice of port by shippers and carriers. The private sector now has more control over the process of Port growth and decline.

## *The Shipping Act of 1984*

The Shipping Act of 1984 addressed problems of the steamship line common carriers that compete with one another in particular foreign trades, but also collude with one another, in arrangements called "conferences." It streamlined the process they follow in dealing with the FMC. For example, agreements among these carriers to fix rates, pool cargo, or establish joint services now go into effect automatically after 45 days, whereas before they became effective only after notice, hearings, and appeals. If agreements are challenged as "substantially uncompetitive," the burden of proof is on the FMC, rather than the carriers to make the case. Antitrust immunity is granted automatically to the carriers in some cases so long as the agreements were properly filed. Further, the carriers may now set intermodal rates for services that include the inland portion of through rates, thus resolving a jurisdictional problem with the Interstate Commerce Commission (ICC), which regulates land-based carriers operating in foreign or interstate commerce. For carriers, the act allows them to react more quickly in a competitive environment to take advantage of intermodalism, and to be protected from FMC interference except in cases of severe abuse.

The act also protects shippers—the consumers of transportation services. A shipper can get a lower freight rate or special service from a carrier even if that carrier is part of a conference agreement that has set rates and services because carriers now have a right of independent action: they can withdraw from conference memberships or set new rates or services with only ten days notice. Shippers are now authorized to enter into service contracts in which shippers agree to commit a minimum cargo within a specified time period, whereas the ocean carrier or conference commits to a rate schedule for the cargo and specific service guarantees. Shippers can now form nonprofit shippers associations to consolidate freight to secure volume discounts or service contracts.

The new flexibility offered ocean carriers and cargo shippers follows the recent trend of deregulation in the U.S. transportation industry. Deregulation in waterborne commerce, however, did not go nearly as far as it did in trucking, rail, and air transportation. Nevertheless, the FMC now has less freedom to interfere with transportation pricing and the private sector has more freedom in the rating and routing of ocean shipments.[7]

## *The Legal Right of a Port to Cargo*

Since the 1920s, some Ports have argued before the FMC that the Shipping Act and Merchant Marine Act provide safeguards against economic discrimination at the hands of the carriers. They seek protection for flows of cargo that allegedly are exclusive or "naturally tributary" to individual ports or port ranges. They are reacting against a long-established carrier practice of drawing cargo into load center ports through artificial financial or routing inducements, such as port "equalization," "absorption," or "transshipment." Under these inducements, the carrier either pays the shipper the additional overland transportation cost necessary to bring cargo to the load center rather than the port nearest to the shipper, or the carriers will transship goods overland at their expense from the port where the goods originally were delivered to the dock where the carrier's ship will call (Payne 1986, 259). In both cases there is a "losing" Port, one that is denied cargo moving business. The winning Ports and their carriers argue that the cost savings and efficiencies improve transportation services and reduce shipper and consumer costs. They point out as well that innovation is stifled if flexibility or change is not permitted. The losing Ports and the carriers serving them argue that they can provide adequate service and that their infrastructure investments (in many cases paid for mostly with federal dollars) and local economy need protection. The losing Port argues also that Section 8 of the Merchant Marine Act requires port protection as a matter of national policy, and that the Shipping Act forbids common carriers from giving undue or unreasonable preference or advantage to any person or locality.

Payne (1986) outlines the various principles the FMC has used to resolve these competing claims. Although the first port discrimination cases were resolved by balancing the interests of shippers, carriers, ports, and the public, the 1941 case *City of Mobile*[8] ruled in favor of the ports and enunciated the principle that ports have a fundamental legal right to cargo traffic that is "naturally tributary" to that port. Mobile successfully argued that the ocean carriers equalization rates designed to funnel Midwestern goods through East Coast ports negated the lower rail rates available to Mobile, thus damaging port capacity of the United States, harming competition, and adversely affecting commerce. In a number of cases over the next thirty years, carrier practices designed to bypass some ports in favor of the load center port were challenged, and the FMC was called upon to interpret the scope of the port's right to naturally tributary cargo. In some cases the competing ports were found to share the same cargo hinterland: a diversion from one to another was, thus, fair competition, not an unlawful diversion. In other cases, carriers were permitted to defend their practice by showing that there was inadequate cargo going to the complaining port. In one case involving a complaint by Portland, Oregon, against Seattle, the FMC determined that the ocean carrier had to call Portland, Oregon, on every other voyage in order to satisfy in part the right of Portland to continue to handle its local cargo.

In 1978 the FMC made a radical departure in its doctrine in what have come to be known as the "minibridge" cases. These cases involved goods flowing by rail and ship from the lower Mississippi River Valley to Europe via the Port of Charleston rather than New Orleans; and the flow of goods from the Far East to the United States Eastern Seaboard via Pacific Coast ports rather than East Coast Ports. The FMC recognized that no port or range of ports could claim a multistate inland region as "naturally tributary" to it and that if transportation innovations and efficiencies were to occur, a balancing of interests between ports and carriers was needed, not a "right" that protects certain ports. The commission noted that the protection of ports is only one of many factors in national policy. The new rules developed by the FMC, articulated first in the *CONASA* minibridge case,[9] have been applied in two important cases since then, one that protected the port, the other upholding the carrier practice.[10] Payne (1986) concludes that the doctrine of naturally tributary cargo may have no vitality left, but that the FMC still has the burden of balancing important and conflicting interests.

> The question therefore becomes one of policy; is there some point at which the rights of carriers frantically funneling cargo by whatever means, at various rate levels into larger and larger ships and load centers must bow to the greater public interest encompassing a port and its community? This must be considered in light of the substantial powers which have been given to steamship lines (both individually and collectively), under the Shipping Act and which can damage ports. Despite the benefits that have accrued from competition [and] innovation . . . there is still an ever-present danger surrounding the control of the shipping outlets for this nation's commerce solely by commercial interests (311).

Taken together, the Shipping Act of 1984 and the new FMC port discrimination policies suggest that Ports will get less special consideration or protection from the FMC, and the ocean carriers will have more freedom to choose and use the port that is most cost-effective to them. Ports will need to depend far more on their own competitive ability in the marketplace. This puts the burden on them to make themselves attractive either through pricing or special services (see Chapter 10).

Some fears have been expressed that the new freedoms will result in further load center concentrations in fewer ports, with severe consequences for some smaller ports and for U.S. port capacity as a whole. Marcus (1986) and Payne (1986) point out, however, that load centers may be proliferating and the fear of continual concentration of cargo in ever fewer ports is unwarranted. Some carriers or commodity shippers will establish load centers in smaller ports because of their location, the availability of space or special services, the possibility of avoiding congestion, and the chance of acquiring a special and dominant role in the smaller Port.

Further, since "ships follow cargo" and some ports can become too large and complacent, aggressive ports can remain competitive through specialization and particular programs that aid the shipper and inland transportation providers. In sum, the retraction of the FMC as a potential ally of the ports may not harm the port industry directly, but it means that each Port must fend for itself in finding a niche in the fast-changing transportation field. However, it is important to bear in mind Payne's observation in the above quote. If Ports ultimately are controlled by private transportation interests, longer term military or developmental interests of the nation may be overlooked. For this reason the FMC must monitor the nation's interests in port development and be ready to act to protect those interests, if necessary.

## NEW COST-SHARING ON FEDERAL NAVIGATION PROJECTS

The fact that Ports will get less help from the FMC in the area of transportation regulation reflects a policy of federal government retrenchment in port affairs. A parallel reduction in the federal role is occurring with respect to federal navigation projects. The Water Resources Development Act of 1986[11] calls for the federal government to pay a smaller share of the costs of these projects than in the past. Local port interests will have to finance a significant portion of the construction and feasibility study efforts, either through their own money or through so-called user fees, which they may be reluctant to impose on their customers. This could have a great effect on the competitive posture of ports—those ports with extensive dredging needs and a weak financial base have a distinct disadvantage compared to others more favorably situated. On the other hand, the new cost-sharing formulas could result in fewer, more justified projects, thus saving public dollars and natural resources.

Traditionally, the U.S. government paid virtually all the costs of deepening and widening of channels, disposing of dredged materials, building harbor works such as jetties and breakwaters, and maintaining the channels and other works. The concepts of cost-sharing and user fees have a lengthy history (Marcus 1976; Hull and Hull 1967), but interest in them accelerated with the growth of deregulation under President Jimmy Carter, and especially under President Ronald Reagan. These presidents withheld their support for new construction projects until the cost-sharing and user-fee issues were resolved. In Reagan's 1982 budget message, a bold cost-sharing and user-fee proposal was set forth to end the federal port construction and maintenance stalemate.

The initial Reagan administration position called for recovery of 100 percent of construction, operation, and maintenance costs from nonfederal sources. The primary nonfederal source would be a federally assessed, nationally uniform set of charges on waterway users.

A lively national debate resulted from the Reagan proposal. High-volume ports worried that if fees were imposed on a national basis, they would be subsidizing competitive ports. Shipping interests argued that they would be paying user fees even if they did not need a deeper channel and that nonpaying indirect users, such as railroads, would benefit. Others argued that since port dredging was in the national interest (by leading to increased exports, hence reduction in balance of trade deficits, increasing employment, and so on), it should continue to be financed out of the general fund (NRC 1985).

U.S. Ports were divided into two factions over this Reagan administration proposal. For years this division in the industry prevented them from effectively bargaining a compromise.

> One faction, which called itself U.S. Port System Advocates, endorsed the concept. These ports tended to be smaller ports with extensive dredging requirements. For their members a nationally uniform charge would function as a form of "cross subsidization," since there would be surplus revenues collected at large traffic and/or low project cost ports. The other faction, called the National Coalition for Port Progress, made up of larger ports with less dependence on federal dredging support, sought instead a system of non-federally set, collected and retained charges at levels sufficient only to pay for their own projects and thus avoid their subsidization of competitive ports (Hammon 1987, 1–2).

The Water Resources Development Act of 1986 reflects a compromise among these contending interests. Under Title I, Section 101, nonfederal sponsors pay varying percentages of the costs of construction of commercial navigation improvement projects during construction according to their depth: 10 percent of the cost up to 20 feet, 25 percent of the cost between 20 and 45 feet, and 50 percent for facilities deeper than 45 feet. In addition, nonfederal sponsors pay for lands, easements, rights of way, relocations and dredged materials disposal areas. They also pay another 10 percent after construction for a period up to 30 years with interest. For deep-draft projects, nonfederal sponsors pay 50 percent of the operation and maintenance costs necessary to keep channels deeper than 45 feet in good repair.

Under Section 204 of the act, the Secretary of the Army can authorize privately constructed and funded navigational improvements if he finds that they are economically justified, environmentally acceptable, and constructed according to accepted engineering standards. If these requirements are satisfied, the government will reimburse the federal share to the nonfederal interest and provide 100 percent of the maintenance to 45 feet, and 50 percent of the maintenance below that depth.

The act authorizes a non-federal interest to collect fees from commercial vessels to finance its share of construction, operation, and maintenance costs of a completed project or project segment. These "Port or Harbor Dues" are heavily regulated in the act. Under Section 208, harbor dues:

> . . . may not be levied . . . in conjunction with a deepening feature of a navigation improvement project on any vessel if that vessel, based on its design draft, could have utilized the project at mean low water before construction . . .

In other words, vessel operators protected themselves by incorporating into the act the principle of "no need, no pay." This concept, although having economic logic in that it asks those gaining the benefits to bear the costs, will put severe limits on the ability of Ports to recoup their share of project costs. As Hammon (1987) notes:

> Many port sponsors will have to absorb such costs, either because fully compensatory port or harbor dues would be so high as to drive ships away or because competing ports already have the depths they need, depths that were fully paid for earlier by the Federal Government and on which port or harbor dues are therefore prohibited. Most public port agencies are not sufficiently endowed with the ability in the long term to pay for channel construction and maintenance (3).

In the end, the Ports' two main arguments against user fees and increased local cost-sharing were not heeded. First, they argued that a percentage of the more than $8.5 billion annual customs receipts collected at the ports (some of which are earmarked to benefit particular programs) should be dedicated to federal channel improvement projects. They argued that the ship using the harbor was only one of many who benefit from port development and that port development advanced a variety of social and economic goals at present and in the future. Thus, in the Port's view, it was unfair to choose any one class of beneficiaries to bear all the costs. Also, they pointed out, the federal treasury is the greatest beneficiary of the ports because of customs receipts collected there, and thus the United States should cover the cost of harbor improvement and maintenance.

Second, the Ports lost the argument that if user fees are to be assessed, they should be widely applicable:

> The ports had lobbied for the right to assess port or harbor dues on as broad a traffic base as possible so as to reduce charges per ship, treat all traffic more or less uniformly, and increase their revenue. Their view was based on the concept that a port constitutes various non-federal investments needed to make it safe and efficient for shipping and that the right to do business in the port carries with it a responsibility to share the cost of such investments. Many European ports recognize this principle. Moreover, even one improved project among several in a port generates broad scale benefits, though not necessarily equal, for all port users. An analogy is a public school system in which

all residents, not just those with school age children, pay school taxes and thus enhance the community in which they live, work, and play. (Hammon 1987, 3.)

Although the Ports lost their two main arguments, the 1986 act authorizes 41 harbor and channel navigation projects, worth $2.2 billion (federal share only), and it allows nonfederal sponsors to initiate planning, design, and construction (with reimbursement for the federal share coming later) with fast-track permit schedules and procedures. Thus, in exchange for the increased financial burdens, the ten-year delay in the authorization and funding of new port projects came to an end.

The role of the federal government in navigation project development is substantially changed. The 1986 act brought about a historic shift in responsibility. In the past, the COE led the way in planning, executing, and financing navigation improvements. With the movement away from total federal control of projects and toward local initiation and implementation, the nonfederal sponsors—port authorities for the most part—are assuming a larger role. Ports with the financial and technical abilities will take the lead and prosper under this new system. In a sense, Ports are in a freer, less confined environment. They are less dependent on the lengthy bureaucratic procedures of the COE. But this new freedom has a cost. The COE will be only one of many entities with which Ports work in achieving navigation improvements. To raise money, alliances with carriers and other commercial interests, as well as local government funding sources, will be necessary. In this arena the Reagan administration has effectively implemented its de-federalization policy.[12]

## PROTECTING THE HARBOR ENVIRONMENT: A NATIONAL PRIORITY

If the federal government is withdrawing from economic regulation and shedding responsibilities for navigation improvement, it is doing the opposite in environmental regulation. In the past two decades, there has been rapid growth in federal law affecting coastal development along shorelands and tidelands and in bays and harbors. Prior to the late 1960s, the COE concerned itself only with the promotion of civil works projects, such as deepening channels in harbors, and with the review of development in U.S. navigable waters to ensure that the navigable capacity of the nation's waterways were not impaired (Power 1977). Since that time, a series of laws and policies have been adopted that broaden significantly the scope of the review performed by the COE and that provide roles for many additional public bodies in the review of coastal development. The regulatory framework today is complex because of the many agencies involved, but coastal policies have emerged that are mutually supportive and articulate a national policy preference in favor of environmental protection in harbors and promotion of public uses.

The first step in this major shift occurred in 1970 when the National Environmental Policy Act (NEPA)[13] was first implemented, and the U.S. Fifth Circuit Court of Appeals decided the case of *Zabel v. Tabb*.[14] NEPA required all federal agencies to consider the environmental consequences of their decisions, to consider alternatives to their actions that may have less damaging consequences to the environment, to consult with other agencies and the public to broaden the base of environmental information, and, when necessary, to prepare an environmental impact statement for those major federal actions that would significantly affect the quality of the human environment. NEPA's effect was especially strong when it resulted in a new interpretation of a regulatory program. This happened to the COE in *Zabel v. Tabb*. In that case the plaintiff land developer was denied a permit by the COE[15] to fill portions of Boca Ciega Bay in Florida for a trailer park because of the adverse environmental consequences of that fill. Studies had shown a direct correlation between landfill and loss of fish resources. The plaintiff claimed that the COE's jurisdiction was limited to a review for the effects of his project on navigation, the traditional function of the COE. The court disagreed with the plaintiff and upheld the COE decision to deny the permit saying that COE's basic law must be read in conjunction with the newer environmental laws now emerging. The court noted that the COE must interpret its law in conjunction with NEPA, which requires considering and weighing environmental consequences, and also said that after consultation with the U.S. Fish and Wildlife Service (FWS) and the National Marine Fisheries Service (NMFS) under authority of the Fish and Wildlife Coordination Act,[16] the COE could recommend denial because of environmental consequences. The court concluded that Congress intended collaboration among agencies and a consideration of diverse factors in decision-making: the COE need not wear "navigational blinders" when reviewing permit applications. Thus, after this case, the COE, the agency with the broadest mandate to review projects affecting navigable waters of the United States, could use a wide range of criteria in reaching a decision. It is through this mechanism that environmental factors have come to play such a large factor in coastal decision-making.

In 1972 Congress adopted two additional laws that substantially enhanced the regulatory framework for coastal development. Section 404 of the Federal Water Pollution Control Act of 1972[17] gave the COE authority to review the discharge of dredged or fill material into the waters of the United States. Since dredge and fill was a traditional COE function, and one that could affect its civil works and existing regulatory functions, applicants for this type of activity would continue to seek permits from the COE as they had done in the past. To ensure that water quality evaluation was built into the review process, the COE had to apply criteria adopted by the Environmental Protection Agency (EPA) with respect to each proposed discharge.[18] As a result, the COE issues the permit, but EPA has a substantial say in the process and can require that an application be denied. Thus, with this act, the COE added water quality to its arsenal of statutory responsibilities, which already included protection of fish and wildlife and consideration of general environmental factors under NEPA.

The Federal Coastal Zone Management Act (CZMA),[19] adopted in 1972, substantially enhanced state and local government regulation of shorelines and added yet another set of factors to the COE decision framework. The CZMA provided financial and legal incentives for state and local governments to develop and administer coastal management programs. These programs would include plans, policies, and regulations to guide land and water uses in the state's coastal zone. Financial incentives included federal grants-in-aid for planning and administering coastal programs. Although the CZMA is voluntary, virtually all states and territories are participating, and coastal management programs are important tools in controlling coastal development (Archer and Knecht 1987). When a state's coastal management program is approved by the National Oceanic and Atmospheric Administration (NOAA), the administering agency, another incentive becomes available to the state—the "federal consistency" provisions.[20] These provisions require that other federal agencies such as the COE, take actions that are consistent with the federally approved coastal management program of the state.[21] Thus, the COE considers the planning and regulatory actions of state and local government in its review process as well as the comments of FWS, NMFS, and EPA.[22]

This broad-based coastal decision framework, involving many laws and agencies at all levels of government, has been elaborated on over the years. The geographical limits, policies, standards, and procedures are more clearly spelled out. For example, the geographical limits of jurisdiction under Section 404, after years of litigation, now include virtually all U.S. surface waters, such as wetlands, bogs, ponds, sloughs, and intermittent streams, as well as navigable waters in general.[23] EPA's guidelines under Section 404 require that any discharge of dredged or fill material minimize adverse impact on the aquatic ecosystem. Also, they prohibit discharges that will cause or contribute to significant degradation of U.S. waters.[24] The COE wetlands policy gives special protection to wetland areas by requiring that the benefits from any alteration of wetlands be shown to outweigh the damage to wetland resources, and that the use to be made of the wetlands area be "water dependent."[25] The COE also has published nationwide permits that provide authorization in advance for certain classes of minor work in navigable waters, such as placement of buoys, nets, and navigational aids, routine repair and replacement, and survey activities.[26] The state CZM programs, which regulate shorelands as well as tidelands and waters, have stressed protection of aquatic resources, public access to the shore, avoidance of coastal hazards, restoration of urban waterfronts, and preference for water-dependent uses.[27]

Not surprisingly, many of the agencies have rules dealing with the same issue or problem. This is so because the authorizing laws refer to the same geographic area and often have the same environmental protection or resource allocation objectives. The result is that similar principles of coastal management are espoused, but they are implemented differently, to the consternation of permit applicants. Most of the agencies, for example, are protecting wetlands, minimizing landfill in water areas, encouraging only necessary water-dependent uses to encroach on the water (e.g., no housing over water because housing is not water-dependent), requiring mitigation where there is an unavoidable adverse impact on aquatic resources, and promoting public access whenever possible. But the agencies will differ on such matters as the definition of wetlands, the justification for fill, the definition of "water-dependency," and the appropriate type and amount of mitigation or public access. (Chapter 12 is devoted to the subject of environmental mitigation in harbors.)

In the late 1980s, Ports and other coastal developers are faced with a maturing environmental regulatory framework, but one that does not yet provide predictability or consensus. Because of the case-by-case review structure, the conflicting standards of review, the differences among coastal environments, and the inadequate links between long-range plans and individual permit applications, the pattern of permit battles on individual projects will continue. Also, new issues arise that must be integrated into the review process. In many older harbors, for example, toxic and hazardous materials buried on port lands or in submerged lands raise problems of containment, disposal, or treatment that will affect the pace, cost, scale, and location of development in the future.

Dealing with the rapid expansion of environmental regulation has resulted in considerable change within Port organizations. Virtually every organization has hired "permit chasers" (sometimes called environmental planners or specialists) who prepare permit applications and environmental impact statements, and who participate with regulatory agencies on joint planning committees dealing with environmental, land use, and neighborhood issues. More significantly, a new set of pressure groups concerned with environmental and neighborhood issues are actively influencing Port policies and the choice of Port commissioners. The result is a new awareness of environmental issues among top staff and commissioners, and a change in procedures that allows this sector of the community greater opportunity to be heard. In the equation of port growth and decline, the management of the natural and human environment has become an important, and in some cases determinative, factor.

It is fair to conclude that federal environmental law has had an enormous influence on the port industry. The sheer number of laws that have been passed since 1970 and the range of subjects they treat show that Congress has given priority attention to the protection of the harbor environment. During this same period of time, few initiatives have been undertaken at the national level to benefit port and maritime industries. Even the Water Resources Development Act of 1986 provides new navigation improvement projects only at a substantial cost to the port sector. With the exception of a "do-it-yourself" option, it has done little to streamline the already complex and lengthy water resources development process. The Reagan Administration has attempted to curtail spending on environmental programs and to impose fast track environmental reviews, but they have had little success. Environmentalism is a powerful political and social force that has restructured the way in which we do any kind of physical development, including development in the harbor.

## FLIRTING WITH PORT REGIONALISM

In the 1970s, trends in the port industry and shifts in federal policy were putting new pressures and costs on Ports. Deregulation of the transportation sector accelerated shifts toward intermodalism and the load-center/feeder port arrangement and raised questions about the viability of some Ports in the future. Federal subsidies for port and navigation projects were being curtailed and Ports were turning to state and local government sources or to the private sector for financing. Environmentalism was imposing new costs and slowing the pace of development,[28] and environmentalists challenged whether all the Port projects proposed were actually needed. In some state legislatures, bills were introduced to consolidate port districts or to impose new regulations on them. Because of these fiscal and environmental issues, Port projects were scrutinized more carefully. Ports realized that in order to gain financing and pass environmental review, they needed stronger justification for their projects. One way to add weight to their argument was to show the need for their project in a regional context. Thus in the mid-1970s groups of Ports began to experiment with various forms of regional port studies and cooperation dealing with the question of facility needs.[29]

These voluntary efforts at cooperation were a defensive reaction to many charges. First, Ports were reacting to charges of overcapacity in container terminals that had been documented in national-level studies (Marcus 1976), and in the concern that Ports not developing container terminals would become obsolete. Second, environmentalists became concerned that redundancy or duplicative services would unnecessarily consume coastal resources such as wetlands and intertidal zones, and they opposed "speculative fills"—projects to fill coastal waters in the hope of attracting a tenant (Hershman 1978). Third, many opposed port expansion if underutilized facilities would result because of the inefficient use of public funds. At this time, virtually all Port projects received some form of governmental subsidy, and a number of commentators were suggesting that the case-by-case evaluation of Port projects was no longer adequate; Port projects should be evaluated against expected regional growth and development (Schenker 1973; NRC 1976; Marcus 1976). Fourth, capital investment costs were quite high and it was becoming harder to raise that money. Not only were self-generated revenues and local tax sources thin, but the federal government was urging a reduced level of support for channel dredging projects and the elimination of tax-exempt status for Port bonds.

Because of these political and economic pressures, Ports agreed to participate with MARAD and other federal agencies in conducting regional port studies. Sixteen such studies were conducted between 1975 and 1983 (Wardwell 1983).[30] The Washington State experience illustrates how the studies were carried out and how the information was used to aid Port development decisions. The Washington example shows that this federally initiated effort resulted in only a very modest step in the direction of greater regional cooperation.

The perceived shortcomings of the local port system in Washington State were particularly controversial in the 1970s and 1980s. Washington has 73 port districts in existence, 15 of which are deep-water ports in western Washington. Six larger ports serve the Puget Sound region (Seattle, Tacoma, Everett, Bellingham, Olympia, and Port Angeles). Concern was expressed over the justification for and manner of investment of public funds in Port facilities. There were charges that duplication of services and inefficient Port operations recurred and that a regional or statewide authority was needed. In 1969, 1971, and 1975, legislation had been introduced by Representative O'Brien from Seattle calling for such a remedy. Similar bills have been introduced into the legislature in the 1980s. The Washington Public Ports Association (WPPA), authorized in 1961 to represent Port interests before the state legislature, initiated efforts to head off externally mandated coordination or statewide control. In 1972, a committee was established to design a study of Northwest Ports. This led to the MARAD-sponsored Port System Study, begun in 1974 and completed in 1975 (Olson 1980).[31]

The WPPA study was to provide a regional analysis as background information for Port development decisions. It was not intended to hinder individual Port development initiatives. The study assessed regional port needs based on existing and anticipated trade patterns, facility capabilities for handling different types of cargo, and technological trends. The study identified the number and types of facilities needed in the region to meet future growth, but it stopped short of saying where those facilities should be located. It emphasized the significant economic role of the public ports, clearly a pitch for the industry's accomplishments. The original 1975 Port System study (WPPA 1975) has been updated at five-year intervals (WPPA 1980, 1985).

The Cooperative Development Committee (CDC) of WPPA was established to use the results of the 1975 Port System Study in the review of Port projects. In this way, Washington's effort was different from most other Port studies in the United States. It went beyond the mere compilation and publication of regional trade and port capacity information and designed an institution to conduct a voluntary "peer" review by the industry of separate Port projects. The CDC reviewed whether the project was needed by comparing its justification to the information in the regional studies. The CDC review also considered the suitability of project siting, design, costs, and environmental factors. Although modest, this was a first for the nation. Never before had competitive port districts looked over each other's shoulders on a regular basis to review project proposals.[32]

By 1982 the CDC had reviewed 34 projects. Only two failed to receive an endorsement, and neither involved cargo handling facilities. The CDC has not reviewed projects since 1983 when the competitive race between Seattle and Tacoma, the two largest Puget Sound ports, heated up. Olson (1980) believes the CDC performed particularly well in sharing information and in preparing updates to the Port System Study. However, he notes two major problems: not all Ports participate—

there are free riders; and there are no sanctions for noncompliance. Levie (1986) concludes that the CDC structure precludes any effective regional planning. First, it is an organ of the WPPA and that organization is fundamentally opposed to any coordination that sacrifices Port autonomy and local initiative. Second, he observes that regional port planning must address multiple use issues and the social opportunity cost of port development, and to do this a broader base of interests and perspectives must be represented. He believes the Washington experiment is not one of regional port planning per se, but local port planning and regional port comment.

The other major effort at regional port planning in the United States, the Seaport Plan for San Francisco Bay (discussed in more detail in Chapter 1) resolves some of the problems plaguing the CDC pointed out by Olson and Levie. All six Ports in the Bay Area participate and the plan is officially adopted by an agency with power of implementation, the Bay Conservation and Development Commission (BCDC). Further, the plan was developed by a broad-based group involving Ports, carriers, environmentalists, and public agencies. The problems of Port growth and competing land use, environmental and fiscal values were addressed directly in negotiating the elements of the plan (Fay 1983). Thus, the Bay Area Plan is an enforceable set of guidelines determining future action, whereas the CDC project review process unofficially tried to influence and improve individual projects.

It is important to keep in mind that the CDC and Seaport Plan for San Francisco Bay are the only two of the sixteen MARAD regional studies that took any steps in the direction of influencing the outcome of Port projects. For the most part, the MARAD regional studies program has been an experiment encouraging port regionalism through financial incentives. It is an important first step, though a modest one. Some new information has been developed and used in project review. The issue of regionalism has been introduced and debated in many locations around the country. But the fundamental structure of the industry as a set of decentralized, autonomous, and independent Ports seems as strong as ever in spite of these regional studies, with the possible exception of the San Francisco Bay example. It is evident that even though environmental, fiscal, and competitive pressures on Ports are growing and regional action could relieve some of the pressures, these developments are not yet strong enough to overcome the independent structure built into the industry.

## THE DECLINING FEDERAL ROLE: BY CONSCIOUS DESIGN

U.S. federal port policy is hard to pinpoint. The federal government's influence comes in discrete sectors of port and harbor activities, such as ocean carrier regulation, navigation improvement, or water quality. Within most sectors, there is variation because federal policy is implemented at regional, state, and local levels, thus involving additional players and policy interpretations. Further, port development often feels the effects of federal policy aimed at others, such as land transportation carriers. Thus, federal port "policy" can vary from harbor to harbor because it is the net effect of many diverse policies. In other words, federal port policy is only known after the fact and in a particular context.

The burden of using, integrating, rationalizing, and influencing these various federal programs is on the local Port manager. In addition to being skillful in the competitive business environment, the successful Port manager must be adept at managing diverse and uncoordinated governmental programs. As noted before, maximizing the subsidies and minimizing the constraints is a substantial part of the job, and one that often can provide the competitive edge.

Even though the United States has no central bureau that aids in the administration and development of ports, as in the cases of Canada and Japan noted at the outset of this chapter, federal regulatory, fiscal, and environmental policy has substantially affected U.S. port development since about the early 1970s. New dredging projects were halted for ten years until a cost-sharing arrangement was worked out and a user-charge system adopted. Port protection against cargo diversion by private companies has virtually ended. A wide range of environmental regulations dealing with land, air, water, wildlife, waste, and chemicals has increased the time and cost of projects. And port regionalism, virtually unheard of heretofore and initially opposed by the industry was undertaken in part because of federal initiative. The fate of ports and harbors is inextricably tied to federal policy even though administrative structure is largely local.

Many other recent changes in federal law and practice have affected Port and harbor development. Not all could be outlined in this chapter. For example, the overhaul of the federal tax code in 1986 results in loss of the use of tax-exempt industrial revenue bonds for some port-related projects and in the forfeiture of arbitrage earnings from money raised by use of tax-exempt bonds. Federal policy imposing personnel cutbacks in the U.S. Customs Service affected the flow of commerce at Ports. Ultimately, user fees to support the Customs Service were established in 1986. The comprehensive trade bill being debated in the Congress in 1987, designed to give U.S. exports a boost in the world marketplace, could affect cargo flows. By administrative action, the Reagan Administration has reduced the port-related activities of MARAD. As a result, regional port studies, university research, compilation of port data, and harbor dredging simulations have virtually ended.

In effect, the national framework provides special-purpose subsidies and specific enumerated constraints, but it lacks leadership—a place in government where diverse agency functions can be coordinated, where problems can be identified and researched, and where public objectives can be formulated. Those agencies with the mandate—COE, MARAD, and FMC—have been restrained fiscally, told to decentralize and interfere minimally in local affairs, and directed to apply user fees and to privatize wherever possible. The primary burden of identifying Port problems and issues, researching their causes and effects, and designing responses has fallen on the shoulders of the American Association of Port Authorities (AAPA), a relatively small industry association that works primarily through the voluntary efforts of its members and a ten-person staff. Other port-related research has been done by the Office of Port and Intermodal Services of the Maritime Administration (1985) and by the Maritime Transportation Research Board of the National Research Council (1976, 1985), now called the Marine Board.

The leadership void within national government needs to be reversed. A focal point for port affairs is needed within the federal agency structure, one that can be effective in coordinating, analyzing, and articulating national port isues and in stimulating action that would lead to their resolution. Ports are an essential national infrastructure for defense and trade, thus national strategic interests of the United States are a factor in many port development decisions. Port affairs cross over many aspects of federal responsibility (commerce, environment, customs, defense); coordination should be a function of the federal government since it is within the power of federal agencies to do something about coordination problems or policy conflicts. Ports are public agencies, and smoother cooperation among all levels of the public sector aids the consuming and tax-paying public.

Leadership in port affairs must deal with more than coordination, however. A number of policy issues have been identified that deserve research, and perhaps action, that could be agenda items for a revitalized port affairs function in the federal government. For example, concerns have been expressed that environmental and recreational goals are given priority in many harbors, that maritime industries such as shipbuilding are being displaced, and that reversion of waterfront lands from recreation *back to* industry at some point in the future when demand increases will be impossible. The federal government has conflicting policies that call for the promotion of water-dependent use of the shorelines and that advance urban recreation and environmental values. Is there a problem from a national perspective and if so should something be done about it? Federal agencies have not responded to this issue even though it has been raised on many occasions, nor has information been gathered about it (see Chapter 13).

Another example relates to the nation's need for deep-water ports—those dredged to a depth of 50 to 55 feet to handle large bulk vessels such as colliers in the export trade. A recent study by the National Research Council (1985) concluded that only one port on each coast is needed for such development and that a choice must be made among contending ports. Given the project-by-project review process, the considerable political influence exerted over navigation projects, and the new water resources law, which gives advantages to wealthier ports that can meet local cost-sharing needs, there is little opportunity for comparison among competing ports and the application of objective criteria for decision making. Some new form of port decision making, other than the single project review, must be conceived and implemented if the NRC recommendation is to be heeded.

Other policy questions affecting Ports await serious consideration. What will be the cumulative effect on Ports of the new user fees for dredging, customs, and trade adjustment assistance; and potential user fees for Coast Guard services and ocean dumping rights? Are the environmental mitigation and permit-processing policies of the COE, EPA, USFWS, and NMFS unduly delaying Port projects and increasing costs? Do the issues surrounding contaminated or hazardous materials and wastes in harbors—inspection, containment, disposal, treatment, liability—require direct action by public ports or can users and waste management firms resolve the problems directly without Port intervention? Should cooperation among Ports within a region be encouraged through federal incentives, building upon the MARAD-funded efforts reported above?

Since 1968 seven major studies analyzing aspects of national port policy have concluded that the federal government needs a better planning and coordinating function with respect to ports and harbors. These studies recommended a variety of approaches, including federal subsidy of port development, regional benefit/cost assessments, and standardization of port data.[33] It is fair to conclude that none of these recommendations urging an enlarged federal role has been implemented, nor have they been given serious consideration. Other domestic priorities have demanded the attention of national policymakers, notably environmental protection, cost recovery, and de-federalization. The suggestion that regional port coordination be advanced (NRC 1976, 127) was implemented by MARAD, but that worthwhile effort was terminated by the Reagan administration in 1982. The decentralized system of port development and administration is alive and well in the United States, and by conscious design, recent changes in federal port policy reinforce that structure. It is time, however, for the federal government to establish a focal point among its many agencies for continuing and responsible attention to the substantive issues of port development.

## *Notes*

1. A number of U.S. Port terminals have been built, operated, or leased by the Department of Defense for miltary transportation purposes since World War II. Military ports and terminals, and DOD policy affecting their development and operation, are important components of the total U.S. port operation, but are not dealt with in this chapter.

2. Subsidies that are not as large as expected, or that involve red tape, or that reduce flexibility may be viewed as constraints. The characterization as subsidy or constraint may depend on who is using the term.

3. There are a variety of ways to classify public agencies relevant to ports, and a variety of criteria for determining relevance. The National Research Council (1976, Appendix C) listed forty-five federal agencies under three categories: federal inspection agencies, federal port-related agencies, and federal independent and regulatory agencies. Marcus, et al. (1976, Appendix A) listed forty federal agencies. Both listings include agencies with only a very minor and tangential connection to the Port, e.g., the Federal Communications Commission, the U.S. Geological Survey, and various local planning assistance programs. Table 5.1 is limited to the primary agencies with which Ports deal.

4. In 1981 the port investigation functions set forth in Section 8 were transferred from the FMC to the Secretary of Transportation. By then, however, the principle of port protection had been built into the regulatory practice of the FMC, as discussed in text below.

5. 46 U.S.C.A. Sect. 1701–1720.

6. 46 U.S.C.A. Sect. 867.

7. Many analysts believe that the Shipping Act of 1984 has ruined the U.S. Shipping Conference System. Shippers have benefited from lower rates, but many carriers are declaring bankruptcy. Further deregulation may occur in the future. The Shipping Act will undergo a major re-evaluation in 1989 at which time further loosening of requirements dealing with tariffs, filings, and antitrust determinations will be considered.

8. *City of Mobile v. Baltimore Insular* 2 U.S.M.C. 474 (1941).

9. *Council of North Atlantic Shipping Associations (CONASA) v. American Mail Lines,* 18 SRR 744 (FMC 1978).

10. The Port was protected in *North Carolina State Ports Authority v. Dart Containerline,* 19 SRR 521 (FMC 1979) *aff'd. sub. nom.* Dart Containerline v. F.M.C. 639 F. 2d 808 (1981); the carrier practice was upheld in *Pacific Westbound Conference—Equalization Rules and Practices,* 22 SRR 946 (FMC 1984).

11. Public Law 99–662; 100 *Stat.* 4089, signed November 17, 1986.

12. In a personal communication to the authors, Al Hammon points out that the Water Resources Development Act of 1986 makes it extremely difficult for private terminal operators to deal directly with the COE. Also, he notes that the law curbs the authority of the nonfederal interest to collect fees because of the "benefits" test, but the federal government receives funds for maintenance dredging through a Harbor Maintenance Tax, which does not require a showing of benefits. To him this unequal treatment is "outrageous."

13. 42 U.S.C.A. Sect. 4321, et seq.

14. 430 F. 2d 199 (5th Cir. 1970).

15. Since the late nineteenth century, the COE has had the power to authorize the construction of works in navigable waters of the United States (Power 1977). This power was codified in Section 10 of the Rivers and Harbors Act of 1899, 33 U.S.C.A. Sect. 403.

16. 16 U.S.C.A. Sect. 662–667.

17. 33 U.S.C.A. Sect. 1344.

18. These criteria are called the 404(b) (1) guidelines found at 40 C.F.R. Part 230.

19. 16 U.S.C.A. Sect. 1451 et seq.

20. 16 U.S.C.A. Sect. 1456.

21. There are five different "federal consistency" provisions covering different types of federal agency actions and different levels of consistency. See Eichenberg and Archer (1987) for a comprehensive treatment of the law of federal consistency under the CZMA.

22. The COE considers many other factors as well, such as historic preservation, effects on marine mammals, endangered species, water and energy conservation and development, property owner interests, etc. See 33 C.F.R. Part 320.

23. 33 C.F.R. Part 323.

24. 40 C.F.R. Part 230.

25. 33 C.F.R. Sect. 320.4 (b).

26. 33 C.F.R. Part 330.

27. 16 U.S.C.A. Sect. 1452(2).

28. The cost to the Port is not insignificant. Additional staff, environmental studies, and mitigation projects are expensive, but the cost of delay can be vital. Although all Ports now have environmental overhead costs, those who are quick and effective in resolving environmental issues can gain community goodwill and a competitive edge. In Chapter 12, Tampa and Long Beach are discussed as examples of Ports with a positive program on environmental matters.

29. Ports have always cooperated in lobbying and public relations, normally through professional associations. However, they have maintained strict autonomy in business dealings and facility development. Even when Ports would share resources to do regional marketing, as in the case of the Ports of Philadelphia or the Columbia/Snake Marketing Group, the object was to promote the general benefits of the region as a destination for cargo and not to arrange particular cargo movements. The latter was a sales function reserved for individual ports.

30. Levie (1986, 53–58) points out that the MARAD studies were preceded by many other federal initiatives of the COE and others to propose or carry out regional or national port studies. These efforts were opposed by the Port industry because they feared too much national interference and loss of local autonomy. By the mid-1970s, the pressures had grown significantly, and the structure proposed for the studies satisfied the Ports. With federal funding coming from MARAD, the federal agency charged to protect and advance Port interests, Ports believed their interests would be adequately protected. As it turned out, local Port interests largely determined the design and content of the studies.

31. This study originally included the Port of Portland, Oregon, but in subsequent updates Oregon was not a participant and ultimately did their own regional study (Oregon 1980).

32. The CDC consists of one representative from each WPPA member Port that chooses to participate. Each CDC member has one vote, regardless of Port size. Member ports are requested to submit "all projects for development of water-related cargo handling facilities to the committee for review and action" (Olson 1980). Three regional committees within the CDC review projects for their respective sub-regions and propose findings and recommendations to the full CDC. Evaluation of projects results in one of four possible judgments: approval, disapproval, take no action, and qualified approval. Approval is extended with a "certificate of endorsement" (Olson 1980, 26). Although the committee's jurisdiction and decision are not legally binding, the incentive to voluntarily subject a project to CDC review is that peer review could strengthen project proposals and could carry some weight in efforts to acquire land use and environmental permits and possible financing (Olson 1980).

33. In 1968 and 1969 studies by the Marine Science Council (a federal interagency coordinating body), the COE and the Stratton Commission (a statutorily created blue ribbon commission) recommended national port and harbor studies that would determine national port requirements and identify ways to improve port response to technological change. (NRC, 1976, 161–164.) In 1976 the National Research Council recommended the establishment of a Federal Aid to Ports Program to pay a significant portion of the capital costs of port projects. Decisions about development would be made at the local level, but would have to conform to defined standards of feasibility, safety, and environmental protection. Marcus et al. (1976) recommended that a comprehensive national port study be undertaken that would analyze the economic impact of federal actions on national Port structure and the competitive relations among Ports. Hershman et al. (1978) urged that the federal Office of Coastal Zone Management study regional and national interests in Port development and marine transportation and assist the states in considering the impacts of their coastal management programs on port development. Finally, the NRC (1985) noted that use of federal funds for 40- to 55-foot

channels should be preceded by a *choice* of which ports to deepen based on four criteria: multicargo capacity, inland transportation connections, costs of construction and maintenance, and environmental consequences. Although the NRC advised against centralized port planning for the United States, they urged an expanded federal role in port research, in the collection, analysis and dissemination of planning data, and in financial assistance for port planning. Further, the panel asked the COE to conduct regional as well as specific cost-benefit analyses to ensure that benefits of navigation projects are not overstated.

## References

Archer, Jack, and Robert Knecht. 1987. The U.S. national coastal zone management program—problems and opportunities in the next phase. *Coastal Management* **15**(2):103–20.

Canada Ports Corporation. 1986. *Ports Canada*. Report to the Minister—1985, Ottawa.

Eichenberg, Tim, and Jack Archer. 1987. The federal consistency doctrine: Coastal zone management and "New Federalism." *Ecology Law Quarterly* **4**(1):9–68.

Fay, Dennis. 1983. Discussion of the San Francisco Bay Area Seaport Plan. In *Coastal zone '83*. New York: American Society of Civil Engineers, 273–86.

Hammon, Alfred. 1987. The Water Resources Development Act of 1986 and ports. *Marine Policy Reports,* University of Delaware **9**(3):1–4.

Hershman, Marc J. et al. 1978. *Under new management: Port growth and emerging coastal management programs*. Seattle: Washington Sea Grant, University of Washington.

Hull, William J., and Robert W. Hull. 1967. *The origin and development of the waterways policy of the United States*. Washington, DC: National Waterway Conference.

Inoue, Satoshi. 1986. Japan's experience in port development. Ministry of Transport, Tokyo.

Levie, Jonathan. 1986. Port regionalism: Cooperation and coordination in port planning and development on the U.S. West Coast. Master's thesis, Institute for Marine Studies, University of Washington.

Marcus, Henry S. 1986. Trends in U.S. liner shipping: Past, present, and future. Paper presented at World Conference on Transport Research, Vancouver, B.C., May.

Marcus, Henry S. et al. 1976. Federal port policy in the United States. Cambridge, MA: MIT Press.

Maritime Administration (MARAD). 1985. Public port financing in the U.S. Dept. of Transportation, Washington, DC.

National Research Council (NRC). 1976. *Port development in the United States*. Washington DC: Government Printing Office.

———. 1985. *Dredging coastal ports*. Washington, DC: National Academy Press.

Olson, David. 1980. Regionalism in port development: The Pacific Northwest experience. Department of Political Science. University of Washington.

Oregon, State of. 1980. *Oregon ports system study 1980*. Salem: Oregon Department of Economic Development.

Payne, Stanley. 1986. The legal right of a port to cargo in the age of containerization: Going, going . . . not quite gone. *Transportation Law Journal* **14** (2):249–325.

Power, Garrett. 1977. The fox in the chicken coop: The regulatory program of the U.S. Army Corps of Engineers. *Virginia Law Review* **63** (4):503–59.

Schenker, Eric and Michael Bunamo. 1973. A study of the Corps of Engineers regional pattern of investments. *The Southern Economic Journal* 39.

Shipping Act of 1984. Public Law 98-237, 98 Stat. 67, 46 U.S.C.A. 1701–1720.

Wardwell, Robert H. 1983. Regional port plans: Government without regulations. *Coastal zone '83*. New York: American Society of Civil Engineers.

Washington Public Ports Association. 1975. Port system study for the public ports of Washington State and Portland, Oregon. Olympia.

———. 1980. Port system study for the public ports of Washington State. Olympia.

———. 1985. Port system study for the public ports of Washington State. Olympia.

Water Resources Development Act of 1986. Public Law 99-662, 100 Stat. 4089.

# Part III

# DIVERSE ACTIVITIES

## Chapter 6

# THE CONTAINERIZATION STORY:

## Meeting the Competition in Trade

*Paul W. Chilcote*

---

During the 1950s the United States was at a crossroad in trade policy. Confronting the nation at the time was the question of whether to return to pre-World War II trade barriers that would safely confine the economy, or to embrace an emerging world economy based on international trade. To turn inward would have been possible because only a small fraction of the nation's economy was dependent upon international trade. Also, postwar Europe and Japan had become robust economies that no longer required active U.S. support. Ultimately, the United States chose to commit to international trade, and the nation has, over time, developed an economy largely dependent on world trade. U.S. seaports, windows on the world economy, have become the critical link in the infrastructure of international land and sea trade.

As international trade rapidly expanded during the 1950s, 1960s, and 1970s, technological advances were transforming the shipping industry. Nearly all marine technological advances over these years had one common result: a dramatic increase in vessel size. Larger vessels, it was found, incurred higher operating and service costs, and because of these higher costs, they called at fewer ports. The resulting changes in steamship scheduling and routes have had a revolutionary impact on Port development in the United States and around the world.

The most far-reaching advance in marine technology development in the past thirty years has been in containerized general cargo shipping. Rapid worldwide transition to container shipping has spawned the construction of major new terminals in virtually every significant U.S. harbor.

Another effect of the new container technology has been to bring the port authority to the forefront of port development. The public port authority, rather than private industry, could best absorb the high costs of container terminal development and assume responsibility for long-range planning. Consequently, containerization has fueled increased competition among port authorities as dozens of them sought to capture new trade by building new terminals. This chapter focuses on the development of container shipping and its impact on ports, the port authority, and connecting land transportation.

## THE TRUCK-SIZE BOX AS A TRANSPORT MODE

Large reusable boxes for quick load/unload of ocean vessels can be traced as far back as Roman times. Early containerization efforts had, as a common goal, the ease of loading and unloading ocean vessels. The attempt was to standardize so that machinery could be used, and to "preload" cargo in order to decrease the length of time that a vessel lingered in port. More recently, the Conex box was a reasonably successful attempt at containerization. This standard 6-foot box was developed by the U.S. Navy during World War II, and over 100,000 were eventually placed into service by the 1950s (Mahoney 1985, 105–20). However, the Navy's Conex box, like many attempts before it, failed to catch on because it did not sufficiently increase shipping productivity. This lack of interest changed rapidly with the development of the modern container, which is essentially a mode in itself—a truck body. The modern container's size and transferability became an advantage for both load/unload and intermodal movement.

Modern containers are truck-size boxes that are typically 8 feet 6 inches or 9 feet 6 inches in height, 8 feet in width, and in lengths of 20, 40, and 45 feet. The capacity of a ship or terminal is often measured in the number of containers, or TEUs, it can handle. (TEU is the acronym for 20-foot equivalent unit, the common denominator of the smallest standard container length.)

The containerization concept is simple. Goods of any kind are packed into a container at any location, usually away from the waterfront. Containers can accommodate a vast array of cargoes; and some are specially designed for specific purposes, such as refrigeration for fresh fruit and frozen meat. The entire container is placed on wheels (a chassis) and moved by truck or rail to the marine terminal. Large gantry cranes lift the container from the marine terminal apron into a ship with slots designed to hold containers in place for the ocean voyage. The process is reversed once the ship reaches port and the containers are off-loaded. Containerization has made obsolete the delivery and storage of goods at waterfront sheds and the loading and packing of diverse small cargo items into the hold of a ship.

The most significant adaptation of a vessel for container use was made in 1956 by Malcomb McLean, creator of Sea-Land Corporation. McLean had truck vans placed on the deck of the vessel *Galveston,* as a substitute for overland truck service between New York and Houston. Being a trucker, McLean understood the advantage of loading an entire truck onto a ship in one step. The detachment of the truck body from the truck wheels (chassis) allowed for a one-time transfer, needing only a set of wheels at the destination for land delivery. The concept soon expanded from the original sixty trucks on a converted T-2 tanker to full containership design. By the early 1960s McLean and several other lines had introduced the container to the major U.S. trade routes, the transatlantic and transpacific (Mahoney 1985, 13–15).

The container brought two major changes to the operation of general cargo shipping—standardization and mechanization. Standardization was necessary to ensure equipment compatibility of a particular line's machinery with vessel hold dimensions. It was natural that shipping lines would eventually agree upon a size standard for containers (much like the railroads had done a century before in track and railcar standards).

Mechanization was a necessary development because of the large dimensions of the truck-size container. The importance of mechanization, however, was in the increase in cargo capacity per lift (over previous methods) and the speed with which a vessel could be unloaded and loaded. Standardization of cargo load, plus the greater speed of loading and unloading, dramatically increased vessel productivity (see Figs. 6.1 and 6.2).

Before the advent of containerization, a general cargo vessel would typically spend approximately half its life in port. Time-consuming steps to hand-stow cargo for the most efficient use of space in the vessel's hold required an enormous amount of port time. While in port, a vessel is pure overhead to the ship owner, since revenues are produced only when cargo is moved to another point, as is true with all forms of transportation. The laborious process of hand-stowing cargo had been somewhat improved by palletization of cargo, which allowed for the use of forklift trucks. However, the amount of time a vessel spent in port still did not change significantly.

High overhead costs resulting from vessel load/unload time in port meant that a precontainer general cargo vessel owner would expand business primarily by purchasing additional ships; not by substituting with a larger ship. The reasoning is as follows: only two options exist for expanding the carrying capacity of a shipping line. One is to build a larger ship to replace an existing ship. The other is to build additional ships of similar size. A larger ship would lower the ocean voyage costs per unit of delivered cargo because a larger ship carries more cargo, requires about the same number of crew (though slightly more fuel), and makes an ocean voyage in less time than that required by a smaller ship making an extra voyage to carry equivalent cargo. A larger ship would cost more to build, but certainly not as much as two ships.

Load/unload costs, however, would actually increase for a larger precontainer breakbulk (general cargo) ship with more decks, since it takes a comparatively longer time and is more difficult to load/unload. Thus, port time would increase at the expense of sea time. Additional costs for load/unload of a larger breakbulk ship are more than additional costs of adding smaller vessels to the fleet. Thus, the most feasible economic decision for the general cargo ship owner was to build additional ships.

**Figure 6.1.** Early twentieth-century forms of cargo handling at dockside include handcart (top) and pallet-loading using ship's boom and pulley (bottom). Both are labor intensive. (Courtesy: top—Museum of History and Industry, Seattle; bottom—Washington State Historical Society, Asahel Curtis)

**Figure 6.2.** 1980s modern container ship (Sea-Land Corp.) at the Port of Tacoma, serviced by gantry cranes that extend over the entire ship. This technology is capital intensive. The combined cost of terminal, ship, containers, and cranes well exceed $100 million. (Courtesy: Port of Tacoma)

As a line increased its number of vessels in the precontainer days, it meant that more ports could be called, and more often. A shipping strategy of more port calls was less costly to the line than using land transportation to collect and distribute cargo at a single port. Breakbulk vessels of the 1960s would typically call at as many as a dozen ports on each side of an ocean, taking up to several months to complete a round-trip circuit. This provided business for ports of all sizes.

Containerization reversed this economic logic by substantially reducing vessel load/unload time and thus reducing the amount of time a ship spends in port. The resulting gain in productivity is quite dramatic. The cargo-handling portion of total costs (of a typical shipping line) dropped from approximately 55 percent for a typical noncontainer breakbulk vessel, to about 21 percent for a typical container operation. By the early 1980s, the cargo-handling percent share had dropped even further to approximately 15 percent of total shipping costs. The enormous increase in vessel productivity, based on the container, greatly decreased the importance of port costs (Gilman 1980).

Sea-Land made an early demonstration of a 40,000 metric ton containership unload/load that required a total of 750 working hours. Sea-Land estimated that 24,000 working hours would have been clocked using breakbulk methods, a productivity gain of 32 times (Fairley 1979). Other estimates vary, usually closer to a five or ten times gain, which is still an enormous improvement in labor productivity. An ocean vessel, instead of spending half its life in port, could now spend a relatively acceptable 10 to 20 percent in port for the cargo handling operation.

Achieving these gains in productivity required substantial capital investment by shipping lines. A breakbulk vessel that cost $3 to $5 million in 1960, became a $7- to $12-million investment for a containership. Additional cost requirements, such as an inventory of containers and other specialized equipment, increased capital investment. Finance interest costs of servicing the large capital acquisition cost of a containership greatly increased as a proportion of overall costs, from a relatively low 17 percent for a breakbulk vessel to 33 percent for an early containership. Capital has maintained about the same share because of the rapid rise in fuel costs during the 1970s. However, with more fuel-efficient propulsion, capital debt servicing has again become the highest single cost component for port-to-port ocean shipping (Gilman 1980; Pearson and Fossey 1983).

In spite of the high acquisition cost of a containership, vessel owners found themselves in a "purchase or perish" situation in the 1960s, because of the inherent productivity gain. North Atlantic and transpacific trades are especially adaptable to container operations because they are highly competitive routes, where an edge over the competition translates into a high revenue potential. These two routes essentially converted to the container within a decade, by the end of the 1970s.

Ocean carriers realized that they could build increasingly larger vessels in order to lower unit container costs (provided, of course, that they could fill the vessel with cargo). The economies of scale were opposite to that of the breakbulk ship and encouraged the continuing increase in vessel size for new containerships. A larger containership does not substantially lengthen the time it takes to unload/load, because additional container cranes can be used as vessel size increases. Therefore, unload/load takes approximately the same time regardless of vessel size. This type of mechanization was not feasible for breakbulk ships because cargo was essentially hand-stowed for each vessel and could not be mechanized.

Increase in containership size since the 1960s has been phenomenal. Owners built vessels as large as they could possibly expect to fill—and usually a little larger, because the incremental addition of capacity did not carry the penalty of a significant increase in operational cost. Average new containerships serving the United States in the mid-1960s were somewhat larger than the largest common breakbulk vessels they were replacing, at approximately 600 to 800 TEU in container capacity. By the early 1970s vessels of 1,000 TEU to 1,100 TEU capacity were becoming common, and increasing in size to approximately 1,800 TEU to 2,000 TEU by the end of the 1970s. This is almost four times the size of the larger breakbulk vessels of the earlier part of the decade.

In the 1980s, vessel size continued to increase, with new vessel additions averaging approximately 3,000 TEU in the mid-1980s. The cost of a large, modern containership now exceeds $60 million. Vessel economies still have not leveled out and larger vessels still provide lower per unit costs. Thus, the economic incentive exists for even larger vessels. A physical barrier, the Panama Canal lock dimensions are temporarily constraining vessel size to approximately 3,000 to 3,500 TEU (however, some vessel designs, such as U.S. Lines 4,400 TEU ships, can fit through the Panama Canal). Economics would suggest that vessels of approximately 5,000 to 6,000 TEU, though unable to transit the Panama Canal, are economically possible by the early 1990s. The early 1990s should also mark a nearly full transition of general cargo to container in the major trade routes.[1]

## IMPACT OF THE CONTAINER ON PORT COMPETITION

A container vessel that has a set itinerary of a few selected ports faces quite different operational criteria than a breakbulk ship. The containership, because of low unload/load cargo-handling costs, can accommodate additional cargo, once the economic break-even point has been reached, with very little increase in costs. It cannot, however, afford to waste time by calling at additional ports for more cargo because of the high capital cost of the vessel and high port costs of servicing the vessel. The containership operation, however, can solicit cargo from a wide surrounding area and transport that cargo to a selected port by using land transport alternatives, and still produce an overall profit.

## The Container Load Center

The higher cost of operating a containership compared to the breakbulk ship means that vessel itinerary selection is far more important to the shipping line. The traditional breakbulk strategy, "go where the cargo is," led to a dozen or so stops along a particular continental seacoast. This strategy resulted in limited revenue sharing or coordination with land transport. In contrast, a containership uses land transport to a single designated port within a region, thus reducing port calls to only two or three per seacoast. Thus began a new pattern of "cargo following the containership" by relying on land transportation to feed the container operation, and this new pattern forced the shipping industry to select those ports where cargo would be concentrated. Furthermore, most containerlines tended to select the same ports because of the complicated range of services that container operations require. The port selected for this concentration of activity became known as the regional container load center.

Most general cargo Ports strive, understandably, to achieve the status of load center port. This status normally results in higher revenues for the Port because of the high volume of valuable cargo passing over the terminal. It also attracts auxiliary services to locate near the port, resulting in the growth of jobs and transport services. Several Ports that are not load centers have converted their terminals to other uses and developed feeder transport services to maximize their new relationship to the load center. Some Ports have found that although they must first provide terminals in order to attract cargo, they still may not obtain containerline services.

As container fleets expanded in the 1960s and 1970s, many port authorities began adding container terminals to their harbor inventory in an effort to serve existing operators or to attract new operators and become a load center. Requirements included paved yard areas to accommodate containers destined to go aboard a vessel, as well as space for those coming off. The typical container terminal is characterized by large cranes and a minimal number of buildings. In the 1970s terminals required approximately 20 to 25 acres, 600 feet of berthing, and none of the warehousing or transit sheds characteristic of older breakbulk terminals. Several contiguous terminals, furthermore, became a desirable feature as containership size increased, to allow for easy expansion of particular operations and for sharing container cranes between terminals.

These public ports[2] invested hundreds of millions of dollars in container terminal development and aggressively sought shippers and steamship lines as customers for their new facilities. Previously, Port terminals were often funded by private interests, especially railroads, terminal operators, shipping lines, and others. The large investment required by a container operation, however, has usually been provided by Port-related government funding in an effort to ensure continued ocean general cargo services within a particular region—and the public port became the financing conduit.

## *Load Center Contenders: The East Coast and the Gulf*

On the East Coast, The Port of New York/New Jersey was quick to anticipate the potential in container terminal development. The Port had taken an enormous leap by the early 1960s, becoming a regional container load center that eclipsed both Boston and Philadelphia. The large Atlantic containerlines concentrated their European and Asian trade in New York, using truck services to other cities instead of additional port calls by ocean vessels. New York, by the early 1970s, had a volume of over 1.8 million TEU, some eight times the combined rival Boston and Philadelphia TEUs, a percentage share that New York still holds (Containerization International Yearbook 1975, 1987).

As New York consolidated the primary port-of-call position in the North Atlantic, other Atlantic Ports vied for second and third port-of-call, as shipowners reduced the number of calls from a dozen ports to three or four. Geographically, mid-Atlantic and South Atlantic Ports had regional advantages over the larger Ports near New York (especially Boston and Philadelphia) because of lower land transport costs to interior hinterlands. Strong local Port rivalry developed in both regions. In the mid-Atlantic the rivalry was between Baltimore and Norfolk (now known as the Virginia Ports Authority, or VPA). In choosing between Baltimore and the VPA, a shipping line had to weigh the comparative advantage of land access versus water access, with Baltimore offering the advantage of closer rail and truck proximity to Midwestern markets, and VPA offering easy ocean access to the containerlines. As containership size and operational costs go up, and associated rail and truck costs go down (due to deregulation), the relative change, by the mid-1980s, had enhanced VPA container service at the expense of Baltimore.

In the South Atlantic there are a half-dozen Ports vying for emerging containerline services. Asian trade using the Panama Canal en route to New York, as well as European trade with the South offered new trade potential. In the early 1970s no individual Port dominated the region, and all South Atlantic port authorities hoped for a favorable competitive outcome in new trade opportunities and changing container technology. Wilmington, North Carolina, Charleston, South Carolina, Savannah, Georgia, and Jacksonville, Florida were the primary competing Ports in each Southeastern state. Charleston and Savannah emerged as the two leading containerports of the South Atlantic seaboard by the mid-1980s. These two have grown to the point of potentially challenging the Chesapeake Bay Ports of Virginia and Baltimore for second port-of-call status (after New York). Charleston and Savannah, in 1986, had nearly an equal amount of container business and each is seeking improved rail and truck connecting service. As is the case in Chesapeake Bay, any change in containerline service is carefully watched by the neighboring competitor. The rivalry between Savannah and Charleston is one of the strongest in North America. Each has made claims and counterclaims in order to lure steamship lines away from its rival. In the early 1970s the Chesapeake Bay Ports of Baltimore and VPA were at almost twice the TEU count of their southeastern Atlantic rivals, Charleston and Savannah. A decade later, they are approaching parity as well-established secondary container load centers to New York (Containerisation International Yearbook 1975, 1987).

Along the Gulf Coast, Houston and New Orleans, strong regional competitors, have been largely bypassed as container load center ports because containers destined for the Mississippi Valley region are off-loaded on the East and West coasts and moved overland to their destinations. In fact, until Latin American trade began to containerize in the late 1970s, there was little that Gulf ports could do to attract container services, even though they had a strong interest in competing for trade. External conditions, especially those of container economics, favored other routing and land service alternatives into the Gulf Coast states. West Coast ports, for example, by the early 1970s had captured most of the Asian trade bound for the Gulf Coast, whereas Atlantic ports garnered much of the Gulf Coast states European trade.

Nonetheless, strong competition developed between Houston and New Orleans for container shipping that was routed through the Gulf of Mexico. Houston, with the larger local market and good rail services, was able to compete against more historically entrenched New Orleans, which not only had excellent rail but also Mississippi River transportation services. Conversion of Latin American trade to containerization has been relatively limited and slow, thus minimizing container traffic in the Gulf.

## *Container Intermodalism: West Coast Ports*

If cargo-handling productivity gain was the original impetus for a ship owner to convert to a container operation, the second aspect of the containership, ease of conveyance between transport modes, became even more important. On the West Coast, a fast-growing trade opportunity related to the railroads was developing. Asian trade was increasingly being diverted from the Panama Canal route and being routed through West Coast ports and on to the eastern United States by rail. Shippers had found that rail transport to points all across the continent from West Coast ports could be economically advantageous. The combined Ports of Los Angeles and Long Beach in Southern California subsequently became the largest container load center on the coast, and soon will be the largest in the nation, partly owing to their enormous local market and partly to intermodalism.

The Ports of Puget Sound in the Pacific Northwest and the San Francisco Bay Ports are now competing for the port-of-call position second to Southern California, similar to the Atlantic coast situation involving Chesapeake, South Atlantic, and New York. The Bay Area, with a larger local market than any individual Pacific Northwest port, was the leading contender for this second place position until the 1980s.

Seattle, in the 1970s, early became the dominant port in the Pacific Northwest, surpassing both Portland, Oregon, and Vancouver, British Columbia, as the center of container activity. As a regional container load center, Seattle developed inland services to the northern Midwest and Atlantic seaboard for Asian import traffic. Increasingly, the rail advantage is favoring the Pacific Northwest at the relative expense of the San Francisco Bay Area, primarily due to superior rail service. Similar to the situation of Philadelphia and Boston with their close proximity to New York, the Bay Area is also feeling the pinch of being too close to a much stronger containerport complex, that of Southern California.

By the mid-1970s as a result of the growing demand for importing Asian goods into the *eastern* half of the United States, the top four West Coast containerports—Los Angeles, Long Beach, Seattle, and Oakland—became four of the leading five containerports in the United States, after New York.

Occasionally a very large containerline will relocate to another port on the same coast, bringing an enormous amount of business with it, and encouraging its required ancillary port services to move as well. Some of these services include rail yards, forwarder support, warehousing, repair services, and local drayage. The emergence of the Port of Tacoma as a leading U.S. containerport demonstrates how rapidly port fortunes can change in today's climate.

A single containerline (Sea-Land) relocated from Seattle to Tacoma, later followed by another line (Maersk) that began a new service in the Pacific Northwest. Additional service made Tacoma the fifth or sixth largest containerport in the country in 1986, nearly matching Seattle. In less than two years, Tacoma became larger than any Atlantic and Gulf containerport, except for New York. The Tacoma story demonstrates that selection of a port as a container load center is not entirely based on geographical location or population base. It is influenced by a port authority's competitive drive as well. When Sea-Land Services relocated their entire Pacific Northwest operation from Seattle to neighboring Tacoma, 35 miles to the south, it altered the regional load center balance in Puget Sound.

Port rivalry for containerline services within a region, such as between Seattle and Tacoma, often is viewed as wasteful competition that is using taxpayer dollars. As yet no easy mechanism exists for resolving the apparent duplication of marine terminals. Seattle views Tacoma as a "spoiler" because of its lower rates charged for marine container terminals, whereas Tacoma views Seattle as a "spoiler" for causing higher regional rail rates (because of Seattle's lack of rail planning). The state of the art in the design of container handling facilities in the 1980s calls for putting the intermodal railyard onto the marine container terminal. Tacoma has incorporated this design in its terminals (see Fig. 6.3), and consequently Seattle is less able to compete with other West Coast Ports, because Seattle has an intermodal gap between rail and marine terminals. In Seattle, the rail yard is separated from the marine terminals by a number of city blocks, requiring extra time in drayage and stops at two gates. Thus the intraregional Port competition is often more complicated than mere duplication of terminals.

## LAND-SIDE EFFECTS OF CONTAINERIZATION

Containerization initiated sweeping changes, reshuffling Port rankings in trade volume, and redefining the shape, size and equipment of the Port terminal. Many larger Ports suffered drastic decreases in their general cargo throughput, compared to their traditional rivals. In contrast, some small Ports have been transformed into significant new traffic centers. Some Ports were successful in attracting cargo, and others were not so successful, in spite of large investments in container terminals. Oakland, Tacoma, Charleston, and Savannah have become significant containerports, for example, whereas Boston, Philadelphia, San Francisco, and Portland (Oregon) have not been able to maintain their traditional share. Concentration of trade at just a few Ports on each coast is having far-reaching effects on land transportation as well, since large containerlines, with their enormous annual carrying capacity, have to rely on truck and rail lines to efficiently move goods to and from "feeder" Ports and to numerous inland distribution centers. Efficient land transport provided further encouragement to shipping lines to increase their carrying capacity by building larger ships. This, in turn, further raised the stakes for competing Ports to maintain or acquire container load center status with a large shipping line.

### *Freight Consolidation Programs*

As general cargo switched to containers, some Ports went further than merely constructing new container terminals. They developed various programs, such as truck and rail freight consolidation, designed to assist the land transportation end of the container business. In the 1960s most U.S. port authorities concentrated on marine terminal development. Some Ports offered additional services, such as a rail belt line that connected various terminals to the major rail carriers over a common, Port-owned rail collector system. Land infrastructure was left entirely to private enterprise for activities associated with the container terminal, such as warehousing, trucking, and the railroads, and to government agencies for developing street and highway access.

The most remarkable success story in obtaining new containerline traffic based on program development for inland container movement took place on the West Coast. Seattle's port authority, starting in the mid-1960s, implemented a new program for consolidating transpacific cargo bound for the Midwest. This program was intentionally designed to compete with California Ports.

At this time the vast California market alone consumed approximately 10 percent of Asian imports and was increasingly becoming the prime destination for the growing transpacific containerline service. These containerlines needed the local California market to support their containership operations. In addition to off-loading California-bound goods, the lines off-loaded cargo destined for the Midwest as well, helping to fill their excess containership capacity. Private freight consolidation programs evolved in California to serve the transshipment of the Midwest cargo.

Because the Pacific Northwest lacked the local market to attract the containerlines, they had to devise other means to entice carriers and shippers to use their ports. To compete, the Port of Seattle freight consolidation program offered better services than those available in California. They obtained competitive low inland transportation rail rates (which are volume based) for each containerline by pooling the cargo of several containerlines in Seattle. Those containerlines serving the Midwest were, therefore, able to obtain competitive rates and decrease their Midwest land transport overhead. As more containerlines entered the transpacific market, they increasingly used the Port of Seattle as their first, and often only, Pacific Northwest port-of-call. California Ports were not able to follow Seattle's lead in developing consolidation programs, because existing freight handling firms did not want public-sponsored competition. Intermodalism, however, has made consolidation a less valuable strategy because of the increased trend toward end delivery of a full container.

## *Intermodal Services*

In the early 1970s Seatrain, a North Atlantic shipping line in U.S./European trade, made an historic agreement with several railroads to provide intermodal container-on-flatcar (COFC) service between Europe and California, using a new rate structure. The railroads would provide a very low-cost rail service between the Atlantic port (in that case New York) and the California portion of the journey, and at a rate below that charged to domestic intermodal users, as long as the containerline promised to carry only international (European) cargo.

Seatrain, using excess capacity of their containerships in the Atlantic crossing, were thus able to carry European cargo to California via New York and the U.S. railroads, thereby avoiding the long and costly voyage through the Panama Canal. The voyage to New York would have already met Seatrain's basic ship overhead costs with local New York traffic, and so, carrying extra containers for California did not cost the containerline very much more. The cost trade-off was that even though the rail cost was low, it was higher than the ocean shipping between New York and California. However, the rail cost was lower between New York and California than the overall ocean route between Europe and California, and so Seatrain still made a profit based on the small net cost difference. The revenue obtained in California, because of the relatively low rail rates, made the overall proposal work.

An additional bonus was that it saved about a week of transit time for Europe/California traffic because railroads are faster than ships using the much longer sea route through the Panama Canal. Seatrain was able to offer the same rate as the carriers using the all-water Panama Canal routing, yet deliver cargo faster via the railroads and use their excess containership capacity on the Atlantic crossing as well.

In effect, the new program was an agreement between the container shipping line and the railroad to use a rail substitute for a portion of an international movement. This new form of transport service became known as mini-landbridge (MLB), or minibridge. This term was derived from the concept of a landbridge, where the United States would be used as a rail landbridge between Asia and Europe, thus avoiding either the Panama or the Suez canals. The economics of the Asia-Europe landbridge have yet to work out. However, the shorter version—for example, between Asia and the U.S. Atlantic seaboard—works extremely well.

MLB works even better in transpacific trade than in the transatlantic. The magnitude of potential traffic is greater than the original Europe-California movement by nearly tenfold. Containerlines serving the West Coast with excess containership capacity obtained a considerable new volume of cargo under the new MLB program (and a subsequent hybrid called microbridge to serve Midwestern distribution centers such as Chicago). For transpacific containerlines that wanted to compete with carriers in the Asia-Atlantic seaboard all-water market (via the Panama Canal), MLB offered an easy solution with no extra vessels needed. There was the necessity of sharing cargo revenue with the railroads, but it was still competitive because, like Seatrain, the basic overhead was met with Pacific Coast traffic. Conceptually, MLB was a rate and rail service mechanism that allowed West Coast containerlines to compete with lines using the Panama Canal by using U.S. railroads. In the early 1970s, transpacific containerlines with recently added fleet capacity needed extra cargo.

The shift was dramatic for the West Coast containerports of Los Angeles, Long Beach, Oakland, and Seattle (and now Tacoma). The diversion of traffic that normally might have used the Panama Canal was significant. By the late 1970s, as much as half the transpacific cargo between Asia and the eastern United States was moving through West Coast ports instead of the Panama Canal, and the diversion was increasing (Port of Seattle 1978).

Meanwhile, the next cycle of containership productivity was being introduced to the transpacific, that of much larger and more fuel-efficient vessels. The largest (at mid-1985) were the 4,400 TEU vessels of the U.S. Lines (owned by Malcomb McLean, the originator, now ex-owner, of Sea-Land Corporation), the largest that can fit through the Panama Canal locks. These vessels have a very low cost per container when the vessel is filled. Even with the longer distance through the Panama Canal, these ships were expected to compete with the MLB rates of the transpacific-Atlantic seaboard trade. But the "expected" did not turn out to be reality because of rail industry changes.

In 1980 the railroad industry was deregulated and the new marketing approach of the railroads vastly increased intermodal productivity, creating high-volume corridors of intermodal unit trains (Fig. 6.3) that cost much less to operate than the early MLB trains. Overall railroad employment in the United States has been cut by over one-third since the 1980 rail deregulation (Statistical Abstract of the United States 1984),[3] and abandonment of low-density traffic rail branch-lines has also greatly lowered overall rail costs.

Recently, the introduction of new technology, in the form of the double-stack intermodal railcar (Fig. 6.4), has further advanced rail productivity for container traffic. Stacking one container on top of another essentially means transporting two containers almost for the cost of one. The corollary, however, is an increase in rail terminal costs, which has further encouraged concentration of container activity at

**Figure 6.3.** A recent technological innovation in cargo handling is double-stack railcars at an intermodal yard that is immediately adjacent to the ship berth. (Courtesy: Port of Tacoma)

**Figure 6.4.** A closer look at terminal yard equipment double-stacking specially built railcars with containers. (Courtesy: Port of Tacoma)

fewer places. Rail technology has adapted to ocean shipping and, as a result, has made rail intermodalism the driving force in ocean shipping with entire trains transported directly and quickly to destination centers.

## *The International Container*

The rapid increase in rail intermodal container traffic, such as MLB, has caused another important change since rail deregulation: the use of the international container to carry internal U.S. domestic freight. An imbalance in the directional flow of international container trade often exists, when imports are significantly greater than exports, for example. This international imbalance of containers, where empty containers are transported back across the continental United States for refilling, is an inefficient use of rail intermodal capacity. To remedy this, empty containers used for international trade are carrying domestic U.S. traffic that is usually imbalanced in the opposite direction to that of international imbalance. The innovative double-stacked container railcar has also expanded the use of international containers for domestic U.S. freight, because of the productivity gain in the use of containers over that of piggyback trucking (truck trailers placed on rail flatcars with wheels attached, called trailer-on-flatcar, or TOFC). Because of the wider use of rail service, load center ports will gain economically as international containers increasingly dominate domestic freight business.

## SHIFT OF FACILITIES WITHIN HARBORS

Some Ports experienced a redistribution of traffic within their physical harbor area as a result of containerization. An excellent example is in San Francisco Bay. San Francisco had been a dominant West Coast port since the gold rush of 1849, until losing the top position to Los Angeles in the 1920s (when it moved to second place). Oakland, across the Bay from San Francisco, had been a relatively minor port since its creation in the 1920s. The Port of Oakland, however, saw the potential of containerized technology early in the 1960s and constructed several major terminals specifically designed for containerships, whereas San Francisco was still building general cargo terminals unsuitable for efficient container operations.

Within a half-decade (almost overnight by port standards), the shipping industry vacated San Francisco for Oakland container terminals as each shipping line shifted to container operations. By the early 1970s Oakland was handling approximately ten times the container traffic of its cross-Bay rival and a substantially greater amount of overall cargo (Containerisation International Yearbook 1977).

Another harbor example where there is more than one port authority is in Southern California, where the Port of Los Angeles and the Port of Long Beach share the harbor of San Pedro Harbor. In this case, there is a nearly even split of container terminal development as each rebuilt older breakbulk terminals. Competition between the two Ports has not changed the overall balance because of nearly equal land and financial resources.

In the New York/New Jersey harbor, which is operated by a bi-state public port authority, the shift in container terminal activity was away from the congested general cargo piers of Manhattan over to New Jersey. The Manhattan pier area was then made available for uses more in line with its proximity to downtown and its extremely high real estate value.

In other cases, such as Los Angeles, Long Beach, Oakland, and Baltimore, conversion of general cargo terminals into container terminals made economic sense because there were no competing land uses and fewer new expansion areas. There are some Ports such as Seattle where redevelopment of general cargo terminals into container terminals took place in close proximity to downtown. In Seattle's case, the Port is now entering a period of land use conflict, especially as container diversions to Tacoma lower the demand for Seattle's terminal. The maturation cycle of Seattle would suggest a change in land use to more urban uses, especially close to the downtown city core.

## EFFECTS ON LABOR

The enormous gain in longshore productivity realized through containerization has decreased labor requirements in handling the nation's commerce, and this is so in spite of the vast increase in general cargo commerce.* In addition, jurisdictional rights between different labor unions have been complicated by containerization. Under breakbulk shipping, the steps involved (unloading of vessel into nearby transit shed, then to nearby warehousing, then to customer) were well defined over the years of union growth. With containerization, in which most containers are moved intact to the customer often hundreds of miles away, jurisdiction among longshore warehousing, trucking, and rail became controversial—especially with the decline in labor requirements due to mechanization of labor.

Evolution of containerization, from ease of vessel unload/load to a true intermodal movement, took well over a decade—essentially the 1970s. During that time, union jurisdictional rights became clouded as quickly as they were resolved because the industry was in a state of constant change. Although there is a greater need than ever for skilled workers to operate the vast and expensive machinery now required by the container operator, the net result has been a greatly diminished labor requirement to move cargo.

## SPECIALIZATION AMONG PORTS

Port competition is increasingly defined by specialties such as container, bulk cargo, and noncontainer (often called neobulk). Containerports, for example, generally do not compete for bulk cargo. The largest of the containerports serve their entire region, with neighboring ports primarily sharing containerline services by land. Some neighboring Ports also have specialized routes and smaller containerline services. Other Ports are specializing in specific commodities or services, such as Portland, Oregon, for grain and auto traffic, or San Diego for recreational and tourist activities, or Miami for the cruise ship business. Specializations usually evolve from an attempt to diversify the port when the loss of general cargo traffic becomes significant. Industrial development noncontainer activities and other economic activities (such as airports) often result from diversification attempts (see Chapter 7).

Specializations in U.S. Port development that promote feeder services to the regional load center, however, are rare. Throughout the nation, Ports are still seeking load center status for container services, in spite of load center economics that suggest poor odds for success. For example, New England feeder services into the New

*The story of how the reduction in the longshore labor force was handled on the coasts of the United States is complex and beyond the scope of this chapter (see Couper 1985).

York load center would provide better opportunities to local shippers than development of additional container terminals designed to attract a large transatlantic containerline. In San Francisco Bay, the development of Oakland, with superior connecting rail service, would provide better regional benefit than continued container terminal development in San Francisco proper.

## THE GROWING IMPORTANCE OF THE RAIL CONNECTION

In the immediate future, containerization will, for all practical purposes, become synonymous with general cargo. Growth will come from a continued conversion to container in remaining world trade routes where significant breakbulk traffic still exists, such as Africa, South America, and South Asia. Intermodality, especially the direct movement from ship to rail transport, will continue to influence North American Ports.

Most of the transatlantic and transpacific trade is now containerized, though a shift to container is still occurring for some commodities that were once considered noncontainerizable (such as scrap steel, lumber, and various agricultural products). Additional trading subregions of the world are also entering the primary trade routes by container through feeder networks. For example, much of Southeast Asia is now converting to container via excellent service at Singapore. Thus, expansion of transatlantic and transpacific trade is still occurring because of the container.

Intermodality, however, is the big change now impacting U.S. Ports. Significant productivity increase over the past five years in land transport, especially rail, has greatly lowered relative land transport costs. The significance is that lower rail costs have been able to maintain competition with ever-increasing containership size (and, therefore, lower cost per container) of traffic using the Panama Canal. The double-stack railcar has maintained an economic balance with the largest containerships using the Canal. If containerships get any larger, which they will need to do in order to maximize economies of scale for lower per unit container costs, then they will no longer be able to fit through the Panama Canal. A containership that is unable to fit through the Panama Canal will no longer be constrained by design dimensions of the locks, and can, therefore, be much larger than those now in use. A larger vessel will cost more and be forced to use double-stack rail services. Therefore, the rail connection will be a primary consideration for a very limited number of port calls.

It is conceivable that very large containerships—of, say, 5,000 to 6,000 TEU size—filling several double-stack rail trains at a time will be calling at a limited number of ports on each coastline. Which ports, and whether all large container lines will call at the same two ports are not known at this time. If the recent past is any indication, the load centers will not necessarily be concentrated at the same ports, and railroad service will essentially determine which ports are selected. The West

Coast will likely evolve into two load centers, most likely Southern California and Puget Sound, whereas the Atlantic Coast may continue with the three load centers now emerging, New York, Chesapeake Bay, and the South Atlantic.

Feeder services will, of necessity, become more important. Those Ports that recognize the new role of feeder services will provide the best international trade opportunities for their local shippers. For economic as well as political reasons, most port authorities will be required to increase diversification activities into noncontainer, and even nonfreight, activities, such as recreation and urban development.

International trade growth of the United States and containerization go hand in hand. The container has greatly increased the productivity of ocean, land, and rail transport, which, in turn, has lowered relative transportation costs for international trade. The container, in essence, has made the world smaller by making the transport of so many goods simpler, faster, and cheaper. The port is a vital link in this transport system, but the land-to-sea transfer cost at the port is a shrinking share of the total manufacturer-to-consumer transportation cost. This suggests that Ports will have a harder time influencing the routing of goods in future international trade.

## *Notes*

1. As of this printing, the first larger-than-Panmax containership has been ordered by American President Lines.
2. Nearly all U.S. public ports have invested in container terminals. The vast majority of investment has been in the top 20 ports, which are, based on traffic and the author's approximation: New York, Los Angeles, Long Beach, Oakland, Seattle, Baltimore, Virginia Ports Authority, Tacoma, Charleston, Savannah, Houston, New Orleans, Portland (Oregon), Anchorage, Honolulu, Miami, Jacksonville, Philadelphia, San Francisco, Boston.
3. Rail employment in class I railroads decreased from 480,000 in 1980 to 274,000 in 1986 (Railway Age 1986).

## *References*

*Containerisation International Yearbook*. 1975, 1977, 1987. World Container Port Traffic League Section, National Magazine House, 72 Broadwich St., London, W1V 2BP, England.

Couper, Alastair. 1985. Social consequences of maritime technological change. Donald McKernan Lectures in Marine Affairs, Seattle, Washington Sea Grant Program in cooperation with the Institute for Marine Studies, University of Washington.

Fairley, Lincoln. 1979. *Facing mechanization: The West Coast longshore plan*. Los Angeles: University of California, Institute of Industrial Relations.

Gilman, Sidney. 1980. *Ship choice in the container age*. Liverpool: University of Liverpool, Marine Transport Centre.

Mahoney, John H. 1985. *Intermodal freight transportation*. Westport, CT: ENO Foundation for Transportation.

Pearson, Roy, and John Fossey. 1983. *World deep-sea container shipping,* and *Containerline performance and service quality*. Liverpool: University of Liverpool, Marine Transport Centre.

Port of Seattle. 1978. *Marine terminal five-year outlook,* April, pp. 11–18.

*Railway Age*. November 1986, 20.

*Statistical Abstract of the United States*. 1984. Bureau of the Census, p. 626.

## Additional Reading

*Industry Journals*

Container News
Containerisation International (London)
Modern Railroads
Railway Age

*Industry Newspapers*

New York Journal of Commerce
Seattle Daily Journal of Commerce

*General Periodicals*

American Shipper
Forbes
Traffic World

*General Publications*

Annual reports (Form 10K) of the BN, Conrail, CNR, CP, CSX, Norfolk Southern, Santa Fe, and Southern Pacific railroads.

Association of American Railroads, CS-54A weekly traffic reports.

Dooley, Frank, Kenneth L. Cassavant, and Arthur C. Scheunemann. 1983. *Transportation needs of Washington agriculture phase I: Transportation usage by commodities and identified problem areas,* prepared for Washington State Department of Agriculture. August.

Gilman, Sidney. 1983. *Ship choice in the container age*. Liverpool: University of Liverpool, Marine Transport Centre, Grower Publishing, Ltd.

Pearson, Dr. Roy. 1979. *Containerline performance and service quality*. Liverpool: University of Liverpool, Marine Transport Centre, Grower Publishing, Ltd.

———, and John Fossey. 1983. *World deep-sea container shipping*. Liverpool: University of Liverpool, Marine Transport Centre, Grower Publishing, Ltd.

# Chapter 7

# FOSTERING ECONOMIC DEVELOPMENT:

## The Noncargo Alternatives

*James E. Randall*

---

The word *seaport* conjures up an image of a fleet of fishing trawlers unloading their daily bounty as gulls wheel screeching overhead, or container cranes standing sentinel over freighters. Although these images are accurate, they reflect only a stereotypical perception of the U.S. port. Instead, the breadth of port activities exceeds the imagination. Public port authorities, as waterfront custodians, engage in a multitude of activities, many of which are only marginally linked to the disposition of freight or passengers. From the Port as real estate agent and developer, to custodian of bridges and tunnels, to coordinator of community festivals, the seemingly disparate roles of the port authority have a unifying thread: improvement in the quality-of-life within the port community. For most U.S. seaports in the 1980s, this has meant economic development.

In this chapter the role of the Port in fostering economic development is examined. The first section pertains to the traditional perspective on economic development at the public port and how it is measured by economic impact studies. The next two sections show the trends that have led port authorities to expand their economic development efforts into the provision of "noncargo" goods and services. These are defined as functions that are intended to generate revenue and/or employment, but are not directly related to the cargo-handling mandate of the port authority. In the final section a number of the new economic tools that have been implemented at U.S. seaports are discussed, with particular attention to recreation, tourism, and communications. These examples show that public ports have adopted a rich diversity of functions with varying levels of success in order to meet the challenges of rapid change at the seaport.

## TRADITIONAL ECONOMIC DEVELOPMENT: ECONOMIC IMPACT STUDIES

Economic development is a necessary function of the Port to maintain fiscal solvency and fulfill expansionary goals (Goss 1979, Chang 1978). Because many public ports derive at least a portion of their revenues from the voting public, it becomes imperative for the Port to convince the community of its contribution. This public relations effort has often included public information or promotional programs and, invariably, economic impact studies. The latter permit the port authority to quantify the impact of seaport expenditures on employment, income, and sales generated within the local community or region.

Although applied at varying levels of sophistication, economic impact studies have some characteristics in common. By using data on the industrial composition of the community, the direct and indirect consequences of port activity are estimated. Direct impact is measured as employment, wage payments, and purchases of goods and services by firms in the community as a result of their direct interaction with the Port. Indirect impacts refer to the multiplier effects of these original purchases in an iterative process as goods and services are consumed to fulfill the demands of those firms and employees. Obviously this regeneration of production cannot continue forever. As ripples from a rock dropped in a pond slowly fade, so do the multiplier effects of port wages and expenditures. At every step in demand and consumption, leakages of two forms occur. First, a portion of wages and profits is not reinvested or consumed immediately but is saved, slowing down the impacts of progressive production. Second, wages and profits are spent on products and services produced outside the local community, thereby contributing little to local output demand. Thus, for every dollar invested by the Port, that dollar plus a portion of another dollar provide benefits to the local community. In input-output analysis,[1] one of the more sophisticated forms of Port economic impact studies, Port functions are matched to the specific industrial categories in order to trace actual impacts in the economy.[2]

As might be imagined, the impact of different Ports on the economy of their communities varies considerably. For instance, Washington State public ports were said to have generated $1.4 billion in wages, $4.5 billion in total sales, and 99,000 jobs in 1976 (Washington Public Ports Association 1978). In Montreal, the Port contributed $438 million in total sales and 16,500 jobs to the city's economy in 1978, but this constituted only 2 percent of local employment and less of the city's total product (Bailly and Weiss-Altaner 1981). At the Port of Southampton, England, the port employment multiplier for the city was 1.66 (Witherick 1981). Finally, using a 1970 national input-output matrix, the U.S. Maritime Administration (MARAD) estimates that seaport industry operations within the United States have a sales multiplier of 1.6 (MARAD 1978).

The use of economic impact studies in the form indicated above has been criticized on a number of grounds.[3] The most significant criticism is that the selection of port-specific industrial categories does not reflect the true functional profile of the Port. What *does* a Port do and how can this bundle of activities be modeled effectively? Even broad inventories of the economic role of the Port have been limited to operational services (pilotage, customs, cargo handling), ancillary services (shipping broker and forwarder, naval outfitter), distribution (warehousing), and maritime industry (water-dependent manufacturing) (Witherick 1981). Despite the apparent breadth of this inventory, all of these functions are traditionally trade- and marine-related, and therefore the taxonomy fails to accommodate specific nonmarine activities carried out by individual Ports.

Even the most ambitious impact studies treat Ports as reactive entities, bounded in function by the stereotypical perception of a set of port activities. Instead, a seaport is a diverse institution that is proactive in nature. Only by examining the projects, investments, and strategic planning of the individual Port can its true contribution to community economic development be revealed. In this light, the contention by MARAD that impact statements ". . . must reflect the port industry's unique mission to move waterborne cargo" becomes meaningless (MARAD 1978, 12).

## THE TRENDS BEHIND DIVERSIFICATION

### *Shifts Away from Water-Dependent Land Use*

One of the first forays of Ports into the use of noncargo development was the promotion and management of sites for water-dependent manufacturing firms. The traditional symbiosis between ports and manufacturing has been explained most adeptly by industrial location analysts (see Hoover 1948). Where goods must be shipped from one mode of transportation to another, terminal and handling costs can be minimized by locating the point of production at the transshipment point rather than some other intermediate point. Figure 7.1 shows that procurement and distribution costs (the two components of total transfer costs) increase in a stepwise fashion as one moves from the market (C) or the raw material source (A), respectively. Total transfer costs are lowest at the transshipment point (B), and therefore the firm will process the product there.

Ports have long considered as part of their responsibility the facilitation of the needs of port-related industries in the harbor. In most instances, Ports have some degree of control over certain aspects of manufacturers' operations, including policing of cargo movements, monitoring waste disposal, and mediating conflicts between manufacturers and other waterfront users. This involvement escalated in those instances where Ports became creators of real estate from dredged landfill, itself often a byproduct of the maintenance or deepening of a navigable harbor. At least

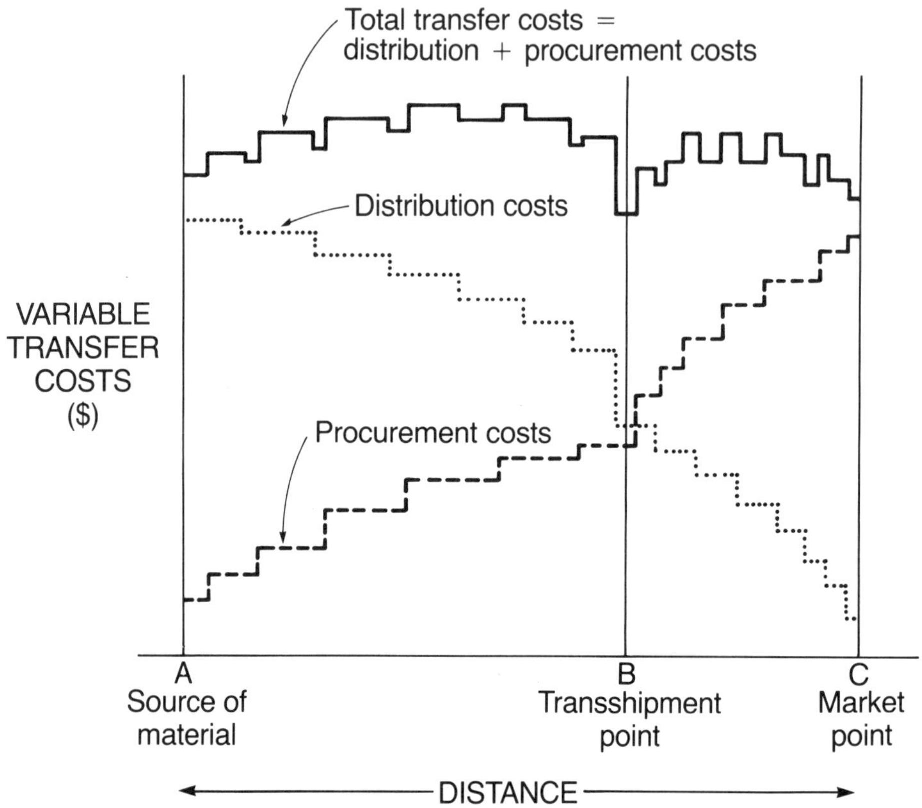


**Figure 7.1.** This conceptual rendering shows that the optimal location for transshipment is the point where combined distribution and procurement costs are lowest. (Courtesy: James Randall)

initially, Ports maintain a commitment to reserve these industrial lands for water-dependent users such as iron and steel and sawmills. However, this commitment is tested when it is recognized that the demand for waterfront land by this narrow range of industrial firms is illusory. Not only are these heavy industry sectors of the U.S. economy subject to cyclical and structural stagnation or decline in output and employment, but the need of these firms for large parcels of inexpensive land is forcing them to move away from the central city waterfront land created by Ports. The need to recoup development costs means that port authorities become more flexible in their qualifying standards for potential tenants, and more amenable to alternate or non-water-dependent users.

Two other trends have contributed to changing land use in ports. Internal structural changes in port technology and needs (for example, capital intensification, increased space requirements, larger and more specialized ships) have resulted in the underuse and availability of smaller, obsolete port facilities such as finger piers (see Chapter 6). Also, communities on central waterfronts adjacent to the downtown business district began to eye underused and cheap Port lands for residential, recreational, and commercial uses. Ports responded to these trends in two ways. First, having become desperate to replace revenue lost through decentralization, deindustrialization, and changing port technologies, Ports searched for alternative activities to fill vacant or underused land and maintain revenue streams. Second, alternate uses were thrust upon the Port because of the changing demands and value systems of the urban community. Regardless of the cause, the breadth of economic projects accommodated by the Port has increased.

## *Stagnation in Trade and Employment*

Port authorities are being exposed to the benefits and costs of increased competition in a number of other forms as well. Following consistent annual increases in aggregate volume of trade handled at U.S. seaports, since 1978 there has been a stagnation or actual decline in marine trade. Table 7.1 reveals the magnitude and the parameters of some of these trends.

Total cargo tonnage decreased by 11.5 percent from a high in 1979 of 2,074 million short tons (Table 7.1). Although both domestic and foreign cargo transfers share in this decline, spatial and functional variations in this phenomenon of decline are masked by these aggregate figures. When disaggregated into the categories of Dry Cargo and Tanker Cargo, the figures in Table 7.1 can be reinterpreted. Almost all the foreign cargo losses are attributable to declines in tanker traffic, specifically the import of petroleum products. Dry cargo shipments have remained fairly constant through the early 1980s. Shifting spatial patterns of international trade have also led to changes in the relative rankings of U.S. Ports in their level of marine trading

**Table 7.1.** U.S. Commercial Waterborne Trade

| Year | Domestic Commerce* | Foreign Commerce* | Total |
|---|---|---|---|
| 1960 | 761 | 339 | 1,100 |
| 1970 | 951 | 581 | 1,532 |
| 1974 | 983 | 764 | 1,747 |
| 1975 | 946 | 749 | 1,695 |
| 1977 | 973 | 935 | 1,908 |
| 1978 | 1,075 | 946 | 2,021 |
| 1979 | 1,080 | 993 | 2.074 |
| 1980 | 1,077 | 922 | 1,999 |
| 1981 | 1,054 | 887 | 1,941 |
| 1982 | 957 | 820 | 1,777 |
| 1983 | 967 | 751 | 1,718 |
| 1984 | 1,033 | 803 | 1,836 |

*Millions of short tons.

*Source:* U.S. Corps of Engineers, *Waterborne Commerce of the United States* (Vol. 5, 1984).

activity. Total foreign commerce has declined by an average annual rate (from 1980 to 1983) of 4.7 percent along the Atlantic Coast, 9.1 percent along the Gulf Coast, 2.7 percent along the Pacific Coast, and 6.1 percent in the Great Lakes ports. Some of this decline can be attributed to periodic recessions, such as those that occurred between 1980 and 1982. But it appears that the era of rapid annual growth has given way to one of instability.

Container traffic at U.S. ports has been growing by leaps and bounds during the 1970s and 1980s (see Table 7.2). This aggregate growth, however, masks some important features. First, container traffic has also been affected by periods of recession, although to a lesser extent because diffusion of the technology was still occurring over this fifteen-year period. In the long term, as containers take an ever-larger share of general cargo, it can be expected that their growth will plateau and eventually track the pattern of change in general cargo shipments. In addition, the employment and revenue benefits accruing to port communities from container traffic is not spread uniformly across the nation's ports. Much is concentrated at a small number of gateway ports. Of the 11 million 20-foot container equivalents passing through U.S. ports in 1984, 6.1 million (55.2 percent) went through the five ports of New York/New Jersey, Seattle, Long Beach, Oakland, and Los Angeles.

Last, and perhaps most indicative of the development problem of the port community, is the decrease in marine employment at seaports. A consistent decline in the number of private shipyard workers nationally suggests that ports may be losing one of those functions that epitomizes the marine setting, shipbuilding and repair. In addition, these relatively labor-intensive functions constitute an integral link be-

**Table 7.2.** Containerized Cargo Handled at U.S. Ports

| Year | TEUs* (in 1,000 units) |
|---|---|
| 1975 | 5,267 |
| 1976 | 5,583 |
| 1977 | 5,477 |
| 1978 | 5,944 |
| 1979 | 7,243 |
| 1980 | 8,567 |
| 1981 | 8,431 |
| 1982 | 8,729 |
| 1983 | 9,559 |
| 1984 | 11,059 |

*TEU = 20-foot equivalent units based on material aggregated from individual container ports in the United States.

*Source:* From 1981, 1983, 1984, 1985 December issues of *Containerisation International*.

tween the urban community and the port. Those familiar with marine operations know that the other mainstay of port employment, longshore jobs, has sunk to low numbers. Nationally, average monthly longshore employment has declined *every* year from 1974 (65,113) to 1983 (34,727). In Table 7.3 total employment declined by 30.1 percent from the high of 179,000 in 1981. These examples of employment loss puts added pressure on port authorities to generate alternate opportunities for employment. Unstable employment and cargo tonnage indicate that the roles traditionally filled by Ports are becoming more difficult to define. No longer can an individual Port expect consistently increasing levels of marine cargo activity. Since revenues must still be generated, Ports are engaging noncargo roles more frequently to maintain their functional and financial solvency.

Ports can experience financial distress for reasons other than an inability to generate consistent increases in cargo-related fees. Operating capital for many Ports depends to a certain extent on local property tax revenues, and industrial restructuring and outmigration has eroded this property tax base. Although the adverse effects of these demographic and economic changes have been felt most severely within small-scale communities in peripheral locations, they are nonetheless apparent in communities of all sizes. For example, the financial crisis that faced New York City in the mid-1970s was not experienced to the same degree by the Port of New York/New Jersey, but it led the Port to reassess its role. The Port recognized that it had a responsibility to maintain the community's economic vitality. This broadened perception of its mandate justified the extension of Port involvement in improving the urban transportation infrastructure and promoting the industrial values of the community. At the other end of the spectrum, small coastal Ports in Washington and Oregon have suffered from the collapse of some elements of the commercial fishing industry and a soft lumber market. Traditional dependence on these two water-ori-

**Table 7.3.** Employment at U.S. Ports

| Year | Private Shipyards Production Workers (in 1,000's) | Longshore Jobs (in 1,000's) |
|---|---|---|
| 1970 | 108.5 | — |
| 1975 | 133.4 | 63.7 |
| 1977 | 139.9 | 56.5 |
| 1978 | 138.4 | 52.1 |
| 1979 | 136.7 | 49.1 |
| 1980 | 141.8 | 48.7 |
| 1981 | 142.2 | 46.2 |
| 1982 | 130.2 | 42.4 |
| 1983 | 108.7 | 34.7 |
| 1984 | 93.8 | — |

*Source:* Shipbuilders Council of America, Washington, D.C., 1985.

ented sectors has left many of these Ports in precarious positions and forced them to diversify their operations.

## Port Market Power

Goss (1979) suggests that a comparison of Ports based on level or mix of import and export commodities, geologic or geographic factors, or economic factors represents a superficial analysis of seaport differences. Most of these factors are not causal, but instead are outcomes of an underlying structural cause of Port activities—the level and extent of power that port authorities can exert upon others. As with any industrial organization, Ports have varying degrees of market power. In addition, because they are a governmental entity entrusted with certain monopoly powers, they have additional influence in local urban economic affairs. As a major provider of transportation services, Ports can influence urban planning decisions that are perceived to have impact on the operations of the Port. The locations and capacity of rail, road, and air connections are obvious extensions of Port influence; but the imprint of Port influence on the urban community is vastly more pervasive and subtle. Port staff are called upon, or call upon themselves, to lend technical and managerial assistance in a wide range of public agencies and private committees. At the senior level, Port commissioners and managers participate on committees that determine urban and regional economic policy. At the junior level, research analysts, engineers, and technicians assist in the implementation of projects for the community. These projects may have only marginal relevance to marine transportation, but, in participating, Ports add a marine perspective to noncargo issues and become more familiar with them.

In a broader context, the duties performed by Port staff and the skills attained are both cause and effect. Port employees have traditionally been drawn from marine backgrounds and their duties matched the skills of the individuals during periods of marine cargo expansion. There was little incentive to venture into projects outside the realm of cargo orientation, and the narrow set of skills contributed to inertia in the entrepreneurial activities of the staff. This can be seen as a reactive phase of Port management. Recently, however, because of government deregulation of transportation carriers, increased competition among alternative waterfront land users, and increased concern to maintain share of cargo volume in a competitive trade market, the composition of Port staff has changed. Along with their knowledge of marine operations, Port decision-makers are called upon to be business managers, accountants, lawyers, public relations promoters, and research analysts. The new skills of the staff facilitate their participation in innovative projects.

A Port's market power is also expressed as a result of its role as guardian of waterfront properties, often among the most valuable industrial property. By default, Ports become real estate managers and may be called upon to repeat this role elsewhere in the community. This expertise, along with their ability to borrow money at favorable rates, makes a formidable combination in the arena of noncargo projects.

## A BROADER CONTEXT: KONDRATIEFF CYCLES AND INVESTMENT STRATEGY

A number of perspectives can be taken to explain the use of noncargo economic development tools by Ports. Hanappe and Savy (1981) adopt a broad structural and macro-approach to explain the changing mix of activities carried out by Ports over the long term. By using the concept of national economic change commonly referred to as "Kondratieff' cycles, they suggest that the pattern as well as the rate of economic growth at ports have undergone fundamental changes.[4] Kondratieff cycles consist of long series of structural change repeated every fifty to sixty years, the initial phase of which is a period of rapid growth (twenty to thirty years) producing high rates of return on investments and rising standards of living. This upswing is followed by a period of stagnation and decline characterized by lower levels of investment and the reconstitution of capital in a number of alternate forms in an attempt to maintain profit levels. These investment changes are themselves a result of underlying structural characteristics of society, including the age of capital stock and international division of labor. The cycles become self-perpetuating because, as the main industrial markets become saturated with capital attempting to secure a portion of the profit within the dominant industrial sectors, production increases, and ultimately market prices, and then profits begin to fall. Capital then flees these sectors and undergoes a period of instability before attaching itself to a new set of dominant sectors.

Although the reliability of this conceptual framework has been questioned by a number of economists, it does provide one explanation of the changing functional profile of the U.S. Port. The period of economic growth during the 1950s and 1960s corresponds to the growth phase of the most recent cycle. Investment was buoyed in the petroleum and iron and steel industries and consequently in marine cargo-handling facilities and harbor improvements. During the economic downturn that has characterized the post-1975 period, capital has been withdrawn from the heavy industrial sectors in response to falling rates of profit, and investors have become more cautious in their behavior, often investing in a wider spectrum of smaller projects on a trial-and-error basis. Investment decisions become more attuned to local labor specialties rather than to the depth of the harbor or quality of port facilities. Port planners are adversely affected by this diffusion of investment away from transport-dependent manufacturing sectors and are also faced with the same investment quandary as private investors. Their response is to become more flexible in their investment portfolio and to increase involvement in a wider range of projects than during the earlier growth phase of the cycle.

Whereas this structural approach may provide insight into the macro-conditions responsible for an aggregate change in Port investment, it does not account for the variations and anomalies that characterize Ports at the micro-level. At this level, port authorities must choose the route that will ensure growth. It is difficult for them to justify investment in cargo-handling facilities for a hinterland that may not yet have generated the volume of trade to warrant this investment. Yet, if port authorities wait until the industrial environment generates sufficient demand to justify marine facilities development, they risk constraining future economic development or losing trade.

In response, Ports may adopt one of three alternative strategies: a supply-led transportation strategy, a demand-led industrial strategy, or an incremental stepwise combination of the two. The transportation strategy involves the development of cargo-handling facilities prior to the present need for these services. Investment in transportation gives present and potential local producers greater shipping flexibility and lower costs. This in turn draws in trading and shipping activity from a wider hinterland in the short term, and in the long term serves as an inducement for the expansion and creation of local enterprises. These Ports will tend to be more cargo-oriented in their functional profile, although during certain periods they will also be peripherally engaged in promoting the industrial advantage of the community.

Alternatively, Ports can break into this cycle of transportation and industrial growth by increasing demand for marine services, by promoting, developing, subsidizing, and generally wooing economic enterprises to their community (that is, by a demand-led strategy). Individually, the new firms may not justify the expansion of cargo facilities, but their collective demand may reach a threshold that is politically and economically justifiable for transportation infrastructure improvement. In fact, in a study of manufacturing activity along the lower Fraser River in metropolitan Vancouver, British Columbia, Randall (1983) observed that many small-scale firms did make use of the central container facilities, even if this use was limited to one or two containers arriving or departing at irregular intervals. The impact of this "grass roots" use of facilities as a stabilizer at sites prone to fluctuations in cargo volumes has not yet been fully examined in the literature.

According to this demand-led strategy, the initial degree of water-orientation of these firms is relatively unimportant. In time, however, these firms are expected to either expand to the scale where water transport of raw materials or finished products is visible, or to attract linked firms that make use of water transportation. Ports that adopt this philosophy of growth would be most likely to specialize in noncargo activities during a significant portion of their growth. Although the provision of marine cargo services remains the essential goal of these Ports, their path to achieve this goal is more wholistic in form, emphasizing the creation of a need before an investment in transportation. The challenge in using the supply-led transportation and demand-led industrial structure is finding the proper balance in supply or demand.

A Port will break into the cycle at some point and undergo changes in its functional profile so that transportation or industrial development creates the conditions for further development. The small Port of Morrow, 250 miles upstream on the Columbia River, took the demand-led route by encouraging the development of several food-processing (especially potato) plants in the community. When the demand for inexpensive shipment of the processed product increased, the Port was able to justify the investment of a $1-million container barge slip and rail-mounted straddle crane. Conversely, the Port of Pasco, Washington, broke into this development spiral via an initial investment in facilities and a traditional marine cargo break-of-bulk orientation, drawing agricultural and wood products from a large hinterland. The well-developed marine terminal attracted and retained industrial users.

The third strategy entails a stepwise approach, making gradual and incremental improvements in the transportation infrastructure as demand or creative genius of Port decision makers dictates. Demand-led and supply-led intervals are likely to alternate. It is uncommon to find this cautious combined strategy publicly advocated at U.S. Ports. Many of the capital investments at these Ports involve a significant commitment in financing and planning. Investment in a $50-million container facility or a $200-million coal or grain terminal is neither incremental nor gradual. However, perhaps where Ports rely on private waterfront firms to develop their own facilities, one would find a situation of incremental change in transport and industrial investment.

This micro-level approach applies to growing ports and communities. However, as indicated by the trends in employment and cargo movement, many U.S. ports are now stagnating or declining. For instance, many of the Great Lakes ports that were established and grew on the basis of iron ore shipments from the Mesabi Range or wheat shipments from the Plains have been bypassed because of the stagnation of heavy manufacturing industries and less costly rail connections to the coasts. On the Gulf Coast, trade at several ports has suffered from a decline in oil imports. How do port authorities react in these more likely scenarios? One solution is to compete more aggressively for the limited trade market and, in so doing, consolidate and focus on the provision of certain cargo services. Although this produces a more efficient operation for the Port, it also implies that employment multipliers for the community decrease as the labor force is trimmed and becomes more productive. This "return to its functional roots" is an attractive option for the Port because the community understands and accepts the Port's traditional role as a provider of cargo-handling services. Thus, risk of failure can be more easily attributed to exogenous economic or political conditions rather than inappropriate decision making or mismanagement by the port authority.

Stagnation or decline could, however, force a Port to choose another strategy, one that will ensure the survival of the Port whether or not the strategy to be used is "transportation-oriented." Faced with a declining industry and neighboring Ports that are determined to maintain their share of trade volumes, port authorities begin to become more adventurous and innovative. Although transportation is always given consideration, the critical question is not whether a new project will lower the cost or improve the efficiency of cargo handling, but rather whether it will generate revenue.[5]

## EXAMPLES OF NONCARGO ECONOMIC ACTIVITY

It appears, then, that Ports have a variety of reasons for entering the arena of noncargo economic development. The unique situation at each Port is governed by the economic and financial conditions facing the Port and its community; the level of control or influence exhibited by the public port over the direction of community development; and the prevailing attitudes, experience, and training within the Port and the community. The range of activities is limited only by finances and imagination, but commonly relate to specific features of the community. For instance, a Port with strong historical ties to passenger transportation may refurbish a sternwheel paddleboat, whereas in a larger city, investment in a convention and trade center complex may be more applicable. A sample of noncargo projects from U.S. Ports gives an indication of this diversity:

International trade center
Foreign trade zones
Development and management of industrial parks
Cruise ship terminals
Marinas
Waste disposal
Pipelines
Bridge and tunnel managers
Urban commuter buses
Intermodal computerized cargo tracking
Teleport
Riverboat operators
Condominium, hotel, and apartment complexes
Public market center
Convention center
Commercial property leasing
International trade liaisons
Whale search and nature watch
Boat shows (Tall Ships)
Public fishing pier
Aquarium
Recreational/festival center
Water taxi service
Rail transit
Hazardous waste management consultants

The depth, character, and motivation for some of these projects are discussed in the following sections. The first part outlines the particular dilemma of small-scale ports and their tendency to embrace recreation and tourism. Then experience at specific ports is scrutinized: the Port of New York/New Jersey, Portland and the Columbia/Snake River system, and Coos Bay, Oregon. These Ports may differ greatly in scale, mandate, and financial resources, but they have in common a rich agenda of non-cargo functions.

## *Recreation and Tourism at Small-Scale Ports*

Cuzan (1983) notes that people's perception of Ports is often biased toward the larger "superports." After all, these large Ports are closest to the major population centers, have the largest aggregate impact when measured in cargo throughput or value of shipments and command the most media attention. The rapid growth in container load centers has given these dozen or so "superports" even more attention because of the sheer volume of cargo moved through them (see Chapter 6). But the hundreds of smaller U.S. Ports deserve special attention because they are intimately tied to the economic health of their communities and frequently have a greater relative impact on the local economy than their larger counterparts. Discussion of economic development at small Ports does not simply involve a shift in scale; distinctive structures and relationships with the urban community are also observed.

Whereas in large port cities there are a number of public institutions, in small port communities there exist fewer and simpler political entities. Port staff assume a position of greater importance. It is not uncommon to find individual Ports that employ a larger number of managers and technicians than the city or county planning departments. As noted earlier, Port managers are considered reservoirs of knowledge on the local economy and are thus called upon to lend assistance in a wide range of community issues, including all forms of economic development. For example, at Newport, Oregon, not only does the Port have a substantial role in marine-related projects, it also has two representatives on the countywide Economic Development Agency, it is involved in a regional consortium to improve the economic health of neighboring towns as well as itself, and the Port plays a lead role in forming new committees charged with specific economic development projects.

The extended role of the public port does not stop at representation on economic development committees. Staff is often called upon to play a local political role and to be the unofficial political representative of the community in county, state, and national forums. This role ultimately translates into economic development because the ability to curry political favor is transformed into receipt of public works contracts, grants and low-interest loans, and inexpensive exposure that may lead to private investment. With an economic base that is so closely tied to the political decision-making process, small Ports have become adept at the arts of political persuasion and grantsmanship.

As general cargo shifted to container load center Ports, small-scale Ports were often relegated to shipping a narrow range of bulk cargoes for local manufacturers or suppliers. The changing structure of the U.S. economy from a producer of goods to a provider of information and services[6] holds special, and potentially catastrophic, significance for smaller Ports and their communities. If a smaller coastal community has any significant manufacturing, it often takes the form of water-dependent industrial firms that have always faced periods of boom and bust. Classic examples include the lumber industry in the Pacific northwest, petroleum products on the Gulf Coast, and the iron and steel industry at Great Lakes ports. This instability in economic and cargo potential is reason enough for port authorities at these sites to investigate alternatives in order to diversify their economic base. The fact that several of these heavy industries have been in a protracted and perhaps permanent bust since the mid-1970s gives additional impetus to search for other economic options.

One of the most popular options is the promotion of recreation and tourism. The mobility of the middle class and their penchant for recreation, combined with the relatively low initial capital investment costs required, make the promotion of community recreational attributes the key to their economic revival. Clark, Wilson, and Binder (1979) suggest that those New England Ports that were bypassed during the era of manufacturing growth are now benefiting because they have retained the ambience that is appealing to the tourist. Fishnets strung at the docks to dry, fresh fish and vegetable markets, sport fishing charters, whale-watching excursions, and clapboard houses now draw people and dollars to port communities.

Some of the more ambitious Ports have taken a larger risk by investing in facilities that are more long term and costly. Often the aquariums, marine museums, hotel and convention center complexes, and steamboat tours have proven to be wise investments in attracting the fickle tourist trade, but there are also many examples where these investments have become financial burdens on the communities and Ports. For instance, at Ports along the Washington and Oregon coast during the late 1970s, the demand for marina slips vastly outstripped supply and two- to three-year waiting lists became the norm. With these conditions, it was understandable that a large number of Ports embarked on ambitious marina-slip expansion plans, hoping to capitalize on the growing demand from sport fishermen and yachters. In a span of several years, the supply of slips expanded rapidly, the commercial fishing industry of the region collapsed, and sport fishing craft and yachts did not grow fast enough to displace this loss. Now there are many ports with marinas facing vacancy rates of greater than 50 percent. As explained in Chapter 9, marina development has special significance for the small-scale Port.

In the broad sense, tourism has become the stimulus for the demand-led philosophy of economic development. Despite occasional setbacks, Ports have used recreation and tourism as a mechanism to achieve industrial development. The recreational industry is used as a key industry or growth pole that pulls in other linked activities.[7] However, in this case the linked industries are not those that provide material inputs or receive outputs referred to commonly by industrial location analysts. Rather, recreational activities attract the attention of entrepreneurs and mark the community as a site with greater potential than other communities. These subtle "cognitive linkages" create the demand for transport facilites by attracting economic activities that may be water-dependent or "water-inclined."[8]

The incorporation of recreational activities into the set of functions coordinated by Ports provides an excellent example of their creativity. The projects that are most successful are those that coordinate with the specific features of the community. However, as with other noncargo roles, Port investment in recreation is not without its problems. Among these are the seasonal nature of recreation and tourism, which perpetuates the boom-bust cycle mentioned above, disenchantment within a community over the impact of tourism on the quality-of-life, and, ironically, difficulty in retaining the ambience that may have been the initial tourist attraction. In Newport, Oregon, the fish processors located at the central waterfront form an integral element in the ambience of the community. But success in attracting recreation-linked investment downtown has driven up land prices and property taxes to the point where it is difficult to justify the fish-processing use at peak value sites without some form of subsidization.

## *New York/New Jersey and the Teleport*

The Port of New York/New Jersey (NY/NJ) is the giant among U.S. Ports. Within the jurisdictional area where the port authority operates, there are over 578 miles of shoreline, including sixty waterfront parks that account for 84 miles of this waterfront and 18.4 miles of beach. Twenty-five percent of the shoreline is either vacant, deserted, or underutilized. In cargo throughput, NY/NJ has consistently ranked first or second in North America, rivaled only by New Orleans. In 1983 NY/NJ handled 149 million short tons of cargo. New Orleans, buoyed by the transshipment of larger volumes of petroleum products, handled 177 million short tons, whereas the next largest U.S. Port (Houston, Texas) handled only 95 million short tons. Perhaps more important than aggregate cargo in terms of employment and sales is the port's ranking in container traffic. Between 1975 and 1984, the annual average of twenty-foot equivalent units of containers handled by NY/NJ was 1,895,000, far more than at any other Port in the nation. There is also a cultural mystique associated with New York's status as the port where millions of immigrants first caught sight of their new home during the nineteenth and twentieth centuries.

New York/New Jersey Port Authority was founded in 1921. The bi-state governing structure was established to avoid the redundancy, antagonism, and inefficiency that would ensue if an economically unified port were divided into two political entities (see Chapter 2). Until the early 1970s, the NY/NJ Port Authority confined most of its functional roles to transportation-oriented projects. In 1978 legislation in the two states was passed that granted the Port a broader mandate: to retain and attract manufacturing jobs through industrial and other forms of economic development. The Port of NY/NJ has now become one of the most prominent actors in urban and regional planning within the metropolitan area. Besides owning and managing seven marine terminals, the Port also operates four airports (including J.F. Kennedy, La Guardia, Teterboro, and Newark), two heliports, four bridges, two tunnels, four bus and truck terminals, the World Trade Center, and several industrial parks (see Fig. 7.2). In 1983 the Port obtained greater financial capability to carry out its broader mandate when the governors of New York and New Jersey established the Bank for Regional Development to be managed by the NY/NJ Port Authority. Increased tolls on bridges and tunnels and rents on Port-owned New York City properties were set aside for transportation and industrial infrastructure improvements.

The NY/NJ Port Authority views its mandate in the broadest sense. Within its Statement of Objectives, the phrases "economic and transportation services," "spur the region's economy," and "capital reconstruction" clearly indicate this broad mandate. Although it is proactive, technically the Port does not view its actions as such. Instead, because the mandate states a responsibility for economic development, the Port *must* become involved in mass transit, promotion of foreign investment, tourism and regional markets, industrial parks, and redevelopment of its waterfront in order to obey this requirement.

Pragmatically, economic development of the area is in part the responsibility of the Port because the viability and efficiency of seaport activities rests on the economic health of the region. This justification is partially derived from the origin-destination hinterland of NY/NJ. Unlike those Ports that are break-of-bulk points with relatively little business drawn from the immediate local community, a large proportion of the cargo now passing through NY/NJ either originated in or is destined for the immediate metropolitan region. The seventeen-county metropolitan region that surrounds the Port had a 1983 population of 15.4 million and a labor force of 7.3 million persons receiving $221 billion in income and generating $86 billion in retail sales.

In addition, the Port is committed to helping the community deal with its fundamental economic and social problems. Port projects are implemented in an area that has experienced high unemployment, industrial and economic outmigration, high crime rates, the threat of financial bankrupty, and poverty. In essence, the Port and community have a symbiotic relationship. Structural forces are at work that are contributing toward a reduction in the manufacturing activity within the region and this has led to a relative loss in cargo traffic to other East Coast ports. This setting helps explain the Port's forays into noncargo projects. The World Trade Center has 9 million square feet of office space and generated $25.6 million in revenues for the Port in 1984. It provides a centralized setting for international public agencies and private firms that are involved in the trade and transportation business. The Fishport project on the Brooklyn waterfront is an attempt by the Port to create a fish harvesting and processing industry in the local area. The projected cost for Fishport is $27

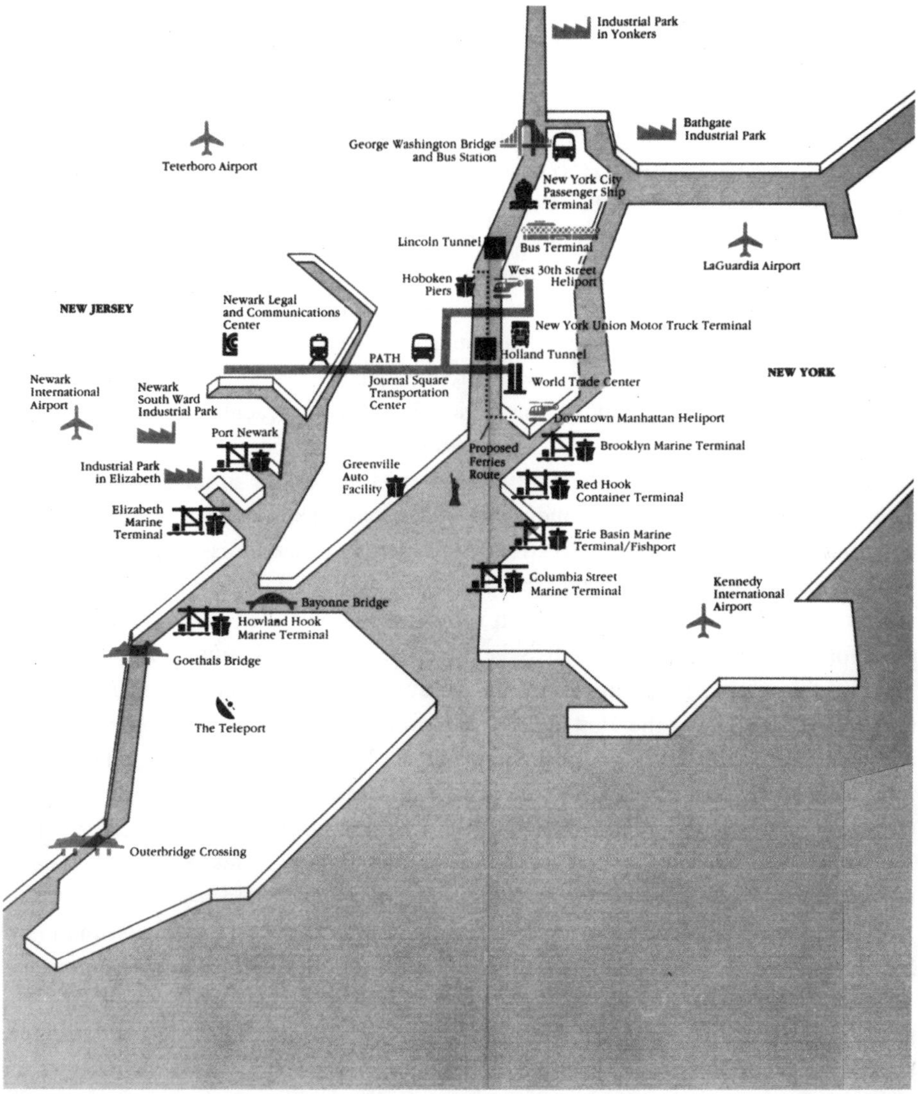


**Figure 7.2.** This map displays the many facilities controlled by the Port Authority of NY/NJ. Aggregate public investment exceeds $5 billion. (Courtesy: Port Authority of NY/NJ)

million (see Chapter 9). The term "resource recovery" is a euphemistic title for a project that converts urban garbage to an energy source through burning. The waste recovery plant is to be located in Newark, New Jersey, and requires a $265-million investment of which the Port of NY/NJ will provide $160 million. It will eventually consume 2,250 tons of the daily 30,000 tons of Essex County garbage and then generate steam for industrial use.

"Homeport" is a project instituted by the U.S. Navy and reflects a policy of decentralization in marine military facilities to a number of locations. Although principally a project coordinated by the city of New York and the Navy, the Port has been willing to make a $15-million investment in marine-related improvements in order to accommodate a seven-ship flotilla at a site on Staten Island. In order to improve the efficiency of urban transit, the Port purchased and gave to the states of New York and New Jersey 2,350 buses at a cost of $400 million. This gift was made with the understanding that the Port would not be responsible for operating or maintenance expenses, nor share in any revenues derived from their operation. A small-scale project that epitomizes the stimulative role of a Port is that of NY/NJ as a venture capitalist. The Port Authority set up a special investment fund of $10 million. It was set up to provide financial assistance to new firms that are committed to creating jobs in the region, but are undercapitalized.

Finally, the Port's Teleport project (see Fig. 7.3) represents a noncargo activity that is not found in other U.S. Ports. It is worthy of discussion because of the unusual management role it permits the Port to assume. Innovations in telecommunications technology have created a new industrial sector that makes extensive use of satellite communications. An interference-free location for the transmission of data, information, radio, and video images becomes an essential locational determinant for these firms, and this requirement both forces and allows them to decentralize from their former urban sites. Ironically, the Teleport project was initiated partially as a result of the success in leasing telecommunications space atop the World Trade Towers. In the late 1970s the Port of NY/NJ could no longer accommodate the telecommunications demand through antennae on the World Trade Towers and perceived that firms would leave the New York area if an alternate source of interference-free transmission were not available within the metropolis. To retain and attract these high-tech firms, the Port proposed the development of a 220-acre interference-free telecommunications site on Staten Island. The site was leased to the Port by New York City on a forty-year basis, and development was financed by the Port (an initial $57 million) and a private partnership of Merrill Lynch Telecommunications and Western Union Communications Systems (a combined $30-million investment). The site on Staten Island ultimately will contain seventeen satellite dishes, a telecenter building that coordinates the transmission and reception of data, and a 100-acre office park. Buried fiber optic cables connect the transmission facilities to the office park tenants and also to customers throughout Staten Island, Brooklyn, Queens, Manhattan, and New Jersey, allowing the creation of nodes of "teleparks" at a distance from the Teleport base. Although Teleport is promoted as a high-technology analogy to an airport or seaport, there are fundamental differences. Whereas an airport or seaport is designed to receive a vehicle transporting a product, a Teleport receives the product directly. In this sense, a Teleport is more analogous to a substation along the route of an oil pipeline.

In terms of returns on investment, net revenues generated from the subleased office properties will be split evenly between the city of New York and the Port. In addition, they are to receive ". . . 25 percent of any excess profits from the com-

**Figure 7.3.** First-phase development of NY/NJ Teleport, a joint public-private venture serving high-volume commercial data transmission needs by satellite for the greater New York area. (Courtesy: Port Authority of NY/NJ)

munications operations, after the private partnership has received a 21 percent return."[9] As of 1987 the first phase of the Teleport is complete and the marketing effort has begun to attract tenants. In order to entice potential users, the Port is granting real estate tax abatements that provide for an initial thirteen-year tax-free period followed by gradual implementation of taxes over the succeeding nine-year period.

The Port became involved in Teleport for a variety of reasons. The Teleport project fits the financial, technical, and managerial expertise of Port staff, and no other private or public bodies were willing to coordinate its development. At the same time, the mandate of the Port Authority allowed projects to be developed that would prevent loss of jobs and capital investment to the community because of outmigration. The eventual success of the project is not guaranteed because the communications industry is subject to rapidly changing technology and heightened competition. However, the project itself illustrates noncargo activities that can be carried out by a large financially solvent Port.[10]

## *Oregon and the Columbia/Snake River System*

Portland is the most significant cargo port in Oregon, accounting for 17.5 million short tons of cargo. Most of Portland's cargo tonnage is in bulk exports, but the port authority can be compared closely to the Port of New York/New Jersey in terms of its range of functions. In a Port of Portland *Values Book,* the first value is stated as follows: "The Port will promote a strong economic base in the community to the fullest extent of its charter with emphasis on marine, aviation, and land development." As with NY/NJ, proposals are evaluated on the basis of their ability to ". . . fill an unmet need in the community's economy." Cargo and passenger-related activities are valued as the base from which the Port derives its mandate, but they are by no means the only functions performed (see Chapter 11).

Revenues from land sales and leasing are rather insignificant when compared to total marine revenues ($2.8 million to well over $40 million in 1983–1984), but this comparison underestimates the involvement of the Port in industrial land development in the city. In addition to the $2.8 million, $14.7 million in industrial development revenue bonds were issued to several major firms to assist them in establishing operations on Port industrial land. The Port has three major industrial parks, Swan Island, Rivergate, and Mocks Landing, which together comprise 3,725 acres and provide jobs for 9,000 Portland area residents.[11]

The remaining thirty-three ports within the Columbia and Snake river systems engage in a variety of tasks (see Fig. 7.4). Several of them (for example, Kennewick, Richland) have no marine cargo facilities, but most have a substantial interest in promoting the economic growth of their communities through the ownership and management of industrial real estate.

Tourism and recreation have been used as the focus for development at several of the smaller port authorities, including Cascade Locks and Hood River. The Port of Cascade Locks, located 50 miles upriver from Portland, financed the construction of a 145-foot replica of a nineteenth-century sternwheeler called the *Columbia Gorge* (see Fig. 7.5). The boat was completed in 1983, holds 600 passengers, and operates year-round tours of the Columbia River Basin. It is also rented for dinners and party cruises for the general public and private groups.

At Hood River an unusual set of topographic and meteorological circumstances has led to the growth of a local $1-million/year industry—sailboarding. Although the Port receives no direct revenues from this activity, it provides promotional and marketing assistance, ensuring that the area's economy will continue to benefit from this water-based sport. Finally, farther upriver at the Tri-Cities area (the confluence of the Columbia and Snake rivers with the three communities of Richland, Pasco, and Kennewick), an annual water sports festival attracts 300,000 visitors. The highlight of the festival is a series of hydroplane speed boat races.

## *Coos Bay, Oregon*

There are conceptual similarities between the large regional Ports, such as New York/New Jersey, and the "microports" that serve a small urban community. With respect

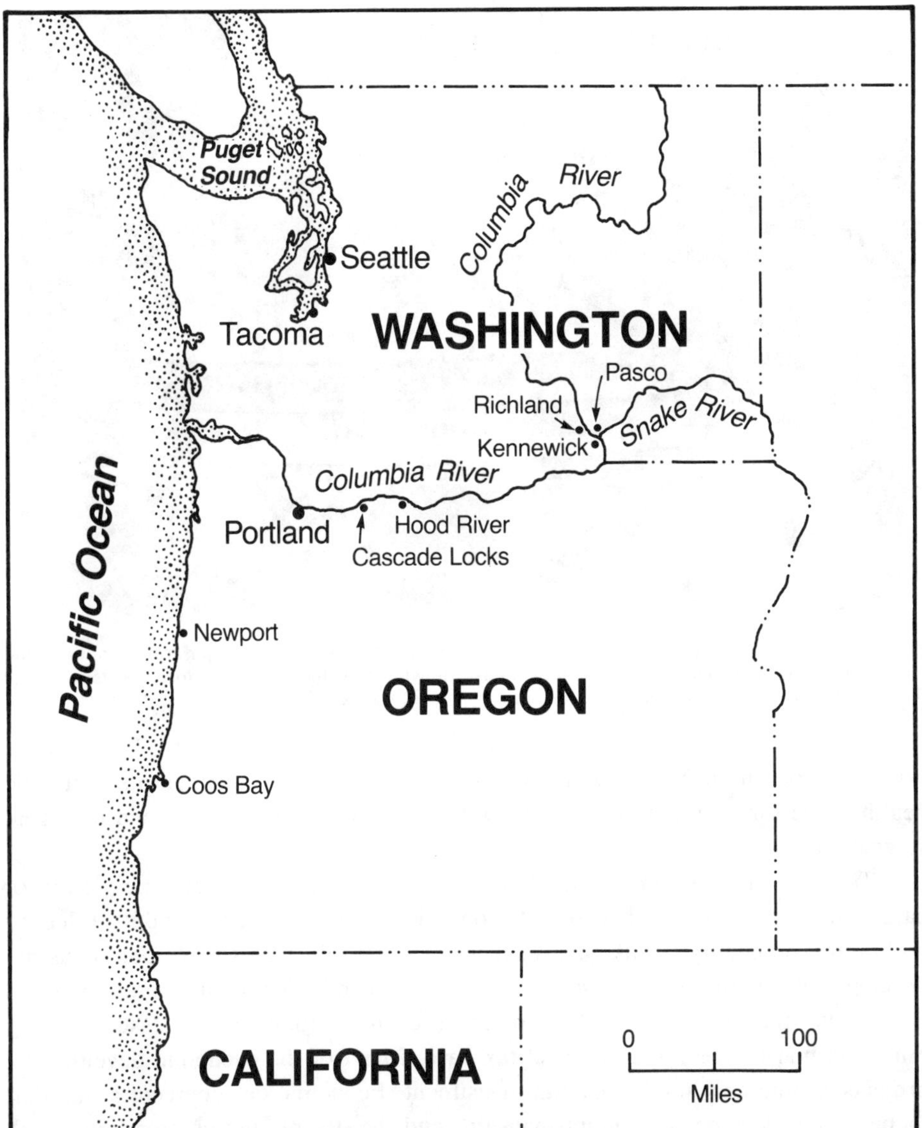


**Figure 7.4.** The Columbia River system between the states of Washington and Oregon contains many ports serving commercial, recreational, and community needs. (Courtesy: James Randall)

to economic development, the latitude given and the roles carried out by the Ports at both scales are often surprisingly similar in character and scope. Coos Bay is a community of 14,400 (1980), located 170 miles south of the Columbia River and 370 miles north of San Francisco Bay. Aggregate freight handled has averaged 6 million tons for many years, much of it wood product exports. As with other lumber-

**Figure 7.5.** For many small port cities, recreation and tourism makes up the major growth industry and ports assist in the development of tourism facilities. Shown here is the *Columbia Gorge,* a sternwheeler financed by the Port of Cascade Locks. (Courtesy: Port of Cascade Locks)

oriented ports on the West Coast, Coos Bay has struggled to maintain its economic health at the same time it has experienced demographic outmigration and industrial stagnation.

The Port of Coos Bay is significant not for its degree of economic vitality or cargo tonnage, but for the diversity of economic activity it promotes and its influence within the community. Unlike many other Ports that rely extensively upon industrial development revenue bonds to sponsor specific economic development projects, Coos Bay derives most of its funds from operating revenues (for example, leased property rents and marine fees) and the local tax base. Many of the economic development projects require only modest capital investment. Economic development at the Port includes site development, grantsmanship, and the use of surplus properties, with site development performed in anticipation of either manufacturing or recreational uses. In order to determine which manufacturing industries would be most appropriate for the community, the Port undertook their own marketing analysis. As a result of this in-house survey, at the end of 1984 the Port prepared a site suitable for the construction of oil industry equipment, including drilling platforms, derricks, and modules. The waterfront manufacturing facilities are not leased exclusively to one firm, but are instead let on contract to any successful bidder. The success of this initiative is tempered somewhat by its highly cyclic character: employment can range from 2,600, when the facility is in use, to zero between orders.

The Port of Coos Bay has long managed a trailer park and marina, becoming involved in the 1950s when the Port was awarded the property by the federal government upon default of the original owners. At that time, commercial fishing trawlers occupied 70 to 80 percent of the marina slips. Since the 1950s the Port has become more active in its role as tourism developer and promoter with mixed results. More than 600 slips are available at the marina for a predominantly sport fishing clientele, but the vacancy rate hovers at 40 percent. Local recreational services (for example, bait shop, hotel/motel, amusement facilities) have been sought to complement the trailer park. The Port has even purchased and installed a television satellite dish within the campgrounds to satisfy recreational vehicle travelers.

Grantsmanship must be classified as a form of economic development because it requires the commitment of Port resources (staff time), and the money generated is invested in the community for the creation of jobs. The Port of Coos Bay makes an effort to seek grants from government agencies and to use the political influence of voters to be awarded public works programs and facilities. Examples include a Pacific Power and Light (utility) award and a grant from the now-defunct Comprehensive Employment and Training Act (CETA). Also, community political efforts resulted in a grant from the Oregon State Economic Development Agency for a barge slip. Although none of these grants would have formed a significant share of the 1984–1985 total budget of $890,000, they represent an important supplement to other unstable sources of revenue.

Utilization of surplus properties, the third tool used by the Port to stimulate economic growth from noncargo sources, indicates the link between economic and political functions. By cooperating with political entities such as the State of Oregon, Coos Bay has developed a screening program to select appropriate merchandise that is made available to communities through the federal government surplus disposition program. Equipment, merchandise, or facilities that may be of use to the Port, to other institutions within the community, or even to other ports is scavenged by Port staff. One acquisition that has created up to seventy local jobs is a 1,000-ton drydock now used frequently for Coast Guard vessel repair work.

As with several of the smaller Ports along the Columbia and Snake rivers, the Port of Coos Bay has been hurt by the economic restructuring within the primary and heavy industry sectors. Communities faced with the debilitating prospects of high unemployment, outmigration, and difficulty in maintaining the quality of public services rely on their local institutions to provide solutions. In both large and small urban seaports, a most venerable institution to handle the task is the public port.

## *Notes*

1. For a detailed discussion of the input/output study, consult William H. Miernyk, *The Elements of Input-Output Analysis*. (New York: Random House, 1965).

2. A multiplier can be derived for employment in the same manner as for sales and wages. For every job that is directly created by the Port, an additional level of employment is created indirectly within a given region by the Port, e.g., addition of 100 port jobs creates demand sufficient for the creation of another 75 jobs within the community.

3. H. C. Davis, in the article "Regional Port Impact Studies: A Critique and Suggested Methodology," *Transportation Journal* **23** (2):61–71, critiques impact studies on three bases. First they lack standardization concerning the range of port industrial categories that determine the size of the multiplier. Second, there are methodological problems involved in the estimation of secondary (indirect and induced) impacts. Last, and as a result of the second problem, evaluation of changes in the level of port services is fraught with the difficulty of separating the marginal impacts of change from the average impacts of change.

4. This particular conceptualization of cyclic change was developed by the Russian economist N. D. Kondratieff and was presented in a 1936 paper entitled, "The Long Waves of Economic Life," *Review of Economics and Statistics* **17**:105–15.

5. In this light, it should be noted that there are many cases where resource-oriented Ports have ceased shipping because there was nothing left to ship (e.g., lumber, coal, iron ore), but it is difficult to find an example among major U.S. Ports where operations have been permanently halted as a result of mismanagement. Unlike a private corporation, and despite government deregulation, the prominence of the Port in the community and the control Ports exercise over public assets such as waterfront real estate ensure their continued existence in some form.

6. For two of many descriptions of postindustrial America, consult Daniel Bell, *The Coming of Post-Industrial Society* (New York: Penguin, 1983), and David Clark *Post-Industrial America* (New York: Methuen, 1984).

7. Perroux crystallized the concept of growth pole theory in a seminal 1955 paper entitled, "Note sur la notion de pole de croissance," *Economie Appliquee,* 7:307–20. Since then it has been adapted by other regional economists and geographers including Boudeville (1966) and Hirschman (1958).

8. Water-dependent industry is a generally accepted phrase implying that a firm requires waterfront access for production or transportation. However, it is too often used as an absolute term. At the margin, there is a larger group of firms and industrial sectors that may be inclined to use water transportation depending on its cost, convenience, stability, etc. vis-à-vis other modes of freight transport. These I have called "water-inclined" economic activities.

9. Many of the preceding statistics are drawn from a July 10, 1984, *New York Times* article by D. H. Charles entitled "New Deadlines for Teleport," on pp. 29 and 43. They are corroborated by Port of New York/New Jersey literature.

10. This description of activities undertaken at the Port of NY/NJ reflects the diversity of Port economic activities. It should be noted, however, that Port intentions and public pronouncements about new ventures are often speculative and require coordination of a large number of groups that control funding, environmental conditions, etc., before they actually occur. Further, political factors often affect the outcome of these noncargo projects. At New York/New Jersey a change in Port administration in October of 1985, combined with an increase in local investment and job creation, produced a policy of retrenchment at the Port. Although the projects noted in the text are still being pursued, the traditional concerns of the Port (marine, air, maintenance of bridges and tunnels) have been reemphasized.

11. This figure of 9,000 employees varies with economic conditions and applies only to the 100 businesses and industries at Swan Island and Rivergate industrial parks. Mocks Landing Industrial Park is expected to provide employment for an additional 2,800 residents.

## *References*

Bailly, A. S., and E. R. Weiss-Altaner. 1981. The Port of Montreal and its future: Economic and spatial consequences for the urban environment. *Cityport industrialization and regional development,* ed. B. S. Hoyle and D. A. Pinder. Oxford: Pergamon Press, pp. 167–80.

Bell, D. 1973. *The coming of post industrial society*. New York: Penguin.

Boudeville, J.R. 1966. *Problems of regional economic planning*. Edinburgh: Edinburgh University Press.

Chang, S. 1978. In defense of port impact studies. *Transportation Journal* **17** (1):1.

Clark D. 1984. *Post-industrial America*. New York: Methuen.

Clark, J., C. Wilson, and G. L. Binder. 1979. *Small seaports*. Washington, DC: The Conservation Foundation.

*Containerisation International*. December 1981, 60–63, December 1983, 57–63, December 1984, 65–67, December 1985, 67–71.

Cuzan, A. G. 1983. The political economy of ports in the United States and Great Britain. *Coastal Zone Management Journal* **11** (3):149–67.

Davis, H. C. 1983. Regional port impact studies: A critique and suggested methodology. *Transportation Journal* **23** (2):61–71.

Goss, R. O. 1979. *A comparative study of seaport management and administration* 1. London: Government Economic Service.

Hanappe, P., and M. Savy. 1981. Industrial port areas and the Kondratieff cycle. In *Cityport industrialization and regional development,* ed. B. S. Hoyle and D. A. Pinder. Oxford: Pergamon Press, pp. 11–21.

Hirschman, A. D. 1958. *The strategy of economic development*. New Haven: Yale University Press.

Hoover, E. M. 1948. *The location of economic activity*. New York: McGraw-Hill.

Kondratieff, N. D. 1936. The long waves of economic life. *Review of Economics and Statistics* **17:**105–15.

Maritime Administration (MARAD). 1978. What U.S. ports mean to the economy. Washington, DC: Government Printing Office.

Miernyk, W. H. 1965. *The elements of input-output analysis*. New York: Random House.

Perroux, F. 1955. Note sur la nation de pole de croissance." *Economie Appliquee* **7:**307–20.

Randall, J. 1983. *Metropolitan Vancouver: Industrial decentralization and port activity on the lower Fraser River*. Working paper no. 12, University of Toronto/York University Joint Program in Transportation.

U.S. Corps of Engineers. 1985. Waterborne Commerce of the United States, Vol. 5.

Washington Public Ports Association (WPPA). 1978. *Washington public ports economic study*. Olympia: Washington Public Ports Association.

Witherick, M. E. 1981. Port developments, port-city linkages and prospects for maritime industry: A case study of Southampton. In *Cityport industrialization and regional development,* ed. B. S. Hoyle and D. A. Pinder. Oxford: Pergamon Press, pp. 113–32.

## *Additional Readings*

Bird, J. H. 1980. Seaports as a subset of gateways for regions. *Progress in Human Geography* **4** (3):360–70.

Hayuth, Y. 1981. Containerization and the load center concept. *Economic Geography* **57:**160–76.

———. 1982. The port-urban interface: An area in transition. *Area* **14** (3):219–24.

Hoyle, B. S., and D. A. Pinder, 1981. Seaports, cities, and transport systems. In *Cityport industrialization and regional development,* ed. B. S. Hoyle and D. A. Pinder. Oxford: Pergamon Press, pp. 1–10.

Kinsey, J. 1981. The economic impact of the Port of Liverpool on the economy of Merseyside—using a multiplier approach. *Geoforum* **12** (4):331–47.

Pinder, D. A., and B. S. Hoyle. 1981. Cityports, technologies and development strategies. In *Cityport industrialization and regional development,* ed. B. S. Hoyle and D. A. Pinder. Oxford: Pergamon Press, pp. 323–38.

Pollock, E. E. 1981. Free ports, free trade zones, export processing zones and economic development. In *Cityport industrialization and regional development,* eds. B. S. Hoyle and D. A. Pinder. Oxford: Pergamon Press, pp. 37–46.

Taylor, M. J. 1974. The impact of New Zealand 'secondary ports' in their associated urban communities. *New Zealand Geographer* **30** (1):35–53.

# Chapter 8

# FISHPORTS:

## Service Centers for a Changing Industry

*Robert F. Goodwin**

---

To the casual waterfront observer, a fishing boat evokes images of a colorful, independent, and adventurous lifestyle; a rugged individualist battling waves and wind while pulling in the rich harvest of the sea; and, perhaps a filet of halibut, salmon, or sole exquisitely garnished and served amid the nautical ambiance of a waterfront seafood restaurant near the fishing boat basin. To the informed observer, that same vessel appears as a large, heavily insured capital investment equipped with space-age navigation and control systems, high technology sonar fish-finding gear, advanced hydraulic equipment for pulling, lifting and trimming nets and lines, on-board refrigeration and fish processing machinery, and life-support systems to sustain a crew at sea for weeks, or perhaps months. That same informed observer will be aware of the intricate but unseen linkages that tie the vessel to the modern urban harbor as securely as the lines that tie its hull to the pier. These linkages are streams of goods and services; services that flow between shore-based establishments and vessels. For convenience, we can conceive of them as packages of marine services: repair and maintenance services, equipment and supplies, consumable provisions; and business, financial, and information services.

A major metropolitan harbor servicing a large and diverse fleet of vessels will likely provide a full competitive range of such services, whereas a rural port harboring a limited number of vessels may supply only the most rudimentary services. Just as a rural household consumer will drive to a major urban shopping center to

---

*Partial financial support from Washington Sea Grant Marine Advisory Program under NOAA Grant #NA86AA-D-SG 044, Project #A/CZ-1 is gratefully acknowledged. Thanks to Mike Giarri, Scott Harrington, Steve Hughes, Dan Huppert, Erling Hulgaard, Neil McKellar, Gerhardt Muller, John Peters, Bob Peterson, Randolph Rossi, Robert Stokes, William Stokes, and B. G. Thompson for interviews or critical review of the manuscript.

find a competitive range of consumer goods, so will fishing vessel skippers put in to a major harbor to gain access to specialized marine services unavailable at the vessel's smaller homeport.

Oceangoing cargo vessels, tugboats, recreational smallcraft, and other floating equipment swell the market demand for similar marine services in a large urban port and the marine service businesses, which developed in response to that market demand, also serve the fishing fleets. For large vessels fishing in distant waters for long periods at sea, the costs incurred in returning to port for repairs and outfitting, together with costs of unproductive downtime while in port, virtually necessitate utilizing a full-service urban harbor to minimize time in port.

## THE MODERN FISHPORT: AN INDUSTRIAL COMPLEX

To illustrate the functions of today's fishport, let us board a large, modern factory trawler (see Fig. 8.1) inbound to its homeport of Seattle, Washington, from Alaska's Bering Sea where the vessel has harvested and processed bottomfish for the last three months.[1] The thirty-six crew members on board are eager to get into port to be reunited with their families and friends for the first time in four months. Families

**Figure 8.1.** A large, modern factory trawler at its homeport, Fisherman's Terminal, Seattle. (Courtesy: Robert F. Goodwin)

and friends will be crowding the dock, equally eager to see and embrace their returning loved ones. Here is the essence of what "homeport" means. Many of these families have long roots in Seattle's Scandinavian fishing community centered in a section known as Ballard, and several generations of a family fish the same grounds, sometimes on the same vessel. Their names reveal their Scandinavian ancestry: Sven, Olaf, Ole, Per, Arne and Johannsen, Jorgensen, Olsen, Perrson, Amundsen.

The twenty crew members who work below decks, processing the catch, will not be required to remain on board, nor will most of the half-dozen deckhands or the two or three cooks. The four engineers, however, will have rotating duties to perform while maintenance and repair work on the vessel is carried out in the shipyard.

Below decks are 1.5 million pounds of frozen filets and blocks of whitefish, already packaged in plastic and cardboard containers. The vessel's first landfall will be at a seafood cold-storage company dock located on an industrial waterway several miles south of Seattle's central urban waterfront where most of the harbor's fresh seafood processing plants once were located. The ship's gear lifts pallets loaded with frozen product onto the dock. Forklift trucks move the pallets into refrigerated warehouses to await shipment by refrigerated rail-car or truck to markets throughout the nation.

Its catch unloaded, the vessel heads for Salmon Bay, a freshwater basin upstream of Ballard Locks. Here fishing vessels ranging from 20-foot salmon gillnetters to 200-foot catcher-processor trawlers crowd into Fisherman's Terminal, a Port of Seattle moorage basin dedicated exclusively to the commercial fishing industry. Well over a month ago, the trawler's company headquarters contacted the Port to schedule dock space and called the shipyard to arrange for repairs and maintenance. Even when lightly loaded, the vessel's 19-foot draft limits the choice of mooring space to the deeper end of the terminal moorage basin. Last year the skipper could not find space here at all and instead tied up at a local tug and barge company's dock.

Within 1,000 feet of the berth, the shorelines are lined with businesses providing all the services necessary to repair, overhaul, service, refuel, and reprovision the vessel in preparation for its return voyage to the Alaskan fishing grounds. An Alaskan king crab boat is almost ready to float out of drydock at a nearby shipyard. The trawler is the next vessel scheduled for that drydock. While the vessel is at dockside, the skipper goes ashore to meet with the company's shoreside personnel and to chat with the staff of the trade association to which he and other trawler skippers belong. Most of the offices he must visit are located in buildings immediately adjacent to the terminal.

The trawler's arrival attracts a small knot of sightseeing tourists who notice the buckled plates and rusty patches on its hull—evidence of an Aleutian storm that halted fishing for two days in April. Some welding, sandblasting, and new coats of paint will restore the vessel to its "first launch" condition. While drydocked, shipyard workers, marine engine specialists, refrigeration and hydraulic engineers, and a host of other subcontractors will turn the vessel into a hive of apparently chaotic activity. A new search radar will be installed; the sonar system will be modified to accommodate a remote sonar unit located at the mouth of the trawl net to monitor the fish actually being caught during a trawl tow; the skipper wants some modifications made to the fish-fileting machine and an additional freezing unit installed; the vessel's 2,600 HP main diesel engine and auxiliary engines will be overhauled. All this work will be completed in three weeks! When the skipper returns from his vacation, new nets will have been stowed; the remaining fuel will have been pumped out, centrifuged to remove water and other contaminants, and returned to the tanks along with 110,000 gallons of fresh fuel, 100 gallons of hydraulic fluid, 200 pounds of freon refrigerant, and ten 40-gallon barrels of lubricating oil; enough food and provisions will be on board to keep the crew fed at sea for four months; and crates of packing materials and cartons for the catch. By the time his vessel has cleared the locks at Ballard, the skipper will have paid out over $250,000 to local firms in Seattle Harbor. Those firms will in turn pay others—employees, suppliers, accountants, lawyers, and, in addition, local, state, and federal taxes. As a result, the impact on local and state economies will have swelled to perhaps twice the original $250,000 taken from the vessel's account. Fishing fleets, then, play a significant role as agents for economic development in urban harbors and the regions in which they are located, a fact of some importance to local public port authorities, and an issue to which we will return. Meanwhile, before the skipper leaves for a well-deserved vacation in Hawaii, let us follow him ashore and observe the importance of information available to him from sources in or close to the harbor.

## *Information Services*

Many levels and bureaus of government play vital roles in the management and regulation of fisheries. The skipper fishes outside U.S. territorial waters in the 3-200 mile Fisheries Conservaton Zone (FCZ), established through the 1976 Magnuson Act.[2] He is one of many U.S. fishermen taking advantage of fish stocks formerly harvested by foreign fishing fleets in the North Pacific. The actions of the Northern Pacific Fisheries Management Council,[3] which is responsible for managing these stocks, are of vital concern for the skipper's livelihood. Changes in quotas of specific species of fish or in the length of seasons in which quotas can be taken are critical factors he must take into account while planning the next season's activities. Professional staff in the trade association to which he belongs keep abreast of any proposed changes in regulations and pass these on to the membership through newsletters and face-to-face contacts at the association's dockside offices.[4] Were the skipper fishing in state waters (0 to 3 miles) of the territorial sea, he could seek out information about state fisheries agencies' regulations in effect in those jurisdictions.

At the Sea Grant Marine Advisory Services office, he could gain further intelligence about fish prices, Coast Guard regulations, new fishing gear developments, vessel stability tests, and a host of other topics germane to his business. While talking with the Sea Grant agent, he would learn of short courses being offered: "Medical Emergencies at Sea," "Troubleshooting Hydraulic Systems," and "Exploiting Underutilized Species." A publication on joint U.S.-foreign fishing ventures catches his eye. In the thirty years this skipper has fished for a living, the changes in technology, in fisheries management and regulations, and in processing and marketing seafood appear to him to have been revolutionary; and maintaining knowledge and skills to adapt to the changes is a constant challenge. The information services available in a modern urban fishing harbor, then, are as important to the fishing industry as the physical services necessary to build, repair, and maintain fishing gear and vessels.

From the foregoing narrative we can define a modern fishport as an *industrial complex*[5] consisting of individual business establishments tied directly or indirectly to the demand for goods, services, and information arising from the presence of fishing fleets in the harbor. Some of these businesses require a waterfront location because of the nature of their work: shipyards, boatyards, marine equipment suppliers. Others gain significant advantage from a waterfront or near-waterfront site because they conduct business with fishers: ships' chandlers, marine banking and insurance offices, marine electronic dealers, fishing industry trade associations, and university marine extension offices. Yet other businesses with strong economic ties to waterfront establishments are scattered throughout the metropolitan area surrounding the urban harbor: marine publishers, subcontractors with mixed marine and nonmarine customers, major banks and insurance companies, and seafood brokers and wholesalers.

## Seafood Processing and Distribution

Although many of the larger urban harbors are no longer the important loci they once were for the seafood processing industry, many of the nation's smaller and middle-size ports play a major role in serving the industry. Recall that the product unloaded from the catcher-processor vessel we followed into port was in processed or semiprocessed frozen form. The shipment and marketing of this product depends on advances made in refrigerated transportation technologies: refrigerated containers, trucks, and railroad cars. Before these technologies were widely adopted, fish were landed at ports close to markets and shipped iced-down, fresh, over short distances. Alternatively, product was canned, cured, or frozen at dockside plants near the fishing grounds for long-term storage and shipment to distant markets. Fresh product can now be landed at ports convenient to the fishing boat owners and moved overland to market by refrigerated carriers, or increasingly by air freight in the case of such high-value species as salmon and lobster. Similarly, product to be shipped frozen is landed at ports near fishing grounds, perhaps partially processed before being frozen locally, then shipped by common carrier to warehouses or reprocessing plants near metropolitan markets. In many cases then, seafood processors have abandoned shoreside facilities in urban ports and relocated to inland sites where land is cheaper and transportation networks less congested. As a result, abandoned canneries and processing plants comprise a large proportion of the vacant, dilapidated structures in some urban waterfronts and have become prime candidates for revitalization and redevelopment schemes discussed in detail in Chapter 13. Gains in efficiency to the seafood industry are losses to the harborfront visitor for whom the sights and smells of fresh fish being unloaded from fishing boats were memorable links to the sea and part of the romantic ambiance of the waterfront.

## CHANGES IN FISHERIES MANAGEMENT: CAUSES AND EFFECTS

We have seen how a factory trawler moored to the docks at Fisherman's Terminal in Seattle placed immediate, complex demands for space, services, and facilities on the port and its marine service firms. Docks and piers designed for 30- to 40-foot salmon boats have become crowded with new classes of vessels built for fisheries whose exploitation on a large scale by U.S. fishermen is a recent phenomenon. Alaskan crab boat fleets, comprising vessels fo 80 to 120 feet in length, and new 160- to 200-foot catcher-processor trawlers fishing the Bering Sea dwarf their older traditional neighbors. Even larger "factory" processor ships cannot be accommodated in the terminal and must moor at private docks.

Behind the effects of impacts and stresses placed on fishports by changing fishing fleets lie complex layers of causes. Some are man-made, some are linked directly to the vagaries of nature acting in the marine environment, and others stem from the interaction between the two.

The marine environment in general and the dynamics of fisheries in particular are imperfectly understood. Complex ecological interactions between marine fish and the environment produce natural perturbations in fish populations. When people harvest stocks of fish, an additional element enters the biological equation, causing further perturbations to a natural system. If not constrained by regulation, fishers would have little incentive to limit their fishing effort and avoid the "tragedy of the commons"[6] at sea. Conservation of the common fish stocks is possible only through the actions of regulatory bodies that enforce catch limits across an entire fleet of fishers. More often than not, the effects of limiting catches lead to inefficient utilization of overbuilt fleets: too many vessels chasing too few fish during too short a season.

Cyclical downturns in or, worse, complete collapse of some fisheries stocks produce boom-bust gyrations for vessel owners and impose unpredictable burdens on fishports. Changes in quotas necessary for the conservation of such fisheries can result in whole fleets being tied up for entire seasons, as was the case recently in both the ocean troll salmon fleet and the Alaskan king crab fleet homeported in northwestern U.S. harbors. Bankruptcies and forced sales of vessels in depressed fisheries often result in uncollectable moorage fees. Because of community pressure and faith in the future, Ports do not evict vessels in default. As a result, space that would normally be available for transient use by active fishing boats is locked up for long-term storage. The opposite problem—large, intractable vacancy rates—besets smaller rural ports used temporarily, under normal conditions, by vessels fishing local waters for short seasons. Where those fisheries are closed or drastically curtailed, the vessels tend to stay in their larger urban homeports.

Since the 1976 Magnuson Act, vast new areas of ocean territory have been reserved for exploitation by U.S. fishing fleets. Once utilized almost exclusively by foreign fishing vessels, the fish stocks in waters up to 200 miles offshore are being caught increasingly by U.S. fishers. But there has been widespread disappointment among many West Coast ports concerning the consequences of extending national jurisdiction over offshore fisheries. The anticipated bonanza for American fishing fleets simply has not materialized.

Although the U.S. fleets, undoubtedly are catching a larger share of the fisheries resources of the 3-200 mile zone, few new vessels have been built to take advantage of the Magnuson Act with the exception of the Alaska groundfish industry, discussed below. Overcapitalized existing fleets have benefited through "joint venture" arrangements with foreign governments, the most common form being a "marriage" between U.S. trawlers and foreign factory ships (Soviet, Japanese, Korean, and East European countries). The product usually is processed and consumed by the foreign nations; the U.S. fishermen retain earnings from sales of their catches to the factory vessels.

The U.S. Alaska ground fishery is a growing exception to the general trend, however, since in this fishery the size of the U.S. harvest has boomed (see Fig. 8.2). The factory trawlers fishing off Alaska are imposing new, severe but brief demands on their homeports.

Further growth in the U.S. catcher-processor fleet vessels is likely to be hampered by biological limitations of the fisheries stocks they exploit—notably cod (NRC 1985). Foreign wage rates, low prices, and little domestic processing capacity have combined to stifle U.S. exploitation of the more plentiful stocks of whiting and pollock. But domestic production of surimi (fish paste) and surimi-based products such as imitation crab legs and other manufactured seafood products such as breaded fish sticks is increasing domestic demand for these lower-value species. As the industry overcomes such problems as spoilage of the product at sea, domestic exploitation of whiting, pollock and other low-value but abundant FCZ fisheries resources will continue to displace foreign fleets.

Beyond conservation of ocean fish stocks and the "Americanization" of the FCZ, the Magnuson Act opened the door to allocation of fisheries for foreign policy pur-

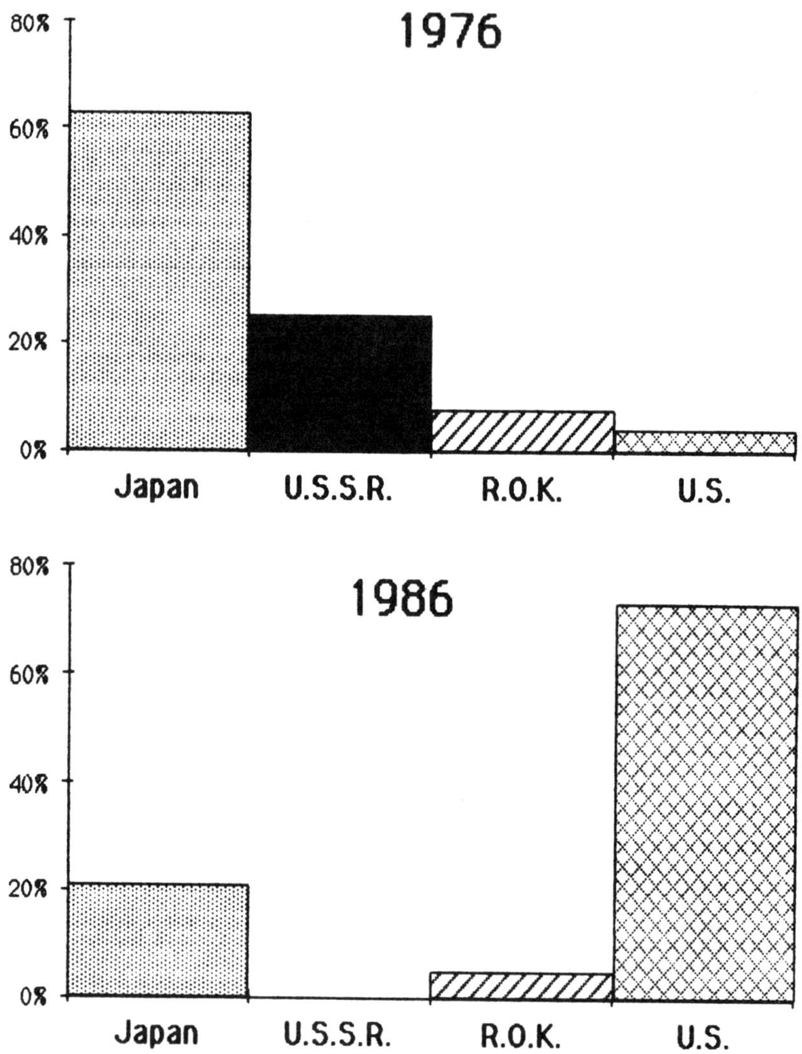


**Figure 8.2.** Percentage of groundfish harvested from the FCZ of the northeast Pacific in 1976 and 1986 by Japan, the U.S.S.R., Republic of Korea, and the United States. (Courtesy: Natural Resources Consultants)

poses: the so-called Fish and Chips policy. The fisheries of the FCZ can be allocated or withdrawn as rewards for or sanctions against the international behavior of foreign nations, especially in trade. As a result, the quotas available to U.S. fishing fleets can shrink or expand for reasons other than conservation of stocks.

The regional Fisheries Management Councils (FMC) also have played a role in implementing controversial judicial decisions. Together with state fisheries agencies, the Pacific FMC has been required to manage salmon runs in a manner that assures treaty Indian tribes the opportunity to harvest 50 percent of the harvestable catch. Severe cutbacks in the ocean troll fisheries off Washington, Oregon, and Northern California were mandated in order to permit salmon to reach the major estuaries where these tribes have customarily fished. One result has been the large number of salmon troll vessels tied up year round at Fisherman's Terminal in Seattle (see Fig. 8.3).

## FORECASTING THE FUTURE

Whereas there are variations among states and municipalities in both enabling legislation and the charters under which public port authorities operate, most Ports perform four key functions: facility development, economic (industrial) development,

**Figure 8.3.** Salmon trolling vessels sitting idle at Fisherman's Terminal, Seattle. Changes in fishing allocations substantially affected the offshore salmon trolling fishery. (Courtesy: Robert F. Goodwin)

land development, and waterborne trade. Each of these functions is enhanced by and, in turn, enhances the development and operation of fishports.[7]

Given the lead time necessary to acquire permits, arrange financing, and construct new or expanded fishport facilities, Port officials need reliable forecasts of future demand for the services they provide. Ideally, such information would be organized around the future size and characteristics of the fishing fleets using the port's facilities, the seasonalities or timing of use, and the kinds of services the vessels would require. However, as we have seen, the uncertainties surrounding the fisheries industry make forecasting difficult. How can Port managers plan for the future? This task is not an easy one, and the consequences of being wrong can be costly. On the one hand, overbuilding in anticipation of growth in fleet size that does not materialize can result in special-purpose facilities being underutilized. Conversely, failing to expand in the face of real growth in fishing fleets can result in overcrowding and loss of business to other ports. Where decline, rather than growth, in fishing fleets is occurring, there still is a need to plan. There may well be alternative uses such as recreational boating or other marine industries competing for space. But before committing to a course of action that would permanently replace underutilized fishport facilities, Port officials must be reasonably certain that the fishing fleet homeported in the harbor will not expand in the foreseeable future. If opportunities to service the fishing fleet are foregone, many of a Port's traditional functions could also suffer. Economic development opportunities, fueled by the fleet and the linked industries it supports, could be lost; those linked marine service firms that would develop on Port land holdings would locate in harbors elsewhere, and shipments of seafood product would move across competing fishports' docks. On what kinds of information, then, do Port managers rely for projecting future facility needs? And from whom can this information be acquired?

Port managers need to be aware of the decisions made by others that affect the manner in which their fishport facilities are likely to be utilized in the future. Decisions made by fisheries management agencies, by fishing vessel owners, by seafood processors and distributors, and by competing fishport managers are among the more critical factors affecting a fishport's future.

## Fisheries Management Decisions

In allocating fisheries stocks among competing fleets, fisheries management agencies and councils make vital decisions affecting fishports. Through their vessel licensing regulations, state fisheries agencies control the number of vessels using a particular kind of gear to harvest certain stocks. Some such form of "limited entry" into fisheries is becoming increasingly common in the United States. Coupled with vessel "buy-back"[8] programs, limited-entry schemes control the size of fishing fleets. Setting quotas for the total allowable harvest of specific stocks in specific areas, state agencies and the regional FMC control the amount of product caught by U.S. and foreign fleets and exert a great deal of influence over the flow of that product into the processing and marketing sectors of the seafood industry. Changes in management strategies applied to certain fleets can result in drastic shifts in the timing and volume of product flows. For example, were the state of Alaska to adopt an individual vessel quota system in its halibut fishery, the current pattern of brief but intense fishing seasons might change to a more uniform but limited year-round fishery on the same halibut stocks (NPFMC 1981). The results would be smoother flows of product into the processing sector and lower peaks and troughs in demand for port facilities by the halibut fleet. Port managers, then, must maintain current knowledge about fisheries management decisions affecting their facilities.

## *Fleet Investment Decisions*

Closely linked to fisheries management decisions are the decisions made by vessel owners to invest in new equipment. An innovation in fishing gear that proves to be successful diffuses rapidly in the industry. The explosive growth in the Washington-based Alaskan king crab fleet during the 1970s is a dramatic example of innovation diffusion in the industry. A slower but equally impressive growth in the U.S. factory trawler fleet operating in the North Pacific FCZ is now underway. Since fisheries management decisions often react to, rather than anticipate, massive investment in new vessels, a Port manager must gather intelligence about the kinds of vessels being built or converted and their logistical requirements while in port.

## *Seafood Processing, Distribution, and Marketing Decisions*

The advent of at-sea processing by U.S. seafood firms is making significant changes in the flow of product from capture to final consumption. Whitefish caught by the factory trawler homeported in Seattle is processed, frozen, and packaged within a few hours of being spilled from the net into the vessel's holds. This rapid processing permits even pollock to be marketed in superior quality. But without adequate land-side cold storage warehouses and refrigerated containers for transshipment, growth in at-sea processing would be stymied. Ports closest to the fishing grounds and served by adequately equipped transportation systems stand to benefit most from this new and growing industry. To capitalize on this growth, Port managers must have a thorough understanding of the logistics involved in moving frozen product and the locational advantages their ports can offer to components of the seafood distribution industry.

## Port Industry Decisions

Public port authorities enjoy a collegial as well as competitive relationship among themselves. That is, while they compete for business, they also share information about common issues, problems, and strategies. For example, Ports, through their membership organizations, attempt to monitor the effects of fisheries management decisions on utilization of their fishport facilities. The Pacific Coast Congress of Harbormasters and Port Managers (PCCHM&PM), which represents Ports with smallcraft harbors along the entire west coast of the United States, including Alaska, has shared information about fishing vessel moorage demand at its regular meetings. More than one of its member Ports has been perilously close to default on bonds as a result of depressed conditions in some fisheries of the Northeast Pacific. Members of the PCCHM&PM participate effectively in public meetings of the Pacific and North Pacific Fisheries Management Councils and report back to the full membership through a newsletter and semiannual meetings. At its semiannual meetings, the PCCHM&PM encourages member Ports to identify their current problems and share information on how they address them. Setting fees for moorage, collecting bad debts, handling hazardous wastes, reducing liability exposure, and other operational issues always figure prominently on the agendas, but Ports also reveal their plans for expanding facilities and providing better service to the fishing industry. As a result, few major surprises concerning the fisheries-related plans of competing Ports encumber a fishport manager who attends PCCHM&PM meetings.

## Tapping Fisheries Expertise in Firms and Agencies

But even if fishport managers are attentive to the decisions being made by fishers, fisheries managers, seafood industry leaders, and competing ports, the intelligence gained is likely to be unsystematic, geographically limited, and of little use beyond the immediate future. For longer range planning purposes, information must be gathered that is systematic, spans local, regional and even international geographic scales, and encompasses all the salient factors affecting the seafood harvesting and processing industries. Furthermore, translating this information into a form that facilitates fishport planning requires the skills of experts in fisheries management, resource economics, seafood science and technology, and industrial siting—skills rarely found among even the largest Ports' planning staffs. It is hardly surprising, then, to find Ports turning to the consulting industry and academic experts for assistance.

Consultants and academic researchers play invaluable roles in compiling fragmented data about the seafood industry into useful, comprehensible new forms, maintaining and updating fisheries databases, and tracking legislative and regulatory policy changes affecting the industry. Because consulting firms are likely to maintain information about the seafood industry at regional and national scales, and the more significant industry trends are likely to affect regional groupings of ports in similar ways, it makes sense for some regional entity to commission studies. On the Pacific Coast, the Washington Public Ports Association (WPPA) contracted with private consultants to study the development of the groundfish industry in the Pacific and North Pacific Fishery Conservation Zones and its implications for ports in Washington. The report (Washington Public Ports Association 1980) presented predictions of expanded U.S. exploitation of the West Coast and Alaska groundfish resources to the year 2000. The report also made specific forecasts by fleet, species, and area, together with the anticipated flows of the catch through Washington's ports. Existing seafood cold storage and processing capacity in Washington's ports was compared with projected expansion of industry capacity necessary to accommodate the growth in groundfish landings.[9]

Perhaps the most valuable service that WPPA's report provided was to deflate exaggerated expectations of an immediate "bonanza" in the West Coast fisheries and the commonly perceived need to vastly expand port facilities to accommodate new vessels and seafood plants to process their burgeoning catch. The cautionary note sounded by the consultants has been validated by the slow growth in both the West Coast trawler fleets and seafood processing capacity in the region.

A later study commissioned by a consortium of fishing firms and industry associations, and published under contract to the NMFS (NRC 1983), examined the economic impacts of West Coast-based trawlers fishing in the Alaska FCZ. The Distant Water Fleet study revealed that the economic impacts and much of the demand for port-related services generated by Washington and Oregon fleets fishing Alaskan waters were concentrated in their mainland U.S. homeports, rather than in Alaska—a fact hitherto obscured by the availability of data only on fish *landings*. Although distant water fleet fishers landed their catch and, hence, most of their *revenue* in Alaska, their *spendings* occurred primarily in and to the benefit of their homeports in Washington and Oregon. The study concluded that of the $1.2 billion total impact of the fishing industry on the economies of the Northwest states in 1981, $0.8 billion (66 percent) was derived from the exploitation of Alaskan fisheries by Northwest mainland states' fleets, and from Northwest labor and capital utilized in Alaskan seafood processing firms (NRC 1983). These findings and those of a more recent report (NRC 1986) have been used effectively by fishing vessel owners and marine support services business groups to pressure the city of Seattle to provide greater protection for the city's marine industrial shorelines threatened by "gentrification" trends. Knowledge of the true economic impact of fisheries on Seattle's economy also lent weight to Port staff recommendations to the Port Commission for expanding and redeveloping Fisherman's Terminal.

Whereas consultants generally turn to their industry contacts for information upon which to base their estimate of future fishing fleet size and characteristics, government agencies at state, regional, and national levels report important information about fish stocks, fish landings, and fleet characteristics in order to support their primary mission of managing fisheries. NMFS prepares an annual statistical report called *Fisheries of the United States,* which compiles fish landing data from federal and state sources as well as fish processing and consumption data, import and export of seafood, and world production and consumption data. Unfortunately, fleet size and characteristics data, although collected and maintained by NMFS' regional fisheries centers, are not included in standard annual reports. A proprietary publisher compiled an extensive compendium of fisheries information, including fleet characteristics and federal FCZ catch data (American Fisheries Directory 1978). This useful compendium has never been updated.

NMFS and the FMCs have commissioned special reports addressing fisheries development in the FCZ, some of which are directly useful to fishport managers and their consultants. For example, the North Pacific FMC commissioned the same consultant who prepared the Distant Water Fleet study to assess future growth in the Alaska groundfish fleet necessary to harvest the entire annual "optimum yield" of the groundfish resource (NRC 1984). The results are expressed as numbers of vessels by region, year, type, and size that are expected to enter the fleet by 1989. Coupled with the information published in the earlier report, these numbers can be useful to specific Ports as they plan new, expanded, or remodeled harbor facilities.

Regional Centers of the NMFS on the West Coast have collaborated in the development of a coastwide database describing fish stocks, fleets, and catches. PACFIN (Pacific Fisheries Information Network) shows promise of becoming useful for port authorities in the three mainland West Coast states (California, Oregon, and Washington). An extensive historical database has been compiled from the three states' fisheries regulatory agencies' files and vessel registration programs. The database can be queried to produce reports on a wide range of factors associated with fishing effort and catch. For example, the number and kind of vessels and their catch of specific species can be tabulated by "principal port" (the port where a vessel's largest share of revenue, fish landings, accrues). Aggregate and average vessel incomes can be tabulated in a similar fashion (Huppert et al. 1984; Huppert 1984).

Research staff are attempting to link the PACFIN database with Alaskan fisheries data. Larger vessels fishing migratory stocks (e.g., salmon or tuna) may land their catch in more than one port in two or more states, imposing unpredictable demands on port authorities' facilities. Inclusion of Alaskan data will be of value because of the strong linkages of Washington- and Oregon-based fleets with Alaskan state and offshore (FCZ) fisheries.

## RECENT FISHPORT PROJECTS: BROOKLYN, BOSTON, SAN FRANCISCO

Public port authorities in several large U.S. port cities are undertaking new fishport developments.[10] The most ambitious of these is the Port of New York/New Jersey's new $27 million Fishport in Brooklyn. According to officials, the Fishport will be "a fully integrated fishing complex" providing moorage and vessel maintenance, and seafood product handling, processing, marketing and distribution.

Although New York City's Fulton Fish Market is the nation's largest wholesale fish marketing center, there are presently no fish-landing facilities in the city; seafood product must be moved overland from fishing harbors in the New England and mid-Atlantic coasts. By building an entirely new fishport, the NY/NJ Port Authority intends to capture a sizable portion of the Northwest Atlantic trawler fleet and the seafood processing, marketing, and distribution industry that it serves. The Fulton Fish Market, located on Manhattan's East River shoreline near the South Street Seaport redevelopment and historic district, is itself under threat from congestion, escalating land values, and the trend toward residential and commercial uses of the urban waterfront. Conceivably, the wholesale functions of Fulton Market could be relocated to Fishport (Fig. 8.4) at some future time, but such a move is presently resisted by the Port and by the city, which owns the market.

Studies conducted by the Port documented that the New York/New Jersey metropolitan market of 15 million residents and 17 million visitors accounts for approximately $1.5 billion in annual seafood sales. According to a Port spokesman, the fishing fleet is dominated by trawlers that are at sea for five to seven days at a time, landing their catch in such ports as New Bedford, Massachusetts. If the fleet

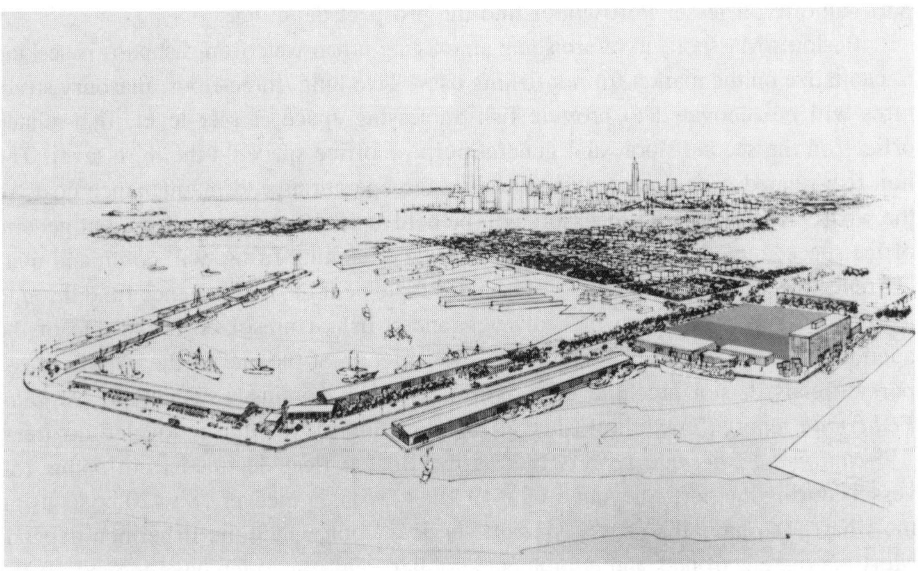

**Figure 8.4.** The Fishport at the former Erie Basin Marine Terminal in Brooklyn will be a waterfront complex for the landing, processing, and distribution of fresh fish and processed seafood products. Construction of the $30-million initial phase by the Port Authority of NY/NJ consists of the rehabilitation and renovation of wharfs, utilities, and buildings in an obsolete marine terminal area. (Courtesy: Port Authority of New York/New Jersey)

can be induced to utilize Fishport, the metropolitan market can be served from a central location. Excellent truck and rail connections provide overnight service to a market of 79 million consumers, and nearby airport terminals can speed high-value product by air to distant U.S. and foreign markets.

The Port will operate berthage, fuel, and ice services, and the fish handling hall where product is sorted, graded, and packaged for the fresh wholesale trade. Processors will lease facilities from the Port for manufacturing frozen and specialty products. A fish auction will be conducted in conjunction with the fish handling hall and should be a drawing card to fishermen, who, in other ports, feel victimized by the lack of competitive bidding on their catches. The presence of an auction will not of itself guarantee that competitive bidding will occur. In Boston, for example, despite the auction held on the Fish Pier, fishermen complain that there are too few buyers, buying in lots too small to expeditiously handle their mixed species catches, and that, furthermore, they must compete with fish brought in by truck from neighboring New England ports. The Brooklyn Fishport, like Boston's Fish Pier, must overcome what the *National Fisherman* characterized as a "catch-22" situation: "In order to attract more buyers, the port needs more boats, but to draw the boats it needs more buyers."[11] Boston is attempting to arrest a long decline of its fishport facilities through a major Fish Pier refurbishing project; in contrast, New York/New Jersey is starting from scratch with no fishport infrastructure or heritage to build on. Clearly then, such a venture poses risks acceptable only to a major, multidimensional port authority: a lesser Port would find the prospect daunting.

Boston's Massport, in refurbishing an existing urban waterfront fishport, is seeking to capitalize on the market for nonfishing uses. Two long, three-floor, masonry structures will be renovated to provide fish-processing space at pier level, fish-related offices on the second floor, and general purpose office space on the third level. The non-fish-related office users will subsidize the operations and maintenance costs of the wharf. Rents for fish processors will be held at $5.50 per square foot, but general office space, rented to architects, insurance, and banking firms, will command market rents of $20 per square foot. Fishermen supplement the maintenance fund through a 1/2 of 1 percent levy on the value of landed fish. Unresolved is the fate of the auction building, a high-ceilinged dramatic structure at the end of the wharf. Massport officials do not preclude the possibility of it becoming a restaurant, *National Fisherman* reports, once the Fish Exchange auction moves into renovated quarters.

*National Fisherman* reports that as the Boston fleet declined from about 150 vessels during the port's heyday in the 1930's to today's meager 18 to 20 boats using the wharf, "so have the nearby support services upon which the fishermen depend. Most marine electronics and equipment suppliers and repairmen have moved off the Boston waterfront."[12] According to the same source, opinions of fishermen, buyers, and processors are divided about the likelihood for success of Massport's bid to halt the systematic decline in fish landings and processing in the face of stiff competition from Gloucester, New Bedford, Portland (Maine), and other New England fishports.

The pressures placed on Boston's Fish Pier by tourist, residential, and office uses have also been felt by West Coast ports. In San Francisco, tourist activities crowd Fishermen's Wharf, congest traffic access, and cause inefficiencies in the commercial fishing vessel operations on the site. In response, the Port of San Francisco has conducted studies and developed a master plan for the wharf that includes some elements similar to New York/New Jersey's Fishport—improved vessel moorage and a fish-handling hall—and other elements drawn from European experience with Fishing Technology Centers in Denmark and Great Britain (discussed below). The Fishermen's Wharf Action Plan calls for classrooms and a second-story walkway providing public viewing access to the fish-handling hall operations. As in Boston, the Port of San Francisco hopes to develop leased facilities for restaurants and related fish markets to provide a "cross subsidy" for the less profitable elements of the Action Plan.

As changing urban land uses encroach on fishports, attrition occurs among firms servicing the fishing fleet. If unchecked, this trend can erode the "critical mass" of marine service establishments upon which these fleets depend, changing the competitive advantages of one port over another. Put another way, the industrial complex, discussed earlier in this chapter, itself is threatened by urban land use change. The Ports of Boston and San Francisco are attempting to halt and then reverse these trends, while the Port of New York/New Jersey is attempting, at the Brooklyn Fishport, to create an entirely new fishing industry complex. Long-established fishports, clearly, have an advantage over wholly new ventures, since the former have expanded and evolved over a long period of time in increments associated with changing demand for services. The complicated interdependencies among fleets, the seafood processing, distribution, and marketing sectors and marine support services present many of the "catch-22's" alluded to in Boston for port authorities attempting ventures like New York/New Jersey's Fishport.

## FISHING TECHNOLOGY CENTERS IN GREAT BRITAIN AND DENMARK

The fishport initiatives underway in Brooklyn, Boston, and San Francisco focus primarily on the physical infrastructure needs of the fishing fleets they are to serve: docks, net sheds, ice and fuel suppliers, fish auctions and wholesale facilities. But, as noted earlier, there are important services that fishing industry personnel use while in port that deal with information, education, and new technologies.

In Western Europe, two nations have developed information and research centers focused on fisheries and fishports that could be instructive for U.S. Ports and fisheries development institutions. The British Seafish Industry Authority[13] (formerly the Whitefish Authority) is a public corporation created by Parliament to administer grant and loan programs for fishing vessels and seafood processing equipment, conduct industry training programs, undertake research in fishing technology, engage in seafood promotion and marketing, conduct research in aquaculture, and provide consultancy services in Britain and overseas. In addition to grant and loan funds from the Ministry of Agriculture, Fisheries and Food, the Seafish Authority is supported by a levy on fish landings and transshipments in Great Britain. At its Hull facilities, the Seafish Authority teaches courses in fish marketing and fishing gear techniques using a flume tank to demonstrate the operation of various fishing nets; a mobile unit takes courses on the road to fishports throughout Great Britain; and staff conduct seafood demonstrations at trade shows. A team of experts in fisheries economics and fishport development provides technical consulting services to port authorities and economic development boards in Great Britain and abroad. In many respects, the Seafish Industry Authority combines in one entity many of the functions provided in the United States by NMFS, the National Sea Grant College Program, the regional Fisheries Development Foundations, and firms in the private consulting sector. The authority's annual reports contain comprehensive data on the British fishing fleets, seafood landings and imports, and household consumption of seafood. Insulated from the regulatory roles of the Ministry and administered by an independent board, the Seafish Authority has established the trust of the seafood industry, port authorities, and others with whom it works.

Denmark's North Sea Center (Nordsøcentret) at Hirtshals on the Danish North Sea coast brings government, the fishing industry, and academic research institutions together at a single center devoted to fisheries development and industry training programs.[14] Close affiliation between Nordsøcentret and Halborg University provides graduate education and research programs at the center; the Danish Institute of Fisheries Technology conducts vessel and gear experiments; the Danish Institute for Fisheries and Marine Research—a branch of the Ministry of Fisheries—conducts research on aquaculture, fisheries stock assessment technology, and other forms of applied research.

The application of the European experiences to the United States is hampered by the scale and diversity of the American fisheries and the complex federal-state-regional fisheries management and development programs found in this country. Unlike Britain and Denmark, no single, central authority is available to coordinate or promote a national fisheries development policy. It is likely, too, that the fiercely autonomous public port authorities in the United States would resist favorable treatment of a few ports, were federally supported fishing technology centers to be proposed at key locations around the nation's coasts. More likely to succeed would be a *single initiative* promoted by an alliance of port, university, state fisheries agency, and industry interests funded from a multiplicity of federal, state, port, and industry sources. This would be a fruitful additional service provided by fishports as they evolve in the United States.

## MODEST CHANGE AND CONSTANT SHIFT

As a result of the "common property" nature of U.S. fisheries discussed earlier, most of the U.S. fishing fleet is "overcapitalized," that is, there are more vessels operating in the fisheries than the number necessary to harvest allowable stocks efficiently. The effect of extending U.S. jurisdiction to 200 miles offshore and enacting policies giving preference to U.S. harvesters and processors has been, with few exceptions, to "soak up" some of the overcapitalized fleet into expanded fishing effort, much of it in "over-the-side" joint ventures between U.S. trawlers and "motherships" of foreign nations harvesting low-value, abundant groundfish resources. Even in the North Pacific FCZ where the greatest remaining foreign exploitation of U.S. fish stocks currently occurs, the existing West Coast and Alaskan trawl fleets could efficiently harvest the entire resource.

Until U.S. processors can compete with their foreign counterparts, either through control of a larger proportion of the global fisheries stocks, or through sharp gains in productivity, most of the low-value groundfish will continue to be processed abroad. Experts contend this is unlikely to change quickly, if at all.

In the foreseeable future, then, fishport managers are unlikely to experience severe increases in demand for their services from either new vessels or new processing plants. What has occurred is a shift in the *mix* of vessels using ports and in the timing and duration of their stay in port. Furthermore, shifts of vessels between ports has been brought about through movement of vessels among different fisheries and through the loss of essential fisheries-related services by gentrification of urban harborfronts. Where the seafood processing industry expands within a Port's jurisdiction, it is likely to be inland, in a general industrial area, rather than on the waterfront.

The vagaries of nature and global fisheries economics will continue to beset the U.S. fisheries, producing uncertainties for management agencies, vessel owners, and fishport managers. "Boom and bust" gyrations cannot be eliminated by even the best informed of management decisions; certain fleets at certain times will be tied to docks, while others reap bonanzas on stocks that are at the top of their cyclical upswings. Fishport managers in the future, then, must caution against expansion of facilities fueled by ill-founded optimism, while planning wisely to avoid erosion of their marine support services that attract the fleet to their port.

### *Notes*

1. Mr. William Stokes, Glacier Fish Company, Inc., Seattle, Washington, in a personal communication with the author, provided detailed descriptions of the activities surrounding the annual visit of a modern factory trawler to its Seattle homeport.

2. Fisheries Conservation and Management Act of 1976 (FCMA), P.L. 94-265.

3. The regional Fisheries Management Councils, established by the FCMA, include the director of each state's fisheries agency, the regional National Marine Fisheries Service director, and a varying number of other members.

4. Usually each principal "gear group" of the fisheries is represented by a separate trade association in its state or region: e.g., the North Pacific Vessel Owners Association (NPVOA) represents the Alaskan crab fleet; the Puget Sound Gillnetters Association represents one part of the regional salmon fleet in Washington State.

5. "An industrial complex may be defined as a set of activities occurring at a given location and belonging to a group (subsystem) of activities which are subject to important production, marketing, or other interpretations." From Walter Isard, *Methods of Regional Analysis: An Introduction to Regional Science* (Cambridge, MA: MIT Press, and New York: John Wiley, 1960), p. 324. A fishing industry complex has been identified by two studies of the Washington State Input-Output Table: Breanda'n O'hUallacha'in, "The Identification of Industrial Complexes," *Annals of the Association of American Geographers,* **74** (3), September 1984. J. Campbell, "Selected Aspects of the Interindustry Structure of the State of Washington, 1967," *Economic Geography,* 50, 1974, 35–46.

6. Garrett Hardin's famous description of the overexploitation of common grazing lands aptly captures the consequences of unregulated access to common property resources: There is no economic incentive for an individual sheep-herder to limit the number of animals he permits to graze on the commons, even though increasing the number will result in overgrazing. While the benefit to this sheep-herder is captured (internalized) by him, the costs of his action (externalities) are borne by all other commons users. By analogy, an individual fisher faces the same problem in an unregulated fishery. Even though the future of the fishery depends on conserving "broodstock," there is no individual incentive to refrain from overharvesting the resource—If I don't catch the last fish, someone else will!

7. Other contributing authors in this volume address port functions and financing tools in full detail. Wide variations exist from state to state and among municipalities in the functions prescribed for port authorities through enabling legislation and specific charters under which individual Ports operate. Most U.S. Ports, however, perform the four functions described in this chapter.

8. In attempts to reduce the number of vessels operating in some fisheries, states have instituted voluntary programs to purchase (buy back) fishboat licenses. Most observers believe the consequences have been to eliminate the least successful vessels from the fleet without making a serious dent in the fleetwide fishing capacity. An example here would be the Washington gillnet salmon fishery in which a significant number of part-time fishers have participated.

9. Two years later (1982) the West Coast Fisheries Development Foundation (Kramer et al. 1982) conducted a similar study, extending its geographic scope to include Oregon and California ports. Similar studies have been done for East Coast ports. For example, a consultant to the New England FMC (Newick et al. 1982) collected detailed information about fishing fleet characteristics by port, growth in the fleet from 1976–80, and financial aspects of vessel mortgages. The mid-Atlantic FMC's consultant produced a three-volume report (Development Sciences, Inc. 1980) describing the fleet fishing waters within the council's jurisdiction. Among the rich data on prices, wages, vessels, and fishing behavior were detailed maps depicting the flow of product by species through specific ports in the mid-Atlantic states. Reports such as these are simply baseline studies however, and provide little guidance to port

managers concerned about the impact of FMC management decisions on the use of their facilities in the future.

10. This information about U.S. fishport projects is drawn from two studies and personal communications with staff of the three ports mentioned: ROMA Architecture and Urban Design, Erik Norgaard and John McMahan Associates, "Fisherman's Wharf Action Plan," prepared for the Port of San Francisco, August 1981, Port of Seattle, Planning and Research Department, "Fisherman's Terminal Redevelopment Study: Public Review Draft," February 1985, communications with Gerhardt Muller, Port Authority of New York/New Jersey.

11. *National Fisherman,* November 1984, p. 102.

12. *National Fisherman,* November 1984, p. 104.

13. Information about Britain's Seafish Industry Authority's programs was extracted from its annual reports and through a personal interview with Neil McKellar, chief economist.

14. Counselor Erling Hulgaard, Consulate General of Denmark, and Dr. Poul Pedersen, director, Nordsøcentret, Hirtshals, Denmark, generously provided information on Denmark's North Sea Center.

## References

American fisheries directory and reference book. 1978. Ed. Burton T. Coffey. Camden, Maine: *National Fisherman.*

Campbell, J. 1974. Selected aspects of the interindustry structure of the state of Washington, 1967. *Economic Geography* **50:**35–46.

Development Sciences, Inc. Marine Group. 1980. Mid-Atlantic fishery conservation zone: Fisheries socio-economic inventory. 3 vols. Mid-Atlantic Fishery Management Council, Dover, Delaware.

Huppert, D. D. 1984. Pacific Coast groundfish management: Evolution and prospects. National Marine Fisheries Service, Southwest Fisheries Center, La Jolla, California.

———, C. J. Thomson, and S. Iacometti. 1984. A report on the 1981 PACFIN research data base. National Marine Fisheries Center, Southwest Fisheries Center, La Jolla, California.

Isard, W. 1960. Methods of regional analysis: An introduction to regional science. Cambridge, MA: MIT Press; New York: John Wiley.

Kramer, Chin and Mayo et al. 1982. System strategy for California, Oregon, and Washington fishing industry and public ports infrastructure needs and assessments. West Coast Fisheries Development Foundation, Portland, Oregon.

*National Fisherman.* 1984. 65 (7). Camden, Maine.

Natural Resources Consultants (NRC). 1983. Distant water study. Seattle.

———. 1984. Properties of domestic fleet and effort required to harvest the Alaska groundfish optimum yield. North Pacific Fishery Management Council, Anchorage, Alaska.

———. 1985. Americanization of the U.S. fishing conservation zone: A discussion paper, Seattle.

———. 1986. Commercial fishing in the state of Washington, a contemporary economic overview of local and distant water commercial fisheries, Seattle.

Newick, C., J. E. Kelly. 1982. The capital sector of the New England fishing fleet: An inventory. National Marine Fisheries Service, Gloucester, Massachusetts.

North Pacific Fisheries Management Council (NPFMC). 1981. The applicability of limited entry to the Alaska halibut fishery, Council Document #14, April.

O'hUallacha'in, Breanda'n. 1984. Identification of industrial complexes. *Annals of the Association of American Geographers,* **74:**3.

Port of Seattle, Planning and Research Department. 1985. Fisherman's terminal redevelopment study: Public review draft.

ROMA Architecture and Urban Design, with Erick Norgaard and John McMahan Associates. 1981. Fisherman's Wharf action plan. San Francisco.

Washington Public Ports Association. 1980. Port system study for the public ports of Washington State, Vol. II, Part 2: Impacts of the extended 200-mile fishing limit. Prepared by CH2M-Hill with Dr. George M. Pigott.

# Chapter 9

# SMALL-BOAT MARINAS:

## The New Professionalism

*Robert F. Goodwin**

---

Personal power, freedom, and the "other worldness" experienced while afloat are qualities of boating that draw North Americans by the millions to buy, rent, charter, or borrow recreational smallcraft, and to pay handsomely for the privilege of doing so. Boaters are as varied as the watercraft in which they play. For some, boating is an end in itself; for others, it is simply a means to an end. Those who fish and hunt use their canoes or outboard-powered aluminum skiffs to pursue their game; those who take their watercraft seriously focus on extracting the last tenth of a knot from polished hulls and taut sails. Some boaters never leave the marina, preferring to sit on deck enjoying conspicuous consumption, cold martinis, and their guests' envious appraisals of their vessel. For many islanders, a boat is transportation to mainland jobs, jury duty, shopping centers. For most, though, boats are the vehicles to escape urban noise, freeways, traffic signals, and yard work. Boats are a means to travel freely: no traffic lanes, no stop signs, no telephones, no neighbor's dogs—just the freedom to be in command of one's own vessel. The meekest accountant, if he chooses, can be as tyrannical a sea captain as an Ahab, and no CEO of a Fortune 500 company enjoys more commanding a role than our weekend smallcraft skipper maneuvering a Chris Craft through the shoals of Deadman's Slough, or negotiating a bar crossing outside a coastal harbor.

Despite congestion, power/sail conflicts, scuba divers, and water skiers, boating is still, incredibly, a largely unregulated pleasure. To be sure, there are registration decals, Coast Guard safety requirements, and speed limits in some waterways, but no license is required to operate a pleasure craft and few restrictions are imposed. Perhaps it is this contrast between our carefully regulated lives ashore and the relative

*Partial financial support from Washington Sea Grant Marine Advisory Program under NOAA Grant #NA86AA-D-SG 044, Project #A/CZ-1 is gratefully acknowledged.

freedom of life afloat that is the most important appeal in owning a boat. It is a tribute to boaters and sailors, power squadrons, yacht clubs, the U.S. Coast Guard, and its auxiliaries that boating activities are conducted with so little loss of life or serious injury.[1]

The popularity of recreational boating has burgeoned since the end of World War II as a result of social and economic changes in our society and technological advances in the boat manufacturing industry. If this trend continues, as experts believe it will, there will be mounting pressure on public and private enterprises to provide the harbor facilities—channels, breakwaters, docks, floats, and such—to accommodate the swelling recreational smallcraft fleet and provide access to the nation's coastal and inland waterways.

But since the early 1970s, new environmental statutes and regulations concerning water quality and shoreline habitat conservation have been put in place. The results has been a collision between the recreational boating industry on the one hand, and environmental activists, planners, and agencies on the other. In this chapter we see, through a series of case studies, how this conflict has shaped the way in which smallcraft facilities are being planned, sited, designed, and managed in the changed regulatory and fiscal realities of the 1980s. Particular attention is paid to the roles of public and private enterprises that develop and operate smallcraft facilities. The chapter includes a discussion of several major trends that will affect the pace and location of harbor facility development serving small craft.

## FINDING MOORAGE AND STORAGE

For all but the canoeist, kayaker, or car-top boat owner, access to the water requires some basic shoreside infrastructure: launch ramps, hoists, mechanical launchers, moorings, or marina berths. Even hand-launched boats require a public shoreline access point with parking nearby. Fuel and supplies must be available at dockside, repair and haul-out facilities nearby. Constructing, maintaining, and operating recreational smallcraft facilities are roles that historically have been shared between public and private entities. For many of the same reasons public port and municipal authorities construct and operate fishboat harbors, recreational marinas are developed and often operated by the same authorities. But a significant proportion of boat moorage and storage in North America is provided by for-profit private enterprise. Even in public smallcraft harbors (discussed in Chapter 8), private businesses operate many boaters services under concession or leasehold arrangements with the public port agency.

A recent nationwide boating facilities survey (Ross 1985) discovered almost 11,000 enterprises providing moorage or storage for recreational smallcraft. Of these, approximately 78 percent (8,300) were managed by private firms and the remaining 22 percent by public agencies: cities, counties, port authorities, state and federal government bodies. Together, these public and private facilities provide over 450,000 in-water berths, 35,000 in-water moorings, 138,000 dry storage spaces, and 3,300 launching ramps.

In Washington and Oregon, and perhaps along the entire west coast, public enterprise plays an even more important role in the provision of moorage facilities. Public marinas in Washington's coastal counties, which comprise less than 20 percent of the 300 facilities, provide almost 50 percent of total slips. Private facilities, although more numerous, are considerably smaller than their public counterparts; private marinas have an average of 50 slips compared to 270 in the average public marina (Goodwin and Stokes 1980). On the West Coast, unlike the East Coast, public port authorities are the dominant force providing public moorage for reasons concerning, among other factors, the physical coastal environments in which marinas must be developed.

Unlike most of the Eastern Seaboard and parts of the Gulf Coast, most West Coast and Great Lakes marinas must be protected from a harsh wave environment. Even in the more sheltered water bodies of Puget Sound and San Francisco Bay, wind-driven waves can be large enough to require substantial breakwaters and jetties to protect smallcraft from storm damage. Barrier islands protecting backshore marshes and lagoons, found extensively on the East and Gulf coasts, occur only in a few areas on the West Coast where river valleys drain the Olympic and Coastal Range mountains of Washington, Oregon, and Northern California or the terraced uplands of Southern California. The costs of providing wave protection have been borne, historically, by federal funds. The private sector is precluded by law from receiving such public subsidies. In general, then, private marinas are to be found only in small, sheltered water bodies, or in estuaries accessible through public navigation projects. Privately funded floating concrete breakwaters, serving double duty as transient moorage docks and public access walkways, have been adequate to protect moorage from storm waves found in these environments.

Regardless of whether public or private enterprises provide them, new boating facilities for moored, trailered, and hand-launched smallcraft will be necessary if the recreational boating population expands at rates forecast by the boating industry, public agencies, and academic experts.

## *Boating for Everyone*

Since the end of World War II, several social and technological changes have brought boating within reach of ordinary American families. Greater personal income, shorter work weeks and longer vacations have removed many social and economic barriers to boat ownership and use, and earlier, longer, and healthier retirement years extend boating further into the aging population.

Where hand-built wooden boats were once dominant, production-line fiberglass and aluminum craft are now standard in the recreational fleet. Cotton, canvas, and manila hemp have been replaced by dacron, nylon, and other more durable synthetics for sails, sheets, and lines. The microchip has brought compact, low-cost communications and navigation systems within the budgets of large boat owners. Bulky inboard engines have been eclipsed by the clip-on technology of outboard motors, or hybrid, unitary, outdrive propulsion systems. Low-speed displacement hulls are reserved only for sailboats; deep-vee and planing hulls, which almost leave the water, have permitted great increases in speed and efficiency in the powerboat fleet. Even the World War II rubber dinghy, from which countless aircrew were rescued after ditching at sea, has been transformed into a high-speed, outboard-powered, inflated craft—a practical vessel in its own right. Further, decreases in weight, coupled with increased hull strength, permit boats to be mounted on trailers and launched by hand, freeing marina berths for the larger nontrailerable vessels and enabling boaters to travel long distances to a choice of destination boating waters.

The National Marine Manufacturers Association (NMMA) estimates that during 1985 over 30 percent of the U.S. population, or 68 million Americans, went boating; that there were over 13 million pleasure boats owned in the United States; and that over $8 billion were spent on recreational boating, which accounted for 6.4 percent of all personal consumption (see Fig. 9.1). By the year 2000, NMMA estimates that the recreational boating fleet will grow to 19 million smallcraft, and that 95 million Americans will participate in some kind of boating activity (NMMA 1985). NMMA adds a cautionary note to this projection, however: "A . . . shortage of marinas, dry storage facilities, and access sites exists on our inland waterways" (NMMA 1985). The implication is that development of shoreside facilities must keep pace with growing demand for water-based recreation.

## *Predicting Moorage Needs*

Deciding when and where to build what quantity of new marina berths or launching ramps requires information that is neither readily available nor, where it is available, necessarily accurate for more than a few years. The information needed is of three kinds: demographic, economic, and spatial-geographic. The forecaster needs to know the size and characteristics of future populations, the future income available to that population, the rate at which households in future populations will choose to participate in boating, and, finally, the shape and extent of the market areas to be served by the moorage or launching facilities. Any one of these factors can change over the years in ways that are unpredictable. In some states, academic researchers, the U.S. Army Corps of Engineers (COE), or consulting firms have built mathematical "models" of the way in which boating fleets grow and change. Typically, the kinds of factors in these "models" that "explain" year-to-year variations in fleet size are: population, per capita income, distance to recreational water bodies, and the con-

**Figure 9.1.** Recreational boating is gaining in popularity as a leisure activity in the U.S. Shown here is Shilshole Bay Marina, Seattle. (Courtesy: Robert F. Goodwin)

straints imposed by the existing supply of facilities. Often these models can replicate historical fleet data quite accurately, but "projecting" the model into the future depends on accurate forecasts of how each of these factors will change in the time period being forecasted. Forecasting socioeconomic trends is far from an exact science. For example, the recession of the early 1980s was not predicted by analysts forecasting economic trends for that period.

Public policies implemented at national or state levels can also affect boating activities. For example, imposition of excise taxes on boat owners, or increased local share of dredging costs can increase the cost of boating either directly to the boater or through increased costs to port authorities passed on to boaters using their facilities.

For a variety of reasons, new facilities generally cost more to construct than old ones: the most easily developed sites close to urban boating populations have in many cases already been developed. New sites for expansion are likely to require extensive dredging, filling, wave protection, and wildlife mitigation. Labor and materials costs have risen with inflation. New facilities, then, must be priced higher than old ones to recover costs and yield a return on invested capital. Although economic theory suggests that operators of older facilities would raise rates to the point at which excess demand (measured by waiting lists) disappears, in fact, older facilities usually charge less for moorage than newer competitors. The effects of variations in price of moorage confound easy market analysis. When transportation expenses are added to the cost of moorage, boaters generally are willing to travel farther to lower-priced moorage than to expensive moorage, resulting in larger market areas being served by the less expensive marinas. Discovering and verifying the shape and extent of boating facility market areas require careful survey of the boating population, or analysis of the home addresses of marina tenants, facility by facility.

The year-round opportunities for boating afforded West Coast, South Atlantic, and Gulf States boaters are denied to boaters in other parts of the nation. On the Great Lakes and the Northeastern Seaboard, harsh winters shrink the boating season to six months or less. Boats must be removed from the water, winterized, and stored on dry land to prevent damage from ice and winter storms. Not only must marina operators recover costs over short seasons, but they must repair severe damage to docks and pilings often caused by storms and ice. In the Great Lakes, thick ice forms easily on the surface, girdling pilings in its grasp. Storm-driven water-level changes called "seiches" lift the ice and the pilings to which it is attached. When the water level falls, the resistance offered by the lake bottom against the pilings is much larger than the friction acting against the same pilings as they are pulled up by the ice sheet. The ice breaks around the pilings, leaving behind a collar, and refreezes at a lower point on the piling; and then the whole process starts again. At the end of the winter, docks can resemble irregular roller coasters. Repairs entail employing a piledriver to reseat the pilings in the lake bottom. Worse yet, storm-driven ice blocks can shear off piles, destroy docks, and require complete rebuilding before the boating season returns in the spring.

Even in the relatively mild coastal states of the Northwest, winters are wet and cold enough to deter all but the most fanatic sailors from venturing out from their year-round moorages. Transient moorage facilities at destination resort areas, awash with visiting summer vessels, are empty and closed during the winter months. Even drastic cuts in moorage rates cannot attract winter seasonal tenants. Planners, then, in addition to estimating future demand for year-round moorage, must factor into their estimates the seasonal peaks and troughs in that demand that are felt in rural resort areas.

In summary, both public and private operators need access to forecasts of market conditions far enough into the future to plan expansion of boating facilities. Given the likelihood that it may take as much as a decade to secure facility permits, arrange financing, and construct the facility, forecasted market conditions can be wildly inaccurate. Port commissions, bankers, private investors, and others involved in the decisions to construct new marinas need to remain flexible and willing to revise the scale of their projects in the light of new market information. The costs of *not* acting on market signals can be bankruptcies in the private sector and unnecessary subsidies from other harbor operations in the public sector.

## RESOLVING MARINA DEVELOPMENT CONFLICTS: THREE CASE STUDIES

It is not unusual for marina developers to expend up to 10 percent of the total development costs of a project on environmental reports, demand studies, and other permit-related tasks before a single pile is driven or a cubic yard of bottom material is dredged at the site. As a result, more than one developer has been forced to withdraw from a project before acquiring all the necessary permits. "It's a good project. Boaters need more moorage in this area, and it'll be built eventually. But I've got to make a living and these permit costs are killing me!" one Seattle marina developer remarked after withdrawing from a saltwater marina project partnership. Or, as a seasoned marina consultant told a conference audience in the Pacific Northwest, "After you get the permits comes the easy part—getting the financing!"

Documentation of the need for a new marina, along with its environmental and social impacts, will be scrutinized during permit hearings held by local or state governmental agencies and, frequently, again at a federal review conducted by the COE. Marinas inevitably alter the shoreline environment through dredging and filling, reducing wave energy, altering natural circulation patterns, and blocking photosynthesis. Chronic and episodic spills of petroleum products, leaching of heavy metals—or worse—from antifouling paints, and poor sanitary practices of some boaters all place stress on water quality in and around marinas. Often just the *presumption* that a marina pollutes is sufficient cause for health agencies to close adjacent beaches and tideflats to shellfish harvesting. Given these adverse effects of marinas, it is clear that those facilities for which a significant need exists are likely to be treated more favorably by natural resource agencies than those for which the need is uncertain or speculative.

Through some examples of recently developed marinas in Washington State*, let us explore ways in which social, economic, and environmental conflicts have arisen and have been addressed and resolved in preconstruction permit hearings and environmental impact documents.

---

*In some of these cases the author has been involved either as an expert witness or advisor to parties supporting or opposing marina development.

## *John Wayne's Legacy: Sequim Bay, Washington*

On the Strait of Juan de Fuca in Washington State, a new marina[2] provides a good example of the role a public port authority plays in recreational boating facility development. This case study illustrates some of the financial, environmental, social, and engineering challenges to be overcome before opening a new marina.

The late John Wayne deeded a segment of shoreline on the west shore of Sequim Bay to the Port of Port Angeles on condition that it be developed as a recreational marina. The adjacent uplands were retained by the Wayne estate for future residential development. The completed marina was formally opened at a dedication ceremony ten years later, on September 15, 1985. Almost a year after its opening, 110 pleasure craft occupied berths in the facility, which has an ultimate capacity for 422 slips.

The Port of Port Angeles is a countywide governmental authority serving Clallam County on Washington's northern coast along the Strait of Juan de Fuca. An existing marina in Port Angeles, the county seat and largest city, harbors a fishing fleet and recreational smallcraft behind the protective natural barrier of Ediz Hook. A strong constituency of boaters moored in the Port Angeles boathaven, but living 17 miles to the east in Sequim, had long advocated development of an additional marina closer to their residences and protected from oceanlike water conditions found in the open strait. Opposing the project were the Washington Environmental Council (a statewide coalition of sports and environmental organizations) and a local group of concerned citizens acting under the rubric Protect the Peninsula's Future.

The John Wayne Marina would be the first significant, permanent moorage facility to be permitted in Sequim Bay, and, therefore, a precedent-setting action by the Clallam County Planning Department, the "lead agency" involved in reviewing permits for the marina. Neither supporting nor opposing the project, the State Department of Social and Health Services made it clear that siting the marina in Sequim Bay would result in the decertification of nearby clam populations for human consumption. This decision was based on a conservative presumption that fecal coliform bacterial contamination from boaters in and around the marina would present a public health risk. Other environmental issues raised included adverse effects on wildfowl and marine life in Sequim Bay due to increased boating activity. But one issue upon which the opposition focused during public hearings was an economic one: namely, was there sufficient demand for moorage in the market area served to justify the cost of development? Or would the taxpayers of Clallam County be burdened by a debt caused by shortfalls in revenue from a partially vacant facility?

In 1982 Washington Sea Grant published a report on the market conditions affecting the marina industry in each of Washington's coastal counties (Goodwin 1982). The report cautioned that adding 422 new moorage berths in Clallam County would satisfy likely future demand for the next decade, and that moorage fees set by the Port could have adverse consequences for competing private marinas in the same region. The Port acted prudently. Only 200 slips have been built in the moorage basin which will ultimately hold 422 vessels, and moorage rates were established at a level that would not encourage an exodus of boaters from private marinas seeking lower cost moorage. Building additional docks and finger piers as the marina fills will conserve port capital and reduce debt service costs during the early years of operation.

To overcome the loss of harvestable clams due to construction of the marina, and the potential losses caused by decertification, the Port agreed to replace the clam beach at another site in the bay and to indemnify the state of Washington for any subsequent clam losses due to fecal coliform contamination. Other actions taken by the Port to minimize unavoidable, adverse environmental effects included protecting an eelgrass herring-spawning habitat, rerouting a small stream, and relocating the entire project to avoid unnecessary dredging and filling of intertidal habitat.

It would have been virtually impossible for a private developer to have constructed a marina on Sequim Bay. At the John Wayne Marina site, waves as high as 5 feet were "hindcasted" from historical wind measurements. Only a rubblemound breakwater can withstand and dissipate such wave energy, and rubblemound breakwaters are costly engineering projects. Only public authorities may be local sponsors of COE public works projects,[3] and only public municipal corporations may issue tax-exempt bonds and levy taxes to service or guarantee the resulting public debt. Without access to such subsidies, private developers would have to charge moorage fees far in excess of market prices in this part of Washington State. Even with access to tax-exempt bond capital, the John Wayne Marina will not be "in the black" for several years.

## *East Bay Marina, Olympia, Washington*

This case study of a public marina development illustrates the importance of conducting careful moorage demand studies and remaining flexible to unforeseen changes in market conditions. East Bay Marina[4] lies at the south end of Budd Inlet in Puget Sound, Washington.

A brochure in glossy color shows hundreds of newly constructed, vacant moorage berths—high-quality, floating concrete-pontoon structures equipped with lighting, water, and power outlets. No boats. The caption reads: TAKE YOUR PICK.

Between 1985, when the photo was taken, and 1987, about 140 of the 600 slips built by the Port of Olympia have been occupied.

Across the bay a recently rebuilt 400-slip private marina has an occupancy rate of about 60 percent. Eighty percent of these boater tenants reside in the immediate urban area of Lacey-Tumwater-Olympia. The private owners of the marina have been able to hold their rates below those charged by the Port, but fear that any downward adjustments in rates by the Port, motivated by the need to fill the facility, could result in a massive exodus of tenants to East Bay Marina and financial disaster to the private partnership.

Whereas no knowledgeable observer doubts that East Bay Marina will fill up *eventually,* the Port faces several lean years during which revenues from other port operations must be used to cover debt service on the marina. To be fair, it must be recognized that development of East Bay occurred during the period when interest rates soared to record highs, personal income fell, and unemployment rose to levels not seen since the Depression. As a consequence, boat sales in Washington State fell 50 percent between 1978 and 1982, and none of these economic changes were predicted. A Corps of Engineers study (COE 1980b) had identified a higher demand for new permanent and transient saltwater moorage in the area than subsequent Sea Grant studies (Goodwin 1982) which had the benefit of substantially deflated state economic forecasts on which to rely. Some cost-containment measures might have been taken, however. For example, fewer slips could have been built; some upland, marina-related improvements could have been placed "on hold"; or the whole project could have been delayed until market conditions showed signs of improvement. But the Port commission had made a political commitment to building the marina and the COE agreed to pay up to 50 percent of the costs of the breakwater and navigation channels. Momentum had gathered to sustain the decision and the marina was built despite mounting evidence of a softening boater market.

The Port of Olympia was not the only port on the West Coast caught between political commitment and a sluggish economy. The Port of Newport on the central Oregon coast developed South Beach Marina during the same recessionary period. The 1,000-slip facility has changed private managers twice since it was opened in 1980. Six years later less than 50 percent of the berths were filled even during the peak summer season.

East Bay Marina faced its share of environmental problems, too. The site was a shallow muddy cove at the south end of a long inlet. The area already had water quality problems—low dissolved oxygen levels during hot summer months, polluted sediments, and poor tidal flushing. Federal and state resource agencies required the Port to dispose of dredged material in confined sites behind protective berms and to install an aeration system to supplement low dissolved oxygen levels in the newly dredged marina basin. The Port is required to monitor dissolved oxygen and to activate the air pumps whenever levels drop below the threshold at which fish kills can be expected. A heavy floating concrete breakwater, rather than an impervious solid structure, was required to minimize disruption of natural water circulation.

## *Point Roberts Marina, Washington*

The third case study illustrates the role played by the private sector in developing a surprisingly large marina. Innovative design and engineering techniques overcame some difficult environmental and physical problems at the site. Point Roberts[5] is a territorial anomaly; it is a peninsula connected to Canada's British Columbian mainland, south of Vancouver, but its southernmost tip is actually within the state of Washington. Residents cross the U.S./Canada border twice to get a haircut! Point Roberts is primarily a recreation and retirement community with its market oriented to Vancouver, rather than to Washington State metropolitan areas.

In the mid-1970s, a Canadian developer proposed to construct a large marina on the point's southern shore. Its tenants would be primarily Canadian vessels seeking moorage close to the Canadian Gulf Islands and the American San Juan Islands. A long fetch across Boundary Bay and the Strait of Georgia creates fierce winter storm waves, which necessitated substantial breakwater and jetty construction. Intertidal and subtidal shore habitats, particularly eelgrass beds, would be destroyed by dredging and filling for a conventional marina constructed along the shoreline. Instead, the developer's consultants designed a moorage basin excavated out of backshore meadows and connected to the sea by a dredged channel. Two short jetties and a short, fixed rubblemound breakwater would be all that was required to protect the entrance channel.

The geometry of the basin and entrance channel was designed to produce a "jet" of incoming tidal water, which would, in turn, cause a "gyre"—circular current—in the marina basin. The resulting circulation during flood tides would provide for mixing seawater in the basin and preventing problems created when water is trapped and static: elevations in temperature, water column stratification, and oxygen depletion, all of which can cause fish kills. Despite being contrary to Washington State Department of Fisheries' guidelines requiring *two* entrances to marina basins, tests conducted at the University of Washington's Harris Hydraulics Laboratory proved that mixing and circulation of water within the marina basin through a single entrance would be superior to that of any other Puget Sound marina studied. The Fisheries Department withdrew their objection.

The 1,000-slip marina was approved, constructed, and opened in 1978. Occupancy has been retarded by the unfavorable strength of the U.S. dollar against its Canadian counterpart and by the recession felt both in British Columbia and Washington State. However, despite some early financial difficulties, the marina is operating profitably at 75-percent occupancy during summer months.

Because Point Roberts Marina is in a rural, undeveloped area, it could be sited in the backshore meadow rather than in tidelands. Such an option is usually unavailable to developers in urban shorelines. In many parts of the United States, locating a marina behind the beach places development in sensitive salt marshes or brackish wetlands—habitats zealously protected by federal and state fish and wildlife agencies. Dredging and filling in these estuarine habitats is approved only when large net public benefits are to be gained, and only then if substantial mitigation measures are undertaken. These important estuarine habitats are common on the Eastern Seaboard and Gulf Coast shorelines of the nation. On the West Coast, points of access and egress to the ocean frequently coincide with river mouths, which drain valuable wetlands and estuaries. Siting marinas in these circumstances must be performed with great care to minimize disruption of fish spawning and rearing areas.

Point Roberts Marina was an unusually large venture for private developers to undertake. But because of its size, the operation gains economies of scale more typical of large public marinas: a 1,000-slip facility costs less than twice the cost of a 500-slip facility. Entrance channels and fairways, which produce no revenue, consume proportionately more of a smaller marina basin than of a larger one. One dock manager is needed for the smallest of facilities, but can as easily manage a larger one, and a small computer can handle accounts, payroll, and inventory for a wide range of scales of operation.

## TRENDS IN MARINA DEVELOPMENT AND MANAGEMENT

From the three cases in the last section and from observation of political and social trends occurring in the 1980s, it is possible to state the major challenges that will face the development and management of marinas to the end of this century. Even though these case studies were of marinas developed in one state—Washington—many of the factors affecting the industry are national in scope and are likely to be felt in other states in similar ways. Environmental laws and regulations are carried out through federal as well as state authorities; tax laws and public works appropriations are the purview of Congress and affect the availability of public as well as private capital nationally; changes in technology diffuse through the industry rapidly, regardless of where innovations originate; and, finally, the marketplace rewards those who operate facilities in an efficient, businesslike manner irrespective of their location on the nation's coasts and waterways.

### *Facility Siting in Urban Harbors*

As the nation's coasts become more crowded and the functioning of marine and freshwater ecosystems better understood, there will be greater scrutiny of development projects that might harm shoreline resources. Development of marinas in pristine or sensitive environments will virtually be prohibited. Wetlands and intertidal areas especially will be declared off-limits to development except where no feasible alternative site exists.

Conversely, within existing urban harbors where the shorelines have been modified already by bulkheading and filling, there will be many opportunities to site recreational boating facilities with minimal adverse environmental consequences.[6] The national trend to redevelop decayed urban waterfronts offers new opportunities for development of recreational boating facilities (see Fig. 9.2). Where industrial marine activities have relocated to new sites away from central city waterfronts, marinas and associated boater services are being seen as appropriate water-dependent uses to take their place. Even where backup land for parking is in short supply, marinas for transient pleasure craft can be accommodated in central urban waterfronts where they add color and vitality to formerly depressed areas. In port cities where marine cargo activity has diminished, pleasure craft facilities may be one of the few water-dependent uses capable of profitable operations on former port industrial waterfronts.

This confluence of trends (industrial abandonment of downtown waterfronts and dramatic growth in recreational boating activity) benefits marina developers who favor sites close to metropolitan populations over more distant rural alternatives. Shoreline habitat has been drastically changed by industrialization of the shore, resulting in fewer adverse environmental impacts arising from marina development; boaters can moor closer to home; and developers can take advantage of various redevelopment subsidies—land cost write-downs, local property tax abatement, community redevelopment financing, and other measures designed to foster revitalization. But higher land costs could preclude development solely for marinas: other more profitable upland enterprises may be required to provide a built-in cross-subsidy to the marina operation. Restaurants, retail shops, boat and accessories sales, and

**Figure 9.2.** Penn's Landing, a redeveloped central waterfront district of Philadelphia, includes transient moorage along with other facilities, such as historic ships and a cultural center. (Courtesy: Marc J. Hershman)

service firms are logical candidates for multiple use of an urban waterfront site, because they benefit directly from their contiguity to a marina.

## Greater Economies of Scale

Economies of scale will encourage marinas to be built as large as possible for reasons alluded to earlier. But unless there exists a demonstrable demand for large numbers of new berths over the short run, the debt service costs on chronically underutilized moorage will consume profits. Prudent developers in uncertain moorage markets, then, will probably phase development to match growth in demand for berths. Eventually the facility will achieve the economies of scale to be gained from being large, at the same time conserving capital invested in docks, finger piers, and utilities.

But even a phased development strategy is far from risk-free. Careful analyses of demand and its origin, occupancy rates and prices at competing facilities, and good intelligence about other new marinas being planned are crucial to designing the appropriately scaled facility. When long delays are encountered in obtaining development permits, the prudent developer will recheck the assumptions made and conclusions drawn in the original feasibility studies before committing to construction.

## Financing Public Marinas

Three factors have affected the financing of recreational boating facilities adjacent to the nation's waterways. Changes to the federal income tax code brought about by the Tax Reform Act of 1986 affect both public and private facilities financing. Publicly financed facilities, however, have been affected by a tightening of appropriations for federal public works projects and a stiffening of the criteria for federal cost sharing in those projects.

The fiscal austerity of the 1980s has resulted in fewer federal dollars being available for public works projects. Navigation improvements and maintenance costs particularly have been scrutinized by federal budget officials and congressional committees. As a result, a higher proportion of the costs of harbor development, waterway maintenance, and navigation aids have fallen on the users of these services. This "user pays" philosophy has permeated local as well as federal decision-making, and recreational boaters are being called upon to pay their full share of the costs incurred by public port authorities in providing smallcraft moorage facilities. As federal funding for recreational boating facilities becomes scarce or severely limited, the costs of developing public marinas will fall increasingly on local authorities; and this at a time when funds for other essential social and physical services are drying up at their federal sources.

The financial instrument most commonly used by local authorities to finance marinas is the tax-exempt municipal bond. Changes to the federal tax code have imposed new restrictions on the use of such bonds where private rather than public entities benefit from the lower interest rates. As a result, the method of financing public marinas, used extensively in California, is now ruled out: the public/private joint venture. Under this arrangement, the public port authority issued bonds to construct harbor improvements, which were then leased to one or several moorage developer/operators. Docks, floats, finger piers, and boater services were built and operated by the lessee, who in turn leased moorage berths to the boater. The lessee also gained the benefit of a federally subsidized navigation improvement (breakwaters, channels, maintenance dredging, bulkheads, and such). The port authority was saved the costs of constructing berths while avoiding some of the risks of uncertain occupancy rates and, hence, revenues.

It is distinctly probable, then, that where new recreational smallcraft facilities are developed by public entities, the entire cost of the facility will be borne by the boaters who use them. Revenue bonds will be the primary source of capital used to construct the facility, and no general tax levy on the district's real property owners for purposes of subsidizing its operation will be tolerated. As a consequence, moorage fees charged to boaters in public harbors will more nearly approach the market rates charged by private marina operators in the same market area. The often huge disparity between public and private moorage rates seen in many areas of the country will all but disappear.

It is not only changes in federal policies that are forcing this issue. Local public demands for accountability in managing public port authorities are often focused on marinas that are frequently characterized as "parking lots for wealthy yachtsmen." Also, private marina operators sharing the same market are wont to cry "Foul!" if they suffer the consequences of "subsidized competition."[7] Some Ports have responded to this issue through formal policies adopted by their commissioners. For example, the Port of Seattle, which operates a 1,200-slip marina on Shilshole Bay (see Fig. 9.3), has stated that it will build additional marina space only if the private sector does not. When Shilshole Bay first opened, there was an exodus of boaters from higher priced, private freshwater moorage on nearby Lake Union to the new saltwater facility on Puget Sound. Local private marina operators have not forgotten the financial impact cause by the Port's pricing policy at Shilshole; and the lengthy waiting list for moorage at Shilshole still attests to the value that boaters place on easy access to saltwater and lower public moorage rates at that marina.

The political clout of wealthy, influential boaters in a port authority's constituency is likely to be diminished by these fiscal and political changes attending marina development. In the past, the views and needs of such groups were felt early and forcefully by elected port officials. If other port functions were declining or undergoing sluggish growth, port commissioners could see the "legitimacy" of their public enterprise eroding. By responding to a maritime need (developing marinas) expressed by community "influentials," those commissioners were able to enhance both their

**Figure 9.3.** Shilshole Bay Marina, Seattle, a 1,200-slip recreational marina operated by the Port of Seattle. (Courtesy: Port of Seattle)

image as leaders of public enterprise and the public's perception of the port's legitimacy. And as long as the federal government picked up half the costs of the facility, there was little political risk in their accommodating boaters demands. Today, however, with this source of subsidy running dry, public ports and their constituents are demanding that boaters pay their own way.

## *Private Financing*

As with other types of real estate, the federal tax codes have encouraged and subsidized marina development. For example, the trend toward "dockominiums" (condominium moorage) has been fueled by the ability of the developer to pass on to the berth purchaser a proportion of such tax benefits as accelerated depreciation schedules and investment tax credits; and by selling berths before and during construction, the developer reduces the amount of necessary long-term capital financing. Closing "loopholes" such as these was part of the motivation behind passage of the Tax Reform Act of 1986. The new tax code may well result in a substantial reduction in demand for dockominium moorage, an attendant increase in development financing costs, and a drying up of the supply of rental moorage produced by nonboating owners of dockominium moorage seeking tax shelters. The dockominium may, simply, have been a "flash in the pan" soon destined to be discarded as a development aid.

Because marinas are rarely viewed as "pure" real estate by lending institutions, private venture capital will continue to play the leading role in financing private marina development in the future. As one banker said, "A marina is a lot like a tennis club. If it goes bankrupt, what can a bank do with it?"

## *Managing Marinas*

There is a trend toward enhanced professionalism in marina management. Newly formed marina trade associations produce newsletters for their members; university Sea Grant programs conduct frequent extension courses and conferences for the industry; public port associations have active marina or smallcraft harbor committees that conduct workshops and undertake commission studies. Computer software packages for marina management have been developed, tested, and adopted by many operators.

The industry has become large enough, also, to spawn research and development in marina design, materials science, and new product development. Regional and national markets exist for precast concrete, aluminum, and plastic floats, marina hardware, boat-hoists, and special pilings.

In short, though frequently characterized as a "Mom and Pop" establishment, the marina is becoming a professionally managed component of a growing recreational industry.[8] Many new private and most new public marinas have over 100 moorage berths, and offer haul-out facilities, upland storage, boat sales and repair shops, restaurants, and many public amenities. In urban areas and destination resorts, the marina is often but one part of an integrated land development scheme that may include residential, hotel, and commercial uses on contiguous upland parcels. Marinas may be constructed more to promote and market the utilization of upland facilities and residential sales than to stand alone as profitable enterprises in their own right. A recently completed shoreside development on Seattle's Lake Union consists of a marina, restaurant, retail, and office space. The marina component was estimated to account for only 4 percent of the gross revenue, but it occupied the largest portion of the site; and had it not been for a city shoreline regulation requiring a water-dependent primary use of the site, it is quite possible the marina would not have been built at all.

The trend, then, in marina management is away from sole proprietorship and toward corporations employing professional management staff and business administration methods. In the recent past, few marinas have had a business plan in place; few knew how much equity they had in their facilities, nor what return they were getting on it; and scarcely any marina managers actively promoted or marketed their services. This passivity is best explained, perhaps, by historically high levels of demand for moorage creating a "seller's market." But the recent recession and overbuilding in some areas have resulted in softened markets and increased competition for tenants. Concurrently, upward pressure on shoreline land costs and submerged land lease rates, and higher insurance premiums are squeezing profits and producing bankruptcies in the industry. Now we find marinas—both public and private—sending representatives to boat shows and using printed advertisement media to promote full occupancy of their facilities.

It would not be surprising if horizontal integration or even franchising of marinas were to occur, as it has in campgrounds and other components of the hospitality industry. Already, on the West Coast, we see one horizontally integrated marina management corporation in place. A boater's tenure at one of the corporation's marinas ensures visiting privileges at all others; a regular newsletter to tenants is also published. And on the East Coast, one company runs nine marinas in three states. Can it be long before we see new membership organizations—Marinas of America (MOA)—emerging to fill boaters' needs for accommodation during cruises in crowded destination areas?

## *Environmental Management*

A proposal to develop a marina, as we have seen, will be scrutinized by public resource agencies, environmental organizations, and citizen groups, and will probably undergo substantial revisions before being approved. Public access and recreational facilities for nonboaters will be required to broaden the constituency served and to achieve affirmative coastal management goals.

Fish and wildlife agencies at the state and federal levels of government are requiring in many cases that losses be "mitigated" where habitat is either modified or destroyed by dredging and filling. Mitigation may be achieved by "enhancing" less productive habitats on or near the site, or by acquiring and deeding to a suitable agency an equal quantity and quality of habitat off the site. Special provisions to protect fish runs may be required, such as requiring development setbacks to provide safe, shallow intertidal habitat through which juvenile fish may transit the site (see Chapter 12).

Because so little data exists about the functioning of marina ecosystems, marina developers are required to bear the cost of monitoring the environment around their facilities. Such monitoring is particularly crucial where habitat enhancement schemes have been put in place. Agencies need to know if these schemes work as planned, so that unworkable solutions can be eliminated in future projects.

Municipal and state coastal management agencies frequently require dedicated public access at the marina: walkways installed on the tops of breakwaters for fishing, public boat launch ramps, picnic areas, and the like. Here again, diminished federal funding for park acquisition and the high costs of maintaining public facilities will add to the pressure on marina developers to provide and maintain public benefits at the marina. These demands will be more forceful in shorelines where little public access currently exists.

## *Alternatives to In-Water Moorage*

The private sector moorage provider—and to an increasing extent the public port authorities, too—respond to a market demand for their services. The public agencies responsible for managing coastal resources ask a different question: Is there a *need* for more moorage facilities? Or can other ways be found to accommodate the need for boat storage? Many boats occupying marina berths are of trailerable size and could easily be stored out of the water, thus freeing wet moorages in marinas for nontrailerable vessels.

Various proprietary and custom-built systems for dry storage of boats have been developed and installed. One commonly used system is a large metal shed, one end of which is cantilevered over water. Racks along each side of a wide center aisle hold motorboats up to approximately 25 feet in length. A traveling gantry crane transfers the boats from their racks through the open end of the building and into the water. The shed may be either built on land or float in its entirety in water.

Open or roofed upland dry-storage-rack systems are designed for two- or three-level installation on dry land. Modified forklift trucks are used to launch, retrieve, and store smaller motorboats. Land storage of larger vessels, including sailboats with fixed keels, has been common practice in regions with cold winters, where in-water winter moorage is precluded by freezing conditions. But one Washington State facility uses open land storage as an alternative to year-round wet moorage. At Dagmar's Landing, large-wheeled "front-loader" vehicles, equipped with extended, adjustable forks, can handle vessels up to 40 feet in length and launch them down a paved ramp (see Fig. 9.4). By careful scheduling, the operator can guarantee that a boat will be in the water moored to a transient dock when the owner wants to use it. Located on a former pasture, this dry-land marina was the first of its kind in the nation and suffered the fate of many innovations: it failed within a few years. A bank assumed management of the facility, which now runs at a profitable level.

But dry-land open storage consumes extensive land area and is, therefore, rarely practical in built-up urban harbors where land costs are high. Within cities, stacked storage systems will be the preferred alternative to in-water moorage for trailerable-size boats. However, the market is a seasonal one; during off-season periods, boat owners tend to store their craft in garages, backyards, or in popular mini-storage rental facilities. The principal users of dry storage will continue to be owners of

**Figure 9.4.** At Dagmar's Landing, a dry-land boat storage facility in Marysville, Washington, huge forklifts launch and retrieve boats from the Snohomish River. (Courtesy: Robert F. Goodwin)

trailerable motorboats who want to avoid the cost of wet moorage (if it is available), but who are willing to pay to avoid the congestion and inconvenience of city streets and public boat ramps in urban areas.

## THE FUTURE: SHARING BOATS AND HARBORS

For the owners of large vessels, moorage space in or close to metropolitan shorelines will be in short supply and higher priced. Rates in excess of $10 per linear foot per month are not uncommon in the crowded harbors of Southern California, and in central Puget Sound, new saltwater moorage is expected to come on line at $6.50 per foot. The owner of a 32-foot sailboat would have to pay $2,500 dollars per year just to store the vessel, let alone use it! It is little wonder, then, that the National Marine Manufacturers Association (NMMA) is funding studies to understand and remove barriers to marina development nationally. NMMA, an organization of boat and engine manufacturers, realizes that the shortage of reasonably priced moorage facilities can retard sales of their products.

Even if the supply of moorage becomes a limiting factor in boat sales, it need not limit the participation of larger numbers of families in boating. The emergence of time-share ownership permits more families to share in the cost of acquiring, maintaining, and mooring larger boats. New commercial enterprises provide leased time-shares to large numbers of boaters, utilizing multivessel fleets. Similarly, nonprofit boating and sailing clubs have been formed to allow shared access to club-owned fleets. Strategies like these have been adopted by groups of general aviation pilots for several decades, providing low-cost opportunities for flying to people who could not otherwise afford access to aircraft. In time-shared arrangements, use of the equipment is greater than it is where the boat or plane is individually owned. When a party in a time-share arrangement schedules use of the boat long in advance, there is an imperative operating not to miss one's turn.

Public port authorities have a role in the development and operation of smallcraft harbors, but there will be changes in the way their projects are financed and in the clientele they serve. The federal government will likely continue to appropriate funds for navigation and harbor improvements only where waterborne commerce benefits can be documented. Harbors designed exclusively for recreational fleets will have the lowest priority.

Where practicable, then, local port authorities will develop future smallcraft harbors to serve both commercial and recreational fleets, rather than developing separate facilities for each. Facilities for the smaller, trailerable boat owner (launching lanes and hoists, dry storage, trailer parking, etc.) and for those who do not own boats (fishing piers, walkways, picnic areas, and public access) will be provided to broaden the constituencies served. Revenues from intensive upland development will be used more so than in the past to defray the nonrecoverable costs of public amenities. And moored boaters will pay their way fully through moorage fees set at, or approaching, market rates. Ports near favored cruising areas will have greater incentive to build marinas because of economic development spinoffs. However, in larger urban harbors where competition from other Port capital development projects is great, the private sector will be the lead developer of new boating facilities.

Boating has been characterized, irreverently, as a lifelong addiction, and a boat as "a hole in the water into which one pours money." Behind these stereotypes lies some truth; but as long as there is water to float on and boats to float in, there will be boaters enjoying their sport. As one Power Squadron Commodore remarked recently, "We will, if necessary, give up our houses, our wives, our firstborn son and our sanity before we'll give up our boat!"

## Notes

1. According to the U.S. Coast Guard, both the number of fatalities and, more importantly, the rate of fatalities (per 100,000 boats) have fallen from 1975–1985. In 1975 there were 1,466 boating deaths, or 12.4 per 100,000 boats. In 1985 these numbers had declined to 1,116 deaths, or 6.9 per 100,000 boats. U.S. Department of Transportation, U.S. Coast Guard, *Boating Statistics 1985*.

2. Much of the information presented in this case study was drawn from U.S. Army Corps of Engineers, Seattle District, Sequim Bay Boat Haven, Final Federal Environmental Impact Statement (NEPA) Section 10/404 Permit Application, February 1983.

3. In this case, the Port decided not to pursue a Corps of Engineers Public Works appropriation for navigation improvements, believing that the application process would slow down the project.

4. Much of the information in this case was drawn from U.S. Army Corps of Engineers, Seattle District, East Bay Marina Final Detailed Project Report and Final Environmental Impact Statement, December 1980a.

5. Much of the information presented in this case study was drawn from an unpublished paper (Heikkala 1978).

6. However, even where substantial urban development crowds the shorelines, the marine environment may be almost undisturbed and will be treated by resource agencies with the same protective attitudes they focus on development proposed in pristine rural shorelines.

7. The 1987 Washington State Legislature considered a bill that would have required public ports to establish moorage rates using the same cost bases as private marinas in their market area. HB 1103 did not pass but will probably be reconsidered.

8. A Sea Grant study reported that the marine recreation industry in Washington State produced an annual impact of half a billion dollars, income of over $90 million, and resulted in the employment of over 8,000 people, statewide (Ellis et al. 1977).

## *References*

Ellis, Roy C., William B. Beyers, Robert L. Stokes, and Darrell D. Brown. 1977. Economics of marine recreation in Washington State—1977. Coastal Resources Program, Institute for Marine Studies, University of Washington.

Goodwin, Robert F. 1982. Recreational boating in Washington's coastal zone: The market for moorage. Seattle, Institute for Marine Studies, University of Washington, May.

———, and Robert L. Stokes. 1980. The moorage industry in Washington's coastal zone. Technical report WSG 80-7, Seattle, Washington Sea Grant.

Heikkala, Susan. 1978. Water quality and marina design: Flushing, internal circulation and the physical hydraulic model. Coastal Resources Program, Institute for Marine Studies, University of Washington. Unpublished.

National Marina Manufacturers Association (NMMA). 1985. A long-term view of boating recreation demand. Chicago.

Ross, Neil W. 1985. Report on recreational boating facilities in the nation to the President's Commission on Americans Outdoors. University of Rhode Island, November 5.

U.S. Army Corps of Engineers, Seattle District. 1980a. East Bay Marina final detailed project report and final environmental impact statement. December.

———. 1980b. Recreational small boat moorage study, Puget Sound and adjacent waters.

———. 1983. Sequim Bay boat haven, final federal environmental impact statement (NEPA). Section 10/404 Permit Application, February.

U.S. Department of Transportation, U.S. Coast Guard. 1986. Boating statistics 1985. Washington, DC: U.S. Government Printing Office.

# Chapter 10

# PORT FINANCES AND OPERATIONS:

## Understanding the Bottom Line

*Thomas J. Dowd**

---

Editor's note: A primary reason why public ports are "autonomous" organs of government is their ability to raise money in the private sector, keep the funds in their own account, and spend it with few restraints. Their receipts do not go into the general fund of state or local government, nor are they dependent on an annual appropriation from a central source. In this sense they operate much like a private business. Many people say the successful Port is one that creates a revenue surplus (i.e., earns a profit) for investment in other public projects. Others claim that the Port must justify itself on a broader basis. They ask what kind of benefits does the community get? How competitive is the Port? Is the Port a good neighbor? As the following interview with Professor Dowd points out, determining the "bottom line" for a public enterprise is a complex task.

**What would you say is the bottom line for a public port agency?**

There are actually two bottom lines. One of them is the financial bottom line—in order to survive, Ports have to make money. But there is another bottom line—the need to keep the taxpayers and citizens happy, or in some cases, to accommodate the governmental entity that runs the Port (like the city, in the case of the Port of Los Angeles). Public ports have to be sure that both bottom lines are positive, because it's possible to make a lot of money and still make people mad.

If the people or the parent governmental agency is mad at the Port, they're going to make things change. There will be fewer and less abrupt changes, if the Port is

*This chapter is an interview with Professor Thomas J. Dowd, Institute for Marine Studies, University of Washington.

making a lot of money—but there will be changes. In order to keep everybody happy, public ports have to keep both bottom lines in mind.

**What are some examples of Port activities that affect the other bottom line?**

Many activities are undertaken in order to attract profitable business. The Port might operate a specialized warehouse facility for apples, for example. It might not pay the Port to operate this cold storage warehouse, but in order to get the apple business to flow through their Port and generate employment, the Port does it.

Just about every Port has activities that it subsidizes in order to encourage regional economic stability. Fishing might be one of them. The Port of Newport, Oregon, is trying to encourage the forest products industry—Ports are doing a lot to encourage their regional economies.

Many Ports are creating parks, bike trails, and these kinds of things—admittedly, in some cases because of agency requirements for mitigation. But often they become involved because it's a way of enhancing that other bottom line. Many people feel that the public are stockholders in Ports. This is especially true when the public does, in fact, provide tax revenues. And they want to get something tangible in return—parks, fishing piers, public access, and such.

The problem with this belief is that it presupposes that the Port somehow is obligated to provide a kind of dividend. If it is doing well as a commercial port, and doing something for the constituents, taxpayers tend to leave the Port alone. If the Port stops doing things for the constituents, or if it is not willing to listen to constituents, then the Port will find a whole bunch of well-intentioned citizens in there "helping make decisions." Ports realize that.

## PORT REVENUES AND PUBLIC SUBSIDIES

**If Ports have to make all these people happy, how do they get the funds to accommodate them?**

Well, there are two kinds of Ports; one is an operating Port and the other is a landlord Port. An operating Port basically operates its own docks and such, and there are fewer and fewer of them. The other kind of Port builds and owns the facilities and leases them to somebody else to run, usually a steamship line or stevedoring company.

A Port that builds the facilities and runs them as an operating Port can make a lot of money in a good year. Tacoma is a good example of that. A number of years ago the Port of Tacoma operated all its docks and made out quite well. During a good year you make a lot of money, but the problem is that in a bad year you don't make so much, and what you end up with is an unstable revenue base. Today Ports that build larger size facilities try to lease them before construction in order to ensure a guaranteed income to pay off the bonds that financed the project in the first place. That's not only prudent for the Port, but it's often required by the investment banking community. You must have a known revenue stream before you can obligate yourself for twenty years. You have got to know where that money is coming from.

**Can you elaborate on how Ports make their money?**

As I mentioned, for the most part, Ports make money by leasing their land and facilities. Ports also offer services for a fee, and that includes dockage, wharfage, storage, and demurrage. Each Port offers a different mix of services and facilities. Once you get beyond services, there are a few Ports that make money from unusual things, for instance, Los Angeles has oil royalty money.

**What are some examples of wharfage, dockage, and demurrage?**

Dockage is the charge for tying the ship up to the dock or terminal. It's a charge that is levied against the ship's owner, and is usually determined by multiplying a rate per foot against the length of the ship or a rate per ton times the gross or net registered tonnage of the ship. In effect, it is a charge for the ship using the terminal or dock facility. Wharfage is a charge against the cargo, and is charged for the privilege of moving the cargo from the ship across the dock. Wharfage is usually charged on a per-ton or a per-container basis. Demurrage is the charge a Port uses to ensure that cargo is expeditiously removed from the dock. When a ship delivers a container, it can sit at the dock only so long before a demurrage fee is charged. In effect, demurrage is a penalty for using the Port's dock facilities as a storage area. At some point in time, it becomes too expensive for the shipper to leave the container on the dock. It's cheaper to find some other place to store it.

**With money from leasing, wharfage, dockage, and demurrage, do Ports operate in the black for the most part?**

Yes! The majority of Ports recognize that in order to continue to be a viable entity, it is necessary to make a profit. They must generate a profit in order to contribute funds to capital projects and pay debt service on revenue bonds.

Even Ports with a substantial tax revenue base realize that they must make a profit. Tax revenues normally will not increase fast enough to pay for new construction, and these days it is almost impossible to get a significant increase in taxes. Thus they need these excess revenues, or profits.

For most large Ports the problem is not making a profit, it is making a sufficient profit to maintain viability and continue to grow. Ports, like any other business, are expected to produce a return on investment (ROI). The level of ROI is a major factor in a Port's ability to remain competitive and to continue to pursue additional opportunities.

**Don't Ports get a lot of their money from tax revenues?**

The majority of Ports do not get tax revenues. For those that do, tax revenue support can be important, especially for smaller Ports, but it is not usually a significant form of income for most larger Ports. There are a few that have some significant tax income, like Tacoma, Seattle, and Houston. Frequently tax revenues are dedicated to capital improvements—they are not used to pay operating expenses. It's a revenue source that goes directly into the capital improvement account.

**Do some Ports have control over the amount or type of tax levies they receive? For example, does the Port of Seattle control the amount of their property tax levy?**

No, it doesn't. In the state of Washington, there is what is called the 106 percent lid law that limits the increase in the amount of property taxes any single entity can receive one year to the next, so the Port of Seattle doesn't control the levy. It could ask for less than the maximum, but there is a lid that determines the maximum. There is usually some form of outside control for all Port-related tax levies.

**Would you say then that it is probably better if a Port doesn't get in the habit of using tax revenues for their operating expenses?**

Absolutely! If Ports take the money and put it into capital improvements, they will have far fewer problems than if they put it into operations. If Ports rely on tax revenues for operating expenses, they are building into their operating budget a source of money over which they have very little control. If Ports use tax money to offset their operating costs, and lose the tax-levying authority or the levy amount stays static or decreases, they are in deep trouble. This sometimes happens with very small, tax-supported Ports.

**But it's said that all Ports are subsidized in one way or another . . . is that true?**

Yes, it's true, because all Ports in the United States are exempt from local government property taxes and their income is exempt from federal income tax. In addition, Ports as governmental units are able to issue tax-exempt bonds, which demand a lower interest rate than if they were to borrow that same money as a nongovernmental corporation. There are some other examples of subsidization. For instance, in California the major Ports were granted the tidelands in front of large cities free of charge. Other forms of subsidization are perhaps more direct, and that is where a parent government directly subsidizes a Port operating unit by either paying some of its expenses or failing to charge the Port for various services that are supplied by the parent government unit.

The subject of subsidies is probably the most misunderstood area in Port management. As I said before, many Ports subsidize activities for the benefit of the regional economy. But a lot of specific functions at Ports are subsidized, because without the specific functions, they would not be able to bring in certain other business that they can make money on. For instance, there are some bulk Ports that subsidize the storage of bulk commodities on Port property in order to get the throughput across the docks of the commodity itself. Again, the word *subsidy* has to be defined. Subsidy is not giving something away. It may mean that the Port is getting less return on investment based on the cost of providing that service than it could "theoretically" get. However, if the subsidized part of Port business helps it acquire or retain the lucrative parts, and the net effect is a profit or other continuing benefit to the community, then the subsidy has worked well. Of course, there are times when a subsidy doesn't work and a loss results, but that is not always true.

**How about borrowing? Don't most Ports borrow money for facility expansion or development?**

Almost all Ports borrow money to pay for capital projects. They don't borrow money in order to pay operating losses or to provide money for noncapital purposes.

There are two types of borrowing that Ports use. One is borrowing money using general obligation bonds that are repaid from the proceeds of tax levies. In effect, Ports are borrowing money on the credit of the taxpayers. The other and the most commonly used method of borrowing is to borrow using revenue bonds, which means that the bonds are repaid from the earnings of the facilities of the Port. Sometimes the payback period on bonds is over an extensive period of time, in excess of 20 years. If the project can earn the money to pay the bonds off from the outset, that is ideal. But often Ports are forced to borrow money for projects that will not earn enough money to pay the bonds off right away. That is why the revenue bonds are supported by a variety of Port operations, including older projects that are now lucrative. If a facility is not going to earn money for a long time (or maybe not at all), then a Port would be forced to borrow using general obligation bonds backed by the taxpayers. The vast majority of Ports do not have a great deal of general obligation bonding authority, and so they rely to a great extent on revenue bonds.

## FINANCIAL GOALS, DISECONOMIES, AND PRIVATIZATION

**Given all the interests to serve and all the ways Ports are able to make money, they must need very clearly defined financial goals.**

I think that many port authorities have rather fragmented financial goals. Most Ports have some feeling for what they would like to see as a return on investment. For example, if the Port is a landlord organization, it would, of course, like to see a return based on market value. But the financial goals of many Ports are not purposefully implemented—they remain in the realm of wishful thinking.

One of the main reasons that Ports fail to attain their financial goals is that they must survive in a very competitive market. Often they cannot charge enough to provide a desirable return on investment because those charges or rates would be noncompetitive. So in order for a Port to get business, it often sacrifices its ability to earn an acceptable return.

A lot of Ports have self-sufficiency as their goal. They often define self-sufficiency to mean that the revenues derived from tariffs and leases plus their income from taxes are equal to, or a little higher than, their expenses.

A problem that a lot of Ports live with is that they have created "diseconomies" (a diseconomy is where you have a negative cash flow). In other words, they have set tariff rates that are too low or lease rates that are too low in order to attract companies and business. And there is an assumption that either taxes will make up the difference, or that somehow other Port users will make up the difference—through cross-subsidization or whatever you want to call it. Often the Port loses sight of the amount of diseconomies it has created, and this can undermine the financial goals of the organization.

**When private businesses are analyzing their operations, they look for diseconomies. They hunt them down and get rid of them. Why do Ports incorporate diseconomies into their operations?**

A lot of Ports eliminate them, too, but diseconomies occur when you are trying to get business. If you get business, it is like a loss leader in a department store. You can sell something for less than cost, but you had better make it up someplace else—and that gets back to our discussion of taxes. A lot of Ports assume that taxes will cover these diseconomies . . . but they may not. Also, a Port can get business that has some profitable aspects, if it will do some things that are not profitable. For instance, a major farm implement firm shipped a lot of its tractors through the Port of Tacoma, primarily because the Port didn't charge any storage fee for those tractors—the tractors just sat there; but they did charge them wharfage when the tractors went on the ship. There was a diseconomy in the storage, but certainly it was made up in the wharfage.

**In dealing with diseconomies—or any other situation—are public port agencies as flexible in their decision making as private corporations?**

No, because Ports are created by government and they usually have two limitations. First of all, they can do only what the law grants them powers to do, and second as a public entity, they tend generally to be more conservative than private business. Ports show a range of styles. On one end of the spectrum a Port can be a very bureaucratically oriented, governmental unit that is terribly conservative. At the other end a Port can be as aggressive as any private business. Take marketing, for example; very aggressive ports like the Ports of Long Beach, Los Angeles, and Port Everglades in Florida market very much like private industry. However, even then there are limits on what they can do. Money is not spent as freely on marketing for Ports as it might be for private organizations. Also, Port marketing people are usually paid a salary, and although there are some incentive bonuses, they are not paid a commission as in a business where the more you sell, the more you make.

**We've been comparing the public port system with the private sector. Do you think that the port system could be privatized?**

From a practical standpoint, I don't feel that it could be privatized, because right now the current level of earnings on invested capital (return on investment) is so low that no one would want to put their money into it.

**Do we need Ports to be public agencies?**

If they weren't public agencies, we would end up with a lot of additional expenses—federal income tax and property tax would apply to a private agency where it does not apply to a public agency. If Ports were all privatized, the price to move a commodity through a Port would of necessity be higher, and this might tip the scale preventing a product's export to another nation, or it might give the Canadian public ports a competitive advantage and result in cargo diversion to Canadian ports.

Also, if Ports weren't public agencies, not only would you have higher expenses, but you wouldn't see any appreciable amount of growth. Private ports would have to borrow money from the private capital markets and this would cost them higher borrowing rates. The projects themselves would need a higher return on investment. Since it's hard to get higher returns because of the amount of competition, it is likely that new investment in Ports would be slow and facilities would be modest.

The public entity provides something that I think would be difficult under a privatized scheme, the ability to achieve economies of scale. Public ports, because of their favorable borrowing capabilities and the use of tax revenues in some cases, are able to build their facilities larger than the present demand requires. In order to have economy-of-scale earnings, you must have a specific size operation. It's not likely that the private port would build large enough facilities because they would want to serve the immediate users at hand where revenue returns would be assured. Also, private operators are unlikely to invest heavily in one location since transportation patterns change so rapidly. Private terminals might be more efficient because there wouldn't be as many and they'd be smaller, but you also wouldn't have the economy of scale in the size of ports, which benefits shippers and consumers through lower costs.

There are other reasons why I think Ports should be publicly owned. The public port has long-term stability and provides single ownership and management of the shoreline. The public port can obtain and manage funds from the federal government. The public port provides a common marketing function in the harbor because it disseminates information on the services available locally and the current rates for doing business, and actively promotes the region. These public functions provide a lot of service to the community that improve the economic climate, even if in indirect ways.

**Yes, it seems as though the idea of privatizing Ports doesn't take into consideration the true economic impact of a public port on its region.**

Well, Ports can have all sorts of goals, one of which is to maximize income. Another one is to maximize employment. Another one might be to stabilize the regional economy—and so, if the Port's *sole* goal were to maximize income, you might argue for privatization.

## MEASURING SUCCESS AT A PORT

**How does a Port know when it is meeting its goals?**

The definition of success is different for each Port. Success to one Port might be breaking even. Success to another might be making money. Success to a third Port might be increasing the number of jobs that are available through Port operations or Port entities, regardless of the balance sheet or profit-and-loss statement.

I don't think it's fair to judge a Port just by its financial statement and whether it makes money or not. A Port might very well not make money and yet be in a situation where it has held the community together because it has stabilized the economy. The Port may have stabilized the economy by subsidizing all sorts of things in order to get people to ship through that Port and make that Port more active during a period when other parts of the economy weren't providing any impetus for growth. The Port of Seattle, for example, helped buffer the effects of the 30,000 layoffs during the Boeing crisis in 1969 when the Supersonic Transport (SST) airplane contract was canceled.

**So you would say that even if a Port is losing money, sometimes it can be considered successful?**

Yes. There is no question about that. And that is why privatization is not such a good idea. A public body can make decisions that are much more long term than a private organization. A private organization is worried about next quarter's results, but a public agency would be more willing (and able) to accept lean years to accomplish something in the long run.

**Who generally makes the financial decisions in a Port?**

Policy decisions as to how much money the Port should earn and the method of return on investment are made by the board of directors or commissioners. In some city-run Ports there may be some overriding policies that are set by the city council or the mayor, but to a great extent, policy decisions, such as the target for rate of return on leases, is set by the board of commissioners.

A problem that comes up is that although these targets are set by the board, they either aren't carried out by the staff or can't be carried out by the staff because of competition problems. Although the board has final say in approving a lease or approving changes in a tariff, the real decision maker quite often is neither the board nor the staff—it is competition.

Let's use the Port of Seattle for an example. If the competing Port of Tacoma drops its tariff on gumdrops, Seattle, in order to stay competitive, drops its gumdrop tariff to the same level. The Seattle Board of Commissioners, trying to be proactive, says, We're going to set our rates so that we cover all our costs. The staff people, living closer to the problem, then say, We are going to have to accept the fact that we can't live up to what the Board wants us to do—we're going to have to set our rates lower so that we can have some business, and get Board approval as well.

Only a few Ports in the country can ignore these competitive pressures. For instance, there are a lot of goods going directly into the Ports of Los Angeles and Long Beach, even though Port costs are higher, because that is where the major market is. And that is why they are consistently earning more money than everybody else—they have a monopoly on a lot of cargo. If you have cargo going to Los Angeles, you would be rather naive to ship that cargo into Seattle or Oakland even though the Port costs are less, because the cost to transport it to Los Angeles would have to be added on, plus the time incentive if it is a valuable cargo, plus the interest on it while it's in transit.

Long Beach and Los Angeles can set their rates, and have, in fact, set their rates, to provide a very good profit for themselves. This type of situation probably exists only at one other place, and that is the Port of New York/New Jersey.

## SHARPENING FINANCIAL DECISIONS

**What are some of the other major problems facing Port managers today?**

The long-run problem is that in order to do financing these days, Ports need to have some kind of known revenue stream, and this situation is very difficult for Ports. It used to be, years ago, that a Port could finance a dock or a facility of some sort, and then go out and market its use. That was when things didn't cost so much. Now they cost a great deal more. Most Ports don't go out and build a new container terminal and buy container cranes unless those large facilities are primarily out on leases, long-term leases with locked-in dollars.

**Does this mean that the lease or some formal commitment for a terminal must be signed before its construction?**

The lease must of necessity come first, because the Port is counting on the money from the lease to pay at least part of the interest and principle on the money borrowed to build the facility. The Port needs a predictable revenue stream. It can't just build the facility on speculation anymore. Admittedly, the very large Ports might be able to build facilities in anticipation of future business—but they would not be willing to take the risk unless that business was clearly in view.

Effectively, what Ports try to do with a container terminal is lease it to a steamship line or terminal operator, preferably a steamship line. That steamship line is obligated to pay, depending on the type of lease, some kind of minimum payment that will at least cover the debt service on the facility, and even some kind of return on investment.

**Why can't the city or county pay for Port projects?**

If they legally could, they might, but Port projects are often at the bottom of the list of their priorities. For example, when the Port of Oakland was first getting underway, the city of Oakland issued general obligation bonds to build the initial port facilities. But there is no way that the city of Oakland is going to build more port facilities with general obligation bonds or tax revenue when it needs, say, a new city hall or civic center or firehouse. The city is totally aware that the taxpayers will accept only so much, so those in power start prioritizing the kinds of things that general obligation bonds and tax money will be used for—and since the Port can generate its own money, the Port projects are often right down at the bottom of the city's list.

**If the Port is dependent on borrowing, does that mean that the investment banking community has a lot of influence on Port decision-making?**

The investment banking community has some impact on the powers of the board of commissioners, or the ability of the board to do certain things. But there are lots of other ways that the board can be limited. Certainly, what a Port has promised its bondholders is a limitation, and what the state statutes allow it to do is a big limitation. Also, the Port's financial capacity is a limiting factor, and its staff and their talents or lack of same are limitations; so there are lots of limiting factors. One of them, and I don't feel that it is a major one in most cases, is what a Port has promised the bondholders.

**How does state law limit what a Port can do?**

Well, in Washington, for instance, Chapter 53 is the part of the state statutes that covers Ports. There is a section in there that delineates powers granted to the Ports. Ports can't own office buildings in the center of downtown nor can they own city halls—there are a lot of things they can't do.

At one time, Ports could not own railroads. Now some of them can. Most Ports can't own convention centers, but some can if the state statutes are adjusted. Statutes define what a Port can do, but this doesn't necessarily cause a Port any problems. When a Port feels that there is a necessary project or function to be handled, it can normally persuade the legislature to alter statutes and provide the authority to do the project.

**If a Port is in a weak financial position, would it normally seek to change statutes in order to do something innovative?**

When a Port is in financial trouble, if it is smart, it does more of the traditional things Ports know how to do rather than go out and look for unique things to do. You have certain talents at a Port, and primarily the thrust should be to do traditional things and make enough money to keep out of trouble.

When a Port is financially well off and has all sorts of money, and someone comes to the Port with a really clever idea, only then should the Port think about becoming involved in something new. For instance, a lot of Ports have been approached to build motels and hotels and restaurants and "boutiquey" shopping centers to lease out to developers. And depending on the state it's in, the Port can do that in some cases and can't in others. But a Port is not usually trying to be unique unless it has an awful lot of money. It's the same thing in private business; in order to be successful, you have to be able to do something well, and you garner talents in order to do that. It's the same thing with a sports team; you have got to do the fundamentals well.

**What can Ports do internally to reduce financial pressures?**

There is a simple answer, and that is cut expenses. There are both fixed and variable expenses. What has happened with most Ports is that the percentage of fixed expenses has increased tremendously in the past few years. So if a Port finds itself in financial trouble, there are fewer choices for cost cutting, because it usually can't cut fixed costs. Fixed costs have to be paid whether you open up shop or not. Variable expenses are primarily people for most Ports. You can always adjust the size of your work force in order to reduce expenses, as a last resort.

**How does a Port assess risk before it undertakes an activity?**

A Port will do a feasibility study, generally through a consultant. You measure risk by going out and asking a lot of questions and trying to understand the business of the activity. Risk is first measured initially by some kind of in-house study, then given to a consultant, somebody who is disinterested. Then the staff presents a recommendation to the commissioners, who can accept or reject it.

Sometimes a commissioner who has been in business a long time says, You know, this is a great idea, except we haven't got the folks to do it, even if we have the money to do it. And so it is rejected. At other times, it is some business interest that asks, What are you doing this for? Shouldn't private industry should be doing this?

Often the kinds of Port projects that are studied are *not* being done by anybody else. Then when the Port initiates the project, private industry says, We want to get into that, too. A good national example is when Ports opened a lot of little self-storage warehouses. It was necessary and required by the community. There was a lot of demand for it. Ports built one or two of these facilities, which soon were completely sold out. At that point private industry decided to get into the self-storage warehouse business itself. It then said to the Port, Why are you in my business?

Sometimes Ports want to work "higher on the cargo chain" and engage in activities other than receiving cargo at a dock. They may want to provide warehousing and distribution services, but this may be looked on as an encroachment by private industry. Therefore most Ports will let private industry try something before they attempt it.

**We talked earlier about privatization. Do you think that going in the opposite direction would make Ports more efficient? In other words, can you envision a way in which the system would operate better with greater state-level control and direction or with more federal financing and attention to port development on a national scale?**

It is hard for me to envision any appreciable positive value in the state or federal government controlling port development. First of all, I don't think that either the state or the federal government would be willing to share the cost of port activities or projects. Second if either the state or the federal government became involved in controlling port operations, it would just put a layer of bureaucracy above what is already in existence. That would be a disastrous situation, since it would make it even harder for the Ports to respond to competition. Third, there is no method that I could see by which either a state or some type of federal organization could force Ports to be more efficient. Competition does this very well.

At this time there are several types of proposed state organizations that would become involved in planning and regulating Port development. The proponents argue that by having a state oversight and planning agency, there would be less potential for duplication of port facilities in a single state. It is possible that new state agencies could rectify this alleged problem, but there is a simpler solution—interport cooperation. Ports should remain independent and autonomous, but they should also cooperate better at a regional or state level. If they worked more closely, they could cut the costs of planning, trade projections, technology monitoring, marketing, and so forth.

I want to point out that some states, like Virginia, North Carolina, South Carolina, and Georgia, have state port authorities that own, operate, and finance all or the majority of ports in their state. These organizations are semiautonomous public ports themselves, not oversight agencies that only plan and regulate. My point is that we don't need the oversight agencies; competition is a pretty fierce watchdog!

## SURVIVING THE RAILROAD/STEAMSHIP SQUEEZE

### What are some of the recent changes in Port operations?

One of the things that has changed the competitive atmosphere of Ports is deregulation of the railroads. Formerly, all the Ports within a port range (this includes Ports within a specific geographic area) had basically the same railroad rates for transit from the Port to an inland location. If a steamship line put a cargo into San Francisco or Los Angeles or Oakland, or Seattle or Portland, they didn't pay exactly the same, but the rate they paid was out of a book that listed rates and charges for services—a tariff—and everybody knew what it was, and it wasn't all that different. The deregulation of the nation's railroads changed all this. Now railroads contract directly with customers, provide time/volume discounted rates, and are not required to adhere to published tariff rates for all customers.

In the old days, the words *hinterland* or *captive cargo* had some meaning—in other words Ports had control of the cargo that was on their docks and destined for their hinterland areas.

That is no longer true today because of a couple of things—primarily railroad deregulation, and also the fact that the steamship lines have what is called a "single bill of lading." They carry something from Yokohama to Chicago on one bill of lading. The shipper consignee pays the steamship line, and the line chooses what port it goes into and what railroad it goes on—the line might even have their own stack-car trains. That causes some real problems for Ports. In fact, the single bill of lading is probably the most serious problem we've got in the port industry. All of a sudden, the Port does not have control of the cargo.

With the single bill of lading, the customer for Port services is no longer the shipper or the consignee—the customer for those services is mainly the steamship line. And the steamship line may not be as concerned about Port costs as about ship utilization or something like that. So it is difficult for a Port to offer price incentives and be sure that this will attract more cargo, because, unlike the shipper or consignee, the steamship line has a whole lot of other things that are also important on its agenda and may not care very much about Port costs. For example, the costs of buying and running the ships are much higher than the Port costs.

### What is the problem confronting Ports as a result of contract rates on railroads?

This is a big problem. Earlier, one of the ways Ports stayed viable was to ensure that the difference in rail rates from one Port to the next was negligible. Preferably, of course, you had a slightly lower rate than somebody else did. And that was in the old days when you could open up a tariff book and figure out what it would cost to move a ton of something from place A to place B, and so you could find out how competitive you were.

Since deregulation, the Staggers Act, the railroads can now provide what they call a "contractual rate agreement," which basically says that they can take into consideration time/volume in their rates. And all they have to do is file those agreements with the Interstate Commerce Commission and they go into effect. The problem is that the information that is actually made public reflects only the subject of the contract and not its dollar value. In other words, it is impossible now for a Port to find out what contract rate is being charged to a shipper to ship cargo from a port to a distribution center, for example, unless the railroad wants to tell—and in most cases, the railroad doesn't. This information is for the most part closed to the Port, so it can lose business and not know what to do in order to get that business back. It has caused a lot of trouble. When there were equity rates, steamship lines tended to be willing to share some cargo among ports like Seattle, Oakland, or Los Angeles. But if a steamship line is shipping "X" amount of goods, and the railroad offers a better rate to ship *all* the goods on their line from Chicago to Oakland, all of a sudden, Seattle dries up and Los Angeles dries up and they don't know how to fight it because they can't find out what the rate is.

**Why did the Port industry let this happen?**

Most Ports didn't realize it was going to happen.

**Weren't they in D.C. lobbying and fighting?**

No. From what I can gather, it was only after passage of the Staggers Act that most Ports realized they were getting their pockets picked.

Most of the Port industry figured that what was happening wouldn't affect them very much, but as it turned out, it was the combined effect of the changes that eventually had a strong impact. The single bill of lading puts the control of the cargo into the hands of the carrier and takes it out of the hands of the consignee/shipper. This creates gigantic volumes that the railroads then go after with time/volume contracts. And the Ports didn't initially see how dramatically those two changes would combine to effectively nullify their ability to market on a selective basis.

**Are the Ports out there pursuing business with steamship lines and railroads then?**

Sure they are. But you must recognize that both the steamship lines and the railroads have their own agendas. The railroads are much more interested in getting long-haul kinds of traffic, so you have a set of economies that are completely apart from the agenda of the Port, and out of control of the Port. The steamship lines are more interested in vessel deployment and trade volumes, and Ports have a hard time influencing these factors.

Another interesting situation is that a Port may be fortunate enough to be served by a very aggressive railroad—then it gets a lot of cargo. If a Port happens to be served by a very unaggressive railroad, it is just out of luck, because this limits what it can do to influence the throughput of cargo.

**Ports appear to be at the mercy of the railroads and shipping lines, if Port rate changes have limited effect on business. How does this affect their mandate?**

It really doesn't have a lot of effect on their mandate. What it does affect is their marketing effort. Ports have had to become more aggressive and professional in their marketing approach. They make every effort to learn as much as they possibly can about a shipper prior to calling on him—how much he ships, where he ships to, on which line he ships, and so on. In order to entice a shipper to use their port, Ports often make joint sales calls with steamship line and/or railroad representatives. Many marketing organizations now use vast computerized data base material that is available from not only private sources, but also public sources in order to do this prescreening.

Admittedly, Port charges have little effect on the various kinds of decisions that the railroads and shipping lines will make. However, rates are not everything. There are various types of services and desired functions that a Port can provide that will make it more attractive. Also, Ports can work with their lessees to adjust various lease provisions in order to encourage a volume increase in business. This is done by providing incentives within the leasing rate structure to move additional cargoes through a terminal or a port. One way to do this is to provide a discount on the wharfage and dockage rates if a specific line or shipper will be willing to guarantee a certain minimum of cargo during a year or perhaps over a five-year period.

**If there are going to be big winners and big losers among Ports in cargo handling, what do you predict will happen to the losers? Should "waterfront park development" agencies replace some public port agencies? Should some Ports go out of business?**

I don't think that we're going to see a lot of the larger Ports disappear. What they may end up doing is becoming more specialized and catering to one or two steamship lines that would use their facilities as a "load center." When the load center concept came out, a lot of people predicted that there would be only a few Ports that would survive. In effect, they predicted that all the steamship lines would choose, let's say, a half-dozen major Ports in the United States to be load centers, and that all the other Ports would shrink on the vine. As it turns out, that hasn't been true. What has happened is that various Ports have been selected by various steamship lines to be load center ports. Some lines will use one Port, some will use another. So there are a number of prospering Ports, even though they are all involved primarily in servicing containerized general cargo liner trades. The Ports that have not been able to attract steamship line load centers have become what is termed a feeder port, which funnels containers into a load center port or a Port that is served by a major liner container service.

The feeder port, such as the Port of Boston, provides a way for containers to go from a specific region, the New England area, to Boston for further transshipment. They are then marshaled in a Boston container terminal and loaded onto a barge, which is sent down to the load center in the New York/New Jersey area where line haul vessels take those containers to Europe or the Far East or wherever. At the same time, the barges pick up containers in NY/NJ that are destined for Boston or New England and carry them back up to Boston for distribution.

These feeder services are just coming on line for many Ports because they are now willing to accept that they probably never will be major lead center ports and an auxiliary role as a feeder port can be very viable. Ports that are neither load centers nor feeders can still participate in maritime trade by becoming special commodity ports for such things as automobiles, lumber products, logs, chips, and various bulk commodities—ore, grain, petroleum, coal, and so on. Ports that can't do even this may have to abandon their marine facilities and become economic development engines for their communities by providing industrial development lands or industrial parks. Some Ports have done well by providing warehousing and storage facilities. A small group of Ports may be able to shift into various kinds of recreational projects, such as marinas and tourist-related facilities, and still remain viable.

An extremely small number of Ports that cannot involve themselves either in load centers, feeder ports, specialty type bulk ports, industrial development ports, or recreational ports will probably become inactive or disband. But I can't see too many Ports doing this. The idea of a Port becoming a waterfront park development agency is probably pretty repugnant to the majority of port commissioners and port managers, and frankly, that would be a last resort for the vast majority of Ports that we have today. Most Ports aren't funded well enough by the taxpayers to do that type of service work. An autonomous park district or a city or county park department would be more effective. I really don't think that there are going to be very many Ports that are unable to find some type of viable function, but those that can't compete will probably become "paper ports"—they will exist on the books, but will perform no Port functions.

# Part IV

# CHANGING TIMES

# Chapter 11

# STRATEGIC PLANNING:

## Defining Port Values

*Daniel Jack Chasan and Thomas J. Dowd**

---

In the "good old days," the period of *On the Waterfront* and earlier, a Port could get by without much planning. It could just stick a pier out into the water, build a long transit shed on it for storing cargo, and invite business. The cargo would come in scattered bits and pieces to be unloaded in slings by gangs of longshoremen. All goods shipped to the people living inland would inevitably come in through the nearest port. Products shipped to other regions or countries by neighboring industries would inevitably go out through the nearest port. Most Ports, of course, did not just sit back and wait for business. They might have to dredge channels—in which case they could call on the U.S. Army Corps of Engineers, which would be glad to do the job with taxpayers' money. They might have to go out and sell themselves to shippers—in which case they might feel compelled to part with a case or two of whiskey, just to get someone's attention. Mostly, though, they were free to pounce—or try to pounce—on any opportunities that came along. Sometimes they got ready for opportunities that never arrived. Ports all along the West Coast lengthened piers and dredged channels to get ready for the flood of business that would follow the opening of the Panama Canal, then found that the opening changed their lives very little. Most of the time, however, the price of guessing wrong was low. People always made plans of one kind or another, but there was little incentive to do it systematically.

*Researched by Tom Dowd and Dan Chasan, written by Dan Chasan. Thanks to the following interviewees at the Port of Portland, Oregon: Lloyd Anderson, Pam Moore, Del Pearson, Glenn Vancelow, Denny West. Partial financial support for this chapter came from Washington Sea Grant, under grant #NA84AA-D-00011.

## THE PLANNING IMPERATIVE

Since the 1960s things have changed. The slings of cargo and the old transit sheds have been replaced by aluminum boxes and blacktopped storage yards. The expectation that a Port will inevitably handle all the cargo to and from its "naturally tributary hinterland" has been shattered by the load center port—the single destination to or from which containers can easily be transported overland, trucked, or carried by rail to an entire region. Many Ports are now bypassed by cargoes that they once would have handled as a kind of geographical right.

There is a premium on speed. No longer are ships sent cruising leisurely down the coast to pick up a little cargo here and a little cargo there. They are turned around as quickly as possible and sent right back out to sea. The operating expenses of big modern ships and the cost of the capital needed to build them are so high that ships must be working most of the time. An old World-War-II-surplus Liberty ship carrying breakbulk cargo could be tied up in port for less than $500 a day. A big cellular container ship loses maybe $50,000 a day in port. The economics have changed.

The economics of developing port facilities to accommodate big ships have changed, too. In the late 1960s, when containerization was new, you could plan and build a new container terminal in a matter of months. Now it takes years. A 50-acre terminal could be built in 1968 for less than $10 million. Now it costs over $50 million. It is no longer cheap to be wrong; and it is no longer financially feasible to do everything port managers can think of doing.

There are other limitations as well. A Port has a finite amount of land: container terminals and auto storage yards require a lot of acreage. They may have to compete for space with other commercial uses (some Port land in downtown areas is more valuable as commercial real estate than as storage or cargo handling areas) and with uses that are not at all commercial. Ports no longer have carte blanche to occupy or alter the shorelines. Two decades ago environmental considerations did not even slow down the dredges and bulldozers. Now a Port may be expected to preserve some sensitive shoreline areas, and it certainly will be expected to provide "mitigation" for its destruction or alteration of others. If it fills in the eelgrass beds here, a Port may have to preserve others in perpetuity. Or it may have to provide money to preserve habitat someplace else (see Chapter 12).

In exchange for filling a 190-acre site in the early 1980s, the Port of Los Angeles had to pay $362,000 to the California Coastal Commission, and the Port's director of environmental planning, Cal Hurst, remarked, "There will probably never again be that inexpensive a compensation. The ante has gone up." Currently, in exchange for the dredging and filling needed to build a new marine oil terminal expected to cost $100 to $150 million, the Port must restore a "degraded" lagoon some 100 miles away. The lagoon lost its connection to the ocean and became brackish, and the Port must dredge its mouth so that it will be open to the tides. The cost is estimated at $15 to $20 million. In the late 1970s, when Sohio was contemplating a huge terminal in Long Beach to handle Alaskan oil, the expense for mitigating air emissions alone was estimated at $85 million. The mitigation would have consisted of placing scrubbers in the stacks of power plants elsewhere in Southern California. Mitigation expenses of that magnitude "throw a whole new wrinkle into port planning," Hurst says.

Traditionally, of course, virtually no one paid a great deal of attention to environmental impacts. Businesses just watched the bottom line. Often Ports did not even do that. They did try to attract cargoes and shippers, but they tended to assume that a busy Port was a prosperous one. If they invested a lot of capital and cut their rates to the bone in order to get business, they did not lose much sleep over the implication that they were basically out there *buying* business, and that the return on their investment was actually negative.

Plenty of Ports still handle cargo at a loss. Generally, their accounting systems are not sophisticated enough to show whether a specific operation or site is really in the black or in the red. But clearly, if, for example, they charge a flat rate that does not cover the capital costs of a terminal and does not enable them to reap greater benefits from an increased throughput of cargo, they are not making money. In order to attract cargo, New Orleans and a number of other Ports lowered the cost of the first level of throughput in the hope that this kind of loss-leader approach would eventually lure more traffic on which higher rates could be charged. Sometimes this kind of pricing works. Sometimes it does not. A Port should probably reap a return on its investment of at least 10 percent—very few actually do that well. During the years from 1981 through 1985, the Ports of Northern California had an operating return on investment (that is, the ratio of their net operating income to their net capital assets) of 8.7 percent. The Ports of Puget Sound did substantially worse, 6.4 percent. The Ports of Southern California—Los Angeles and Long Beach, which have a monopoly on serving a huge population, earned a respectable but hardly spectacular 12.6 percent.

In recent years, increased public scrutiny has forced Ports to look more critically at their return on investment and to confront a Port's built-in conflict between its own bottom line and its responsibility to provide public benefits that may not appear in that bottom line. Ports are virtually always expected to promote economic development in their communities. They are often expected to provide waterfront parks and other amenities, too. Some of their activities have to subsidize others, which means that some have to be very healthy indeed (see Chapter 10).

To cope with conflicting demands, high capital costs, a world in which hinterlands are no longer captive, and an economic environment in which cargo does not necessarily equal profit, a Port can no longer rely on serendipity. What is a Port to do? To some degree and at some time during the past decade, virtually every major U.S. Port has answered "strategic planning." The Port Authority of New York/New Jersey started strategic planning in the late 1970s. The Port of Los Angeles came out with its first strategic plan in August 1986. Some Ports are still feeling their way through the planning process. Many others have tried it and, for one reason or another, given it up.

In countries other than the United States, the national government develops planning guidelines for all the Ports. In many Far Eastern nations, the financing of any Port project goes through the national government, and so a Port builds what the government wants it to build. In Japan the Ports require government financing, and the Ministry of Transport must approve all their plans. The Port of Singapore Authority is actually an arm of the national government. Ports in Canada are more independent, but individual Port projects must comply with the National Ports Policy. If a Port wants to do something that is not in the national plans, it must persuade the federal government to alter its policy.

The United States is perhaps the only major shipping nation that lacks some kind of national port planning. The U.S. Constitution, which says that "no preference shall be given by any regulation of commerce or revenue to the ports of one state over those of another," seems to make centralized planning impossible. Duplication of facilities is expected. Rate cutting (often below the break-even point) to win cargo away from neighboring ports has been common. Del Pearson, senior manager for business planning at the Port of Portland, Oregon, observes:

> I guess people in communities see port authorities as a way of stimulating economic growth, and that's a wonderful political motive, but in the Pacific Northwest right now there are twelve ports within about a 250-mile stretch. [This is like] building four gas stations at an intersection even though there is a limited amount of traffic. It has created an intensely competitive situation that I personally feel gets a little bit out of control.

Ports have planned for the future, but seldom systematically. At best, the glass through which Ports have peered toward the future has tended to have a short focus and a narrow field of vision. In the late 1970s the Port of Boston built a 31-acre container terminal that did not begin operations for two and a half years after completion because of trouble over a union labor contract. Some of the shipping lines that served Boston were unhappy with the situation there and approached the Port of Providence, Rhode Island, with a proposition. If Providence would provide a place for them to unload containers, they would take their business to Rhode Island. The Port of Providence jumped at the opportunity. It bought two container cranes and developed an existing pier to fit the carriers' needs. But it made the long-term capital investments without getting any long-term commitments. Soon after the Boston terminal opened, the shipping lines took their business back there. The Providence container terminal has stood virtually unused ever since.

## MATCHING STRENGTHS WITH OPPORTUNITIES

Strategic planning is supposed to prevent situations like that of the Port of Providence. Basically, it involves matching an institution's internal strengths and weaknesses to the full range of its external threats and opportunities. In the words of an ancient Chinese military sage, "Know self. Know enemy. One hundred battles. One hundred victories."

The first step is simply to prepare a "mission statement" to answer the basic questions: What is our business and where are we going? The next step is to analyze the external environment. Then one sets goals, puts them in order of priority, and devises strategies for reaching them. Once the strategies have been developed, one must allocate the people and resources needed to carry them out, and figure out how progress, or the lack of it, can be measured. In principle, this is all pretty simple. But it can be done badly or well. And once a plan has been created, it can be followed or ignored.

Strategic planning became a hot item—and a classical buzzword—among big American corporations in the late 1960s and early 1970s. "The famed Boston Consulting Group . . . is widely considered the parent of strategic planning . . . " a *BusinessWeek* cover story observed in 1984. "Another important force was General Electric Company, which, along with consultant McKinsey and Company, added . . . seminal concepts. These enable diversified companies to focus more intensely on the market outlooks and competitive factors for each of their operations."

The reason for making strategic planning a formal process, wrote Frederick Gluck, Stephen Kaufman, and A. Steven Walleck in the *Journal of Business Strategy* in 1982, is that "most large enterprises . . . are too complex to be managed with only an implicit strategy. The number of products and markets served, the sophistication of technology, and the complex economic systems involved far exceed most managers' knowledge base or capacity for intuitive decision making. The shoe usually pinches first in financial planning." A company responds by making forecasts based on prior experience, but, as the article continues:

> . . . Sooner or later all extrapolative models fail. Worse, they mislead management into a false sense of security; they imply that the future is predictable from the past. In fact, many companies find that the changes likely to have the greatest impact on their fortunes do not result from continuous and predictable changes, but from unexpected environmental shifts that are only obvious after the fact. . . .
> At this point in the evolution of the planning processes, a creative spark stirs the imagination of the planners and the first true strategic planning is born. Planners suddenly realize that their responsibility is not so much to chart the future as it is to lay out for management decision the key issues that face the company . . . it is often more important to raise key issues than it is to forecast future trends to second decimal place accuracy . . . Frustrated with volumes of irrelevant paper work . . . strategic planners deemphasize the importance of forecasting and begin to focus their attention on understanding the basic marketplace phenomena that are driving the pace of change.

As a good example, Gluck, Kaufman, and Walleck told the following story:

> A strategy team for a heavy equipment manufacturer spent nine man-months reverse engineering [their] competitor's product, reconstructing his manufacturing facilities on

> paper, and estimating his manufacturing cost. This effort revealed that design improvements had produced a substantially lower-cost product for the competition and that no achievable amount of cost reduction could manufacture the company's product as cheaply. Strategic advantage, therefore, was not likely to be won in a price-oriented bid for market share . . . [However] it was discovered that lower maintenance and fuel costs [gave] the company's customers a distinct life-cycle cost benefit. Their sales force was therefore trained to emphasize life-cycle cost advantages. The result was a 30 percent gain in market share and doubling of net profit in three years.

For every inspiring success story, though, there seems to be at least one failure. "To put it bluntly," the *BusinessWeek* cover story declares, "it has become obvious that very few strategies seem to succeed." Indeed, a reassessment of thirty-three strategies described in *BusinessWeek* in 1979 and 1980 found that nineteen failed, ran into trouble, or were abandoned, whereas only fourteen could be deemed successful. The main point of the *BusinessWeek* story was that U.S. corporations had become disillusioned, not so much with strategic planning as with strategic planners. Part of the problem was that strategic planning performed by an "isolated bureaucracy" of planners tended to be too much of an ivory-tower operation. Strategies based on assumptions about the real world, with no real understanding, were not corporate passports to success. For example, according to *BusinessWeek,* at General Electric:

> . . . when data indicated that houses and families were shrinking . . . planners concluded that smaller appliances were the wave of the future. Because planners had little contact with homebuilders or retailers, they did not realize that the kitchen and bathroom were the two rooms that were not shrinking and that working women wanted big refrigerators to cut down on trips to the supermarket. Moreover, top management, which also lacked contact with the market, did not perceive that the planners' data failed to reveal the true story. The result: GE wasted a lot of time developing smaller appliances.

It has also turned out that groups of planners, working more or less on their own, produce plans that no one else wants to use. The psychology is simple: if a consultant or a corporate planning department comes up with a strategic plan, the rest of the organization—the people who actually have to make it work—view it as *their* plan, not *our* plan, and leave it on the shelf.

As a rule of thumb, the more a plan is perceived as someone else's handiwork, the less likely it is to be implemented. Professional planners can help the process along, but they should never be allowed to dominate it. The people who will carry out a plan should have a crucial role in putting it together. The mission statement, goals, and strategies should all have the active support of top management; and everyone in the organization should be familiar with the basic values and strategies.

## REINVENTING THE WHEEL FOR PORTS

Ports, coming to strategic planning later than private industry, have by and large reinvented the wheel for themselves. The port department of the Port Authority of

New York/New Jersey, for example, had its first plan put together by a corporate office of strategic planning. People throughout the port department regarded it largely as a planners' document and did not pay it much mind. The port department had to go through the planning process again, involving more members of its staff, before it had a plan that people were willing to follow. The New York planners also found that strategic planning runs across the grain of the way in which generations of Port officials have preferred to operate. As E. G. McKeever, manager of the NY/NJ Authority's office of strategic planning, comments:

> Any organization should certainly examine its mission, objectives, and goals as it is confronted by changes in the external environment. Like many activities in planning, however, this is easier said than done. The difficulty often lies in the general reluctance of management to explicitly define the organizational mission, objectives, and goals because of a belief that by doing so, the ability to take advantage of certain unforeseen opportunities may be foreclosed (McKeever 1983).

The NY/NJ Port Authority did not give up on the basic idea of strategic planning, however. "How the Port Authority of New York/New Jersey got started in the strategic planning process is rooted in some recent history," explained McKeever. He recounts:

> After a period of severe economic decline and fiscal hardship in the midseventies in the New York-northern New Jersey region, the Port Authority, under new executive leadership, initiated the Committee on the Future in 1978. This was an intensive yearlong examination of trends and needs in the New York-New Jersey region and an assessment of where the Authority, in partnership with public and private sector interests, might have a 'comparative advantage' to intervene effectively with constructive development programs. . . . The potential areas for Port Authority involvement coming out of this effort included non-traditional roles such as the critical need for infrastructure maintenance and renewal; commercial mixed-use waterfront redevelopments and their job opportunities; the need for energy diversification with its important national regional benefits; and development of a program to create new urban industrial jobs. These new ventures challenged management to integrate them with the Port Authority's traditional transportation, port, and trade activities, and to develop a broad program that would better contribute to the region's economic health (McKeever 1983).

If the Port Authority of New York/New Jersey was drawn—or pushed—into strategic planning by the economic problems of its region, the Port of Los Angeles was put in the same spot by an embarrassing abundance of opportunities and demands. As the introduction to the Port's strategic plan explains:

> The development of the Port of Los Angeles Strategic Plan occurred at a time of potentially great change, both externally and internally. . . .
> For example, substantial increases in cargo throughput are anticipated at the Port of Los Angeles during the next 20 to 30 years with the emergence of the Pacific Basin

> as the major trading area in the world. At the same time, shipping line overcapacity and considerable competition for this Transpacific trade are anticipated to continue for at least the balance of this decade. . . .
> In addition, dredging, landfill and energy relocation projects on a scale greater than any development previously undertaken by the Port of Los Angeles are planned to take place within the next five to ten years. . . .
> Add to this the implications of the many current changes in U.S. shipping and tax laws and the changing business practices of domestic and foreign shipping companies, and you have some idea of the environment the Port must operate in and why strategic planning and strategic management are so important (Port of Los Angeles 1986).

The Port of Los Angeles came up with a list of 15 goals, some tied to specific projects and some extremely general. On a very concrete level, the Port resolved to "create a remote landfill and deep-water access south of Terminal Island by 1992 or sooner" and to "develop a plan to locate hazardous liquid bulk cargo facilities that are inconsistent with the Port's Master Plan to a new landfill site south of Terminal Island when it is constructed." It also decided to "increase annual revenue derived from the rental of Port property to partially fund the Capital Improvement Program and support debt service obligations" and "develop a consistent rate of return policy that will be utilized in negotiating all Port agreements." At the same time, on a more abstract plane, it resolved to "increase Department's credibility with employees, tenants, customers, labor organizations and the general public," and to "improve the quality and frequency of management communication and develop an awareness of the need for mutual support."

Perhaps the most striking part of the strategic plan was the mission statement that declared, "the Port is a SERVICE-ORIENTED ORGANIZATION." The *Daily Shipping News* reported that "Under the plan, the Los Angeles Harbor Commission decided the Port is not in business to maximize profits but rather to be a service-oriented organization." The mission statement itself said that "the extent of the Port's profit motive is to earn sufficient revenues to be self-supporting and maintain a prudent level of reserves for future development." The *Daily Shipping News* also noted that "for the first time, the Port has stated in writing that it will not just develop cargo terminals but also will promote shipyards, marinas, and other maritime-related activities."

## THE EVOLUTION OF PLANNING AT THE PORT OF PORTLAND, OREGON

Because so many U.S. Ports have decided that strategic planning is the way to go, and because the process carries no guarantee of success, it is worth looking at a U.S. Port that has done it well. This is not the biggest Port, nor the most successful, but a medium-size, deeply troubled Port with limited prospects. Its claim to fame is not that strategic planning has worked miracles, but that, probably more than any other Port, it has successfully combined the precepts of its strategic plan with the day-to-day functioning of the entire organization. It is the Port of Portland, Oregon.†

Portland lies some 100 miles from the Pacific Ocean, but, served by ships that make their way up the Columbia and Willamette rivers, it has always been a port. As ships have increased in size, and as speed has increased in importance, the distance from the ocean and the narrowness of the channel beyond the Columbia River have become more of a barrier to trade. The Port owns a great deal of land, but its developed terminal sites are widely scattered. The two oldest, terminals 1 and 2, are relatively small facilities on the downtown waterfront along the west bank of the Willamette River. The first serves breakbulk cargo and anything that requires a heavy crane. The second handles a mixture of forest products, containers, and perishable items. Farther out toward the mouth of the river, terminal 4 has a huge grain elevator, and accommodates a variety of other products, including steel that comes in from the Far East, raw logs, and automobiles. Still farther down the river, terminal 5 also handles grain. Finally, terminal 6, the newest site, which is not on the Willamette at all but on the Columbia, handles containers and automobiles (see Fig. 11.1). The Port also owns the Portland Ship Repair Yard, which includes the biggest drydock in the Northwest, on Swan Island in the Willamette. And it operates the Portland International Airport.

The airport is very successful, but legally its revenues cannot be used to subsidize other Port operations; they have to be plowed back into the airport. Therefore the marine operations have to stand alone. They have had their problems. Grain has been a mainstay of the Port for a century, but U.S. wheat exports have been depressed for years. The shipyard makes a good deal of money "laying up" ships that are being taken out of service for extended periods, and it has the only drydock north of California that can serve very large vessels. But U.S. shipbuilding is all but defunct, and high labor costs in the Northwest tend to drive even government business elsewhere. The people passed a bond issue to enable the Port to build a drydock and expand and modernize the shipyard, with the assumption that the revenues would pay off the bonds. That looks at this point like rather wishful thinking.

The container terminal on the Columbia River makes money, but its prospects are strictly limited. The center of the northwestern container trade is not Portland, it is Puget Sound. Ships that take on cargo in Seattle or Tacoma may or may not leave room for containers from Portland. They may or may not call at Portland at all. A lot of cargo going to or from locations in Oregon can go through Puget Sound as cheaply as through Portland. The shipping lines, competing for business, are willing to "eat the dray"—that is, absorb the additional cost of trucking the containers

†The Portland, Oregon case study was researched and written in the summer and fall of 1986, and the strategic planning process at the Port continues to evolve.

**Figure 11.1** Terminal 6 is the newest terminal at the Port of Portland, Oregon, designed to handle containers and automobiles. Mt. Hood is seen in the distance. (Courtesy: Port of Portland, Jim Douglas)

the extra distance. Consequently, many trucks hauling Oregon containers drive right past Portland's terminal 6 on their way north to Puget Sound.

In the absence of strategic planning, Portland missed the boat on containers. During the early and mid-1960s, the Port of Portland was far more successful than the Puget Sound ports. Ships carried the bulk of the wheat crop from eastern Oregon and eastern Washington to the Far East. Many of the same ships came back loaded with Japanese steel. It was a neat little closed circle. Seattle could not break in. Besides, Seattle was short on backup storage space for cargo behind its piers. Portland was holding most of the cards.

In the mid-1960s, though, Seattle, having nothing to lose, gambled on the future of containerization by building container terminals in the hope of attracting the steamship lines. Portland, with little incentive to change, brought a lawsuit before the Federal Maritime Commission (FMC) asking that the Japanese lines be forbidden from taking Portland's container business elsewhere, based on the "naturally tributary" rule, an early FMC rule now virtually abandoned, recognizing a port's right to handle cargo that is "naturally tributary" to it (see Chapter 5). But Portland was only partially successful. The end of the story was predictable. Seattle made itself the premier container port in the Northwest, and the business eventually left Portland for the regional load center in Puget Sound.

Faced with this less-than-promising competitive environment, and, although no one knew it yet, on the brink of a serious recession, the Port started in 1980 to try looking ahead twenty years. In an attempt to budget for eighteen-month periods instead of twelve, the Port brought in a consultant who found that given the Port's available data, it was not possible to look ahead even three to five years! Obviously, the Port needed better planning tools.

The Port really started down the road toward strategic planning, according to Denny West, deputy executive director, when Port managers decided to explore data processing capabilities:

> We brought in a consulting firm that requested each of our departments to make a business case for the things they wanted to do with the computer, and we discovered people couldn't do that very well. We then brought a consultant out from Harvard Business School [who] did a workshop on strategic planning. It was all stars and dogs and cows and chickens . . . the generic idea was to determine what's going on in the world and what you have to do to be competitive. Once you make the commitment to try to answer questions about your competitive environment in an intelligent fashion, you just keep working away at it. One of the first things you realize is that maybe you're not very good at answering the questions . . . and it ends up being an iterative process. We focused for a couple of years on just trying to analyze, competitively, the industries we're in.

The Port's management decided to make the marine department the "guinea pig" for strategic planning. Del Pearson, the department manager, was primarily responsible for the effort. According to Pearson:

> They asked if anybody knew anything about strategic planning and when I said that I had read a *Reader's Digest* article about it, they kind of put me in charge . . . My feeling is that our first attempt was not really successful. Because it was a first-time effort, we were a little bit too concerned with doing things by the book, and a little too concerned with how we were going to write up a document instead of focusing on the thinking that went into it. I would have to say that our first attempt did not include what I would call a whole lot of strategic thinking . . . As we went through subsequent revisions over the next two or three years, a lot more strategy and analysis went into the business plan than there was in the original document.

Although the Port is still learning about the competitive environment, Denny West observes:

> We've built up to the point where we have confidence in our analysis. Now the question is, How do we manage the results? All of a sudden you realize what you need to do to compete [and then you must ask], Does your capital program fit with it? Does your employee development program fit with it? Have you got a [data processing] program that fits? Have you got values that fit? What about your labor strategy? What are the priorities? You can't do everything at once. And, having figured out where we ought to be going, [we need to look at] how we're going to get there . . . And about the time we get that figured out, where we need to go will change, and we'll have to do it all over again . . . I tend to visualize it like an atom—where everything is always in motion.

That realization is crucial. And it is one of Portland's strengths. Just as planning cannot be left to planners, so, too, it cannot be considered a one-shot process, or the plan will end up gathering dust.

A strategic planning process should help a Port figure out how to find a niche in the marketplace, but it should also help a Port determine what underlying assumptions it should make to set its goals and to guide its organizational behavior as it tries to reach those goals. Glenn Vancelow, Portland's research manager, comments:

> First we had the strategic planning process and that led to a distinction between business objectives and public mission objectives . . . We felt that one of our biggest roles as an organization and as individuals in the organization was to balance those two. The two are definitely separate, [and] sometimes opposing objectives in any project or long-term process. We're not just going for profit; and we're not just going for public good. We have to weigh each [project and operation]—which ones do we need to make profitable? Do we sacrifice profit, perhaps, or have some degree of subsidy to get a double benefit?

Questions like that led to the concept of "values-based management." The Port of Portland decided that it should base everything on six fundamental "values": build a strong economy, maintain public confidence, assure customer satisfaction, demand competitive financial performance, insist on excellent leadership and teamwork, and recognize that employees are the Port's key resource.

Executive Director Lloyd Anderson comments on what happened when the Port set about identifying these basic assumptions:

> I think there were quite a number of surprises, [We found that] there are values that are not shared. And some of the middle management people who had controlled their particular departments were seriously threatened by [the new requirement to clearly set out] a value system so that employees two, three, or five layers down suddenly knew that the most important thing to this institution is how we treat our customers. When that kind of value is out on the table—and you have a person who has for years been mistreating a customer—you have a situation where someone who works for this person can say "That doesn't represent the values you've been talking about around this institution. How come you're treating people like that?" And this is applicable to the police at the airport, the marine security, and a lot of other places where people began to question whether what we were in fact doing corresponded to what we claimed our values to be. This threatened and challenged a lot that was going on. And in the process of doing that, the change elicited huge support.

The Port's values coordinator, Pam Moore, acts as an information conduit and is an ombudsman for staff. She monitors staff and management perception of how well the values are used in the decision-making process. Denny West summarizes, "The process of finding words to describe and define values that everyone can agree to doesn't really end. It accomplishes more than just shaking up middle management—port officials believe that it makes decision-making easier, enabling people to focus more sharply on the choices at hand."

As Glenn Vancelow explains it:

> In order to discuss a specific project or issue, or compare a variety of issues, you define a number of 'values.' A strong economy is our number-one value. Public confidence is another, and then we have four others. We can then look at each project on the basis of these values. It helps to resolve the issues, as opposed to everybody just sort of stating their opinion or agreeing with the last guy. It helps to bring up different perspectives on the same issue.

A sharper decision-making process, however, cannot make opportunities materialize out of thin air. And Portland has not, by a long shot, figured out how to overcome the handicaps of geography, world markets, and effects of decisions made some twenty years ago. It has managed to resist the temptation of getting into projects it is not equipped to handle. It said "no thanks" to managing the Portland area's bus system and its new light rail system. Despite Executive Director Anderson's personal interest, it decided not to get involved in developing a new Portland convention center.

Within the Port's traditional lines of work, the very process of planning has led to a more logical organization of the marine operation. Previously, the whole marine operation had been treated as a single unit. When strategic planning started, Del Pearson explains:

> We saw some merit in doing planning in . . . three separate areas. We separated Far East container trade as one group, and we sorted out the second group as the lesser [uncontainerized] carriers. The third group was the charter business that we do. We saw some natural breakdowns in the business, and we started that first planning process by looking at those [areas] as three separate activities. As we went along through that, it was finally decided to restructure the department along those three lines.

Reorganization was not the only change that came up as the marine department went through that first strategic planning process. The department, under Del Pearson, also found itself rethinking the way it went after business:

> A major change in our strategic thinking took place as a result of the first planning session. This was the most important thing that happened. Prior to that first session, cargo marketing involved a group of salespeople going out and talking to cargo accounts, whether they were in Chicago, New York, Los Angeles, Portland, or wherever. The steamship marketing activity [was] separate—there were only one or two people assigned to steamship marketing.

The biggest single change that the Port made was to combine cargo marketing and steamship marketing:

> When we created, for example, the Far East container group, they became responsible for the cargo marketing *and* the steamship marketing . . . the weakness that we saw in the previous setup was the lack of synchronization. What we were seeing in the market was increased cargo control by the steamship carriers, [so it was illogical to have] these as separate activities within the department. The department was reorganized along functional lines . . . so that the business unit that had the responsibility for the container area, for example, had all account coverage, all steamship coverage, all pricing procedures, all inland transportation decisions centered in that one group.

The result, according to Pearson, was not merely new job descriptions:

> The biggest single impact was the increase of the percentage of the allocation of resources that went into the carrier side of the equation as opposed to the cargo marketing side of the equation. Much more resources were devoted to trying to attract carriers, and significantly [fewer] resources were [allocated] to going out and calling on cargo.

Rational decisions alone cannot attract prosperity, however, and the Port had to make a painful decision to cut back its staff by thirty-eight people and its budget by $5 million in fiscal year 1986–1987. Although strategic planning could not alter basic competitive facts, it did enable the Port to make the best of a bad situation. Regarding the cutback, Anderson comments:

> We were very careful how we treated the people that were in the institution, including the ones that were asked to leave . . . now they weren't happy, but nevertheless, the fact that we had the toughness to do it and the sensitivity [to be aware of] how we treated those people reflected well on the organization and on the planning process that had enabled us to focus on what was most important.

And Denny West recounts:

> Part of the reason it worked is that we did it in a manner that was consistent with what we said our values were. [Everyone had] a basic amount of information about what the competitive realities are that we face. And so even though nobody was happy about having to go through [the cutback], there was an acceptance of the fact that it was not random, that it was done in a rational and humane way—and that we needed to do it. As Lloyd Anderson has said, I think we came out of the process with morale stronger in some cases than before. And that's a real payoff.

If strategic planning can help a Port come through major layoffs with high morale, it must have some merit. That may sound like faint praise, but, in fact, strategic planning is not a kind of Aladdin's lamp that will make a Port's wishes come true. Ports seldom lack opportunities. And they seldom lack people ready to give port commissioners a sales pitch for just about every one of them. The trick is in choosing the right opportunities and turning down the rest. Strategic planning provides a tool for distinguishing between solid opportunities and latent disasters, and a rationale for saying no. It cannot transform geography or demographics, but it can help a Port to make the most of them. And as the Port of Portland discovered, it can help an organization develop the internal strength to survive difficult times.

## References

*BusinessWeek*. 1984. The new breed of strategic planner. September 17.

Gluck, Frederick, Stephen Kaufman, and A. Steven Walleck. 1982. The four phases of strategic management. *Journal of Business Strategy*, Winter.

McKeever, E. G., Carl S. Selinger, and Patricia C. Comer. 1983. Implementing strategic planning at a diversified public agency. *Transitions*. Winter.

Port of Los Angeles. 1986. *Strategic plan*.

Port of Portland. 1985. *Values*, Portland, Oregon, April.

# Chapter 12

# MITIGATION:

## Compensating the Environment for Unavoidable Harm

*Ann E. Wessel and Marc J. Hershman**

---

> We like the little bird and we like to protect it and see it do well, but we think it would do much better somewhere else rather than in the middle of our industrial-commercial port with trucks and cars and ships and planes whizzing back and forth . . .
>
> Michael Ferrall, Port of Los Angeles†

With the advent of major federal and state environmental legislation in the early 1970s, the urban waterfront took on a whole new meaning to the community of users and regulatory agencies that surround it. It suddenly became necessary to consider the needs of economic development within an expanded public interest (Neupert 1982, 19–20). The public was no longer concerned only with commerce and economic prosperity. Amenities such as fish and wildlife, clean air and water, access to the waterfront, and view protection took on greater importance, and the regulatory framework established by the National Environmental Policy Act (NEPA), the Coastal Zone Management Act (CZMA), and other statutes reflected this change (see Chapter 5).

As a prominent developer on the waterfront, the public port authority has no

---

*Researched by Marc Hershman, Ray Maddux, Richard Morse, and Ann Wessel. This chapter is a slightly modified version of A. E. Wessel's Master's thesis (1986). Master's theses of R. Morse and R. Maddux, cited in the References, formed the basis for the case studies of Commencement Bay and the Columbia River estuary, respectively. Professor Hershman chaired the supervisory committee for each of the three Master's theses. Partial financial support was provided by the William and Flora Hewlett Foundation.

†From M. Ferrall's testimony before the Subcommittee on Fisheries and Wildlife Conservation and the Environment of the Committee on Merchant Marine and Fisheries, House of Representatives, 1984 (see Figs. 12.4 and 12.5).

choice but to respond to this reorientation in public policy (Perry 1984). In order to carry on its business of providing facilities and services for waterborne commerce, a Port must frequently propose new construction projects. New development must comply with state or local coastal management programs, state and federal fish and wildlife regulations, and water quality standards. It is no longer possible for Ports to assume that governmental approval of projects will not include special provisions to address environmental concerns. The environmental requirements imposed on a particular project can range from fairly simple design modifications or construction sequencing for preventing adverse impacts, to elaborate and costly habitat construction or restoration projects to compensate for unavoidable impacts to fish and wildlife and their habitat. A consequence of this situation is that Ports are being forced to incorporate environmental considerations into their planning and management functions in order to avoid additional costs or timing delays. In the highly competitive world of commercial shipping, Ports cannot afford to overlook factors that might cause marine trade to go elsewhere.

There are a variety of ways that Ports are taking on this new set of concerns. In the United States, the most comprehensive approach has been that of the state of California. Under the provisions of the California Coastal Act of 1976, the four Southern California Ports are required to prepare a Port master plan that includes, in addition to proposed uses of land and water areas, an estimate of the environmental effects of the planned uses, together with a mitigation plan for unavoidable impacts. Upon approval of the master plan by the California Coastal Commission, the Port is then delegated authority for implementation of the plan. Under this approach, the role of the Port has been expanded to encompass all facets of environmental management (see Chapter 1).

Another example of Ports engaging in activity that goes beyond generating commerce is found in Japan, where the role of the Port is to fulfill the function of a marine transportation terminal and to conduct the overall management of the port area, which may encompass the entire urban shoreline. The management aspects of this role include: long-term port area planning, development of various urban facilities as well as land reclamation, regulation of land and water uses, environmental monitoring and improvement, coastal protection (from storms and such) and survey and research (Inoue 1984). The Japanese example goes far beyond environmental concerns, but it does provide some insight into the types of responsibilities that Ports can assume under a different structure of authority.

For United States Ports, the situation found in California is the exception rather than the rule. Nevertheless, most Ports are slowly taking on elements of this expanded role. One element that seems to be gaining acceptance is the planning and implementation of environmental mitigation projects. Mitigation is a loosely defined management practice that has been developed by environmental planners and regulators as a means of allowing resource exchanges as an alternative to denying government permits. The Council on Environmental Quality (CEQ) regulations (at 40

CFR section 1508.20) define mitigation as avoiding, minimizing, rectifying, reducing, and compensating for project-induced resource losses. However, surveys of current practices in coastal zone development indicate that there is no consistent definition of what mitigation is or any consensus as to how it should be applied (Race and Christie 1982, 317–28).

Regardless of this uncertainty, mitigation is being applied as a management strategy, and projects to offset or compensate for environmental impacts are being implemented. It is easy to see why a system that allows for resource exchanges as an alternative to denying government permits is so attractive to resource managers and development interests: " . . . such a system appears to provide the potential to create a 'win-win' situation in which the developer goes forward and the overall status quo of wetland resources is maintained" (Baker 1984, 2).

When considering the relationship between public ports and environmental mitigation, it is important to keep in mind the type of projects in which Ports are involved. First, Port development is generally recognized to be in the public interest. The direct and indirect benefits to the community from stimulation of trade, increased employment, higher land values, and enhancement of governmental income are the reasons that Ports were established as public enterprises, instead of being left in the hands of private developers (Neupert 1982). Second, the dredging and filling of water areas are common Port construction activities necessary to build or maintain channels and terminals. These types of developments are by their very nature water-dependent and cannot be located away from the shore. In addition, ports are generally located in sheltered embayments that are frequently estuarine and may contain salt marsh, shallow intertidal mudflats, seagrass beds, or other habitats that are valuable to fish and wildlife. As a result there are usually unavoidable impacts to environmental resources associated with development proposals. These features are what make Port projects prime candidates for mitigation from a regulatory standpoint. As a management strategy, mitigation can be thought of as a way to "fit" necessary development into the natural environment (Ashe 1982, 3).

This chapter reviews six case studies of environmental mitigation efforts that have taken place in port and harbor areas.[1] The case studies are not intended to be a comprehensive representation of all types of port-related mitigation projects. These studies were selected based on availability of information, geographic diversity, and differences in approach.

The case studies provide useful examples of the major problems and policy issues that are typical of mitigation projects. In some instances they also provide some indication of the ways to solve these problems. The cases also document the variety of ways that Ports are responding to the challenge of planning and implementing mitigation projects. Not all the examples presented directly involve the port authority. However, they are all connected with the harbor environment and demonstrate current approaches to mitigation in harbor areas.

The case studies are organized into two groups. In the first group of projects, mitigation was proposed in response to a specific project plan. The second group looks at the mechanisms utilized by Ports and other entities to incorporate mitigation into a comprehensive planning process, and does not deal with specific projects, as in the first group. Inferences are drawn from a comparison of these approaches to consider the proper level of public port involvement in mitigation planning and implementation. An argument is made in favor of Ports taking an active role in resource assessment and mitigation in order to decrease permit delays and uncertainty over government authorization.

## PROJECT-SPECIFIC MITIGATION: THREE CASE STUDIES—MIAMI, TACOMA, ASTORIA

### *Biscayne Bay and the Port of Miami, Florida*

The Port of Miami is located on a man-made island in Biscayne Bay in southeast Florida. The Port serves the principal tourist and commerce markets of the Caribbean and Latin America. In the late 1970s the Port undertook planning a major expansion project to accommodate projected increases in passenger and cargo volumes (Perez et al. 1983). The key features of the expansion project included enlargement of the existing Dodge Island location of the Port, the filling and expansion of adjacent Lummus Island, the dredging of a turning basin, and the deepening of channels (Gaby and Langley 1985). The dredging and filling associated with the project would cause the destruction of approximately 6 acres of mangroves, 82 acres of seagrass, and 162 acres of shallow water fishery habitat, for a total of approximately 250 acres impacted.

At the outset of project review, mitigation for these impacts was requested by the state and federal regulatory agencies responsible for environmental protection. Mitigation was also proposed by the environmental groups that were members of a standing environmental review committee that was organized by the Port director at the early stages of project planning. The Port was very concerned about denial or excessive delay of permits because of environmental considerations and, therefore, agreed to mitigate as a means of expediting permit approval.

Subsequently, a mitigation plan was included in the Department of the Army permit issued for the project on October 6, 1980. As stipulated in the permit, the mitigation plan called for planting 251 acres of unvegetated Biscayne Bay bottom with seagrass and at least 5.6 acres of intertidal areas with mangroves within four years from the date of issuance of the permit. In addition the feasibility of filling existing dredge holes to the depth of surrounding bottom to mitigate for destroyed shallow water habitat was to be investigated by the Port.

The mangrove planting project was apparently implemented with little controversy in 1984. The dredge hole filling project was not found to be feasible for economic reasons, and was dropped as a requirement. Of particular interest here is the proposal for large-scale planting of seagrass, the largest project of its kind ever attempted. The 251 acres to be planted correspond to the total number of acres impacted by the dredging and filling. According to Gary Nelson of the National Marine Fisheries Service (NMFS), the original agency proposal to the Port was to transplant the existing 82 acres of seagrass to another area. The Port, however, did not feel there was time to conduct the transplanting operation prior to project construction and accepted the 251-acre project as a sort of compromise. In requiring the Port to plant 251 acres, the U.S. Army Corps of Engineers (COE) permit also stipulated a 70-percent survival rate for the plantings.

Phase I of the seagrass mitigation was initiated in 1982. This first phase was to be experimental, testing different species and planting methods, and a 70-percent survival rate was *expected* for the project. A total of 38 acres was planted, a 25-acre plot in an area where high survival was expected, and thirteen 1-acre test plots. The Phase I planting resulted in only 6 acres of surviving seagrass after a year and a half, and the average survival rate for all plots was only 12 percent. The test program, having been designed with expectations of high survival rates, was inadequate to determine the cause of failure, because failure was not anticipated. As a result there was insufficient data to explain the survival patterns and apply them to future planting. The cost of Phase I: $781,245.

There was no additional planting in 1983, since the results for the 1982 planting were not yet in. In 1984 the Phase II planting covered another 20 acres, bringing the total cost to the Port to over $1.1 million. At this point both the Port and the regulatory agencies began to express an interest in reassessing the entire project. Prior to the 1984 planting, the Port requested that it be given credit for the entire 38 acres planted in Phase I, and that it be credited for 3.34 acres for every acre planted in Phase II (using a denser planting technique) regardless of the survival rate for Phase II. At that time these requests were rejected by the COE.

In the latter part of 1984, after the Phase II planting and an interagency meeting held in Jacksonville, Florida, the COE formally proposed that the Port conduct additional experimental planting in the spring of 1985. The purpose would be to identify sites for future planting. The results of the 1984 and 1985 plantings would go into a reassessment report to be prepared by the Port in the fall of 1985. At that time the COE would work with the port authority and appropriate resource agencies to establish what additional measures would be needed to complete the Port mitigation requirements. It was also proposed that the Port transplant seagrasses from an area on Key Biscayne that was to be smothered in a beach nourishment project.

Finally, in April of 1985, an agreement was signed by all parties that established the Port's commitment to transplant the Key Biscayne seagrasses subject to certain conditions. Among other things, these conditions gave the Port full credit for all acres previously planted (60 acres) and eliminated the survival criteria for the transplanting effort. But the Port agreed to make efforts to maximize the chances of seagrass survival by implementing a quality-assurance program during the transplanting and by careful recipient site selection.

The transplanting effort was conducted in the fall of 1985. Approximately 13 acres of grasses were harvested from Key Biscayne. These grasses were then planted over 63 acres in Biscayne Bay, leaving 118 acres under the Port's obligation. There are more grasses available on Key Biscayne for transplanting; however, the estimated cost of planting the remaining acreage is $1.2 million. This is based on the current cost of $10,000 per acre for planting. When the project was first proposed, the cost was estimated to be less than $2,000 per acre. If the remaining acreage is planted, the total project cost would probably rise to $4 million. It is important to keep in mind that even a cost as high as $4 million is small in relation to the total project cost of $500 million. What has become a significant problem for the Port, however, is budgeting for costs that vary widely from the original estimate.

Another factor affecting continuation of the seagrass planting project is the opinion held by some of the biologists involved that there may not be 251 acres in Biscayne Bay that are suitable for seagrasses. It is becoming apparent that if seagrasses do not exist in certain areas, there may be reasons other than the lack of seedlings that prevent them from growing there.

In July 1986 the Port was preparing a proposal for alternative mitigation to fulfill the rest of their obligation. Given the project delays and planting failures, it is likely that the COE will consider alternatives.

This case study demonstrates the importance of calculating for the possibility of failure when planning a mitigation program. The high expectations for plant survival were based on previous studies. However, no project of this magnitude had ever been undertaken before, and none of the parties involved appears to have fully appreciated the difficulties to be overcome. Whereas it may appear as if the Port was being unfairly required to continue paying the escalating costs of the mitigation attempts, it is important to keep in mind that the Port had already completed its dredging and that 82 acres of seagrass had been destroyed. The resource agencies were obligated to enforce the provisions of the permit in order to prevent continued losses of environmental resources in Biscayne Bay.

The uncertainty associated with the technical aspects of mitigation is a key aspect of the Biscayne Bay case study. Even with the results of previous tests available, there were still enough unforseeable problems to severely hinder the project's success and substantially increase its costs.

## *Commencement Bay and the Port of Tacoma, Washington*

The Port of Tacoma and its surrounding industrial area are located at the mouth of the Puyallup River in Commencement Bay in central Puget Sound. The land and waterways that comprise the Port were created by the incremental dredging and filling of extensive tideflats and wetlands that formerly extended across the eastern portion of the bay. These types of habitats are important sources of food and shelter to migrating juvenile and adult salmon. As a result resource managers try to protect the few remaining acres of shallow intertidal and wetland habitat in the Puyallup River delta. This case study considers two mitigation projects in Commencement Bay. The first involves the filling of a shallow intertidal area by Paxport Mills Inc., a commercial hardwood lumber mill on the northeastern shore of Tacoma's Middle Waterway. The second discusses the filling of a freshwater wetland adjacent to the Puyallup River by the Port of Tacoma. Both projects involved lengthy and sometimes confrontational exchanges between the project proponents and federal, state, and Puyallup Indian Tribe resource managers.

**Paxport Mills**

In July 1979 Paxport Mills applied for a permit required under the Washington State Shoreline Management Act. The application described the proposed project as the construction of a seawall with a total of 22,000 square feet of tideland to be filled along a total length of 540 feet. This permit was issued approximately four months later without controversy and with only minor input from state resource agencies. The Puyallup Indian Tribe requested that additional environmental evaluation be conducted prior to any permit approval, but was not strenuous in its objections. In May 1980 subsequent to receiving the state permit, Paxport Mills applied for Section 10 and 404 permits from the COE. The complications in processing this set of permits could not have been predicted from the speedy approval of the state permit.

For the first year after filing the federal permit application, there was considerable confusion surrounding the project. Information on the exact purpose of the project was contradictory. This was an important issue, since water-dependency and the lack of any alternatives must be established in order to gain approval of a COE permit. There were conflicting opinions among agencies over the biological significance of the site and the potential impact of the project. Some agencies recommended approval of the project, some denial; most recommended the permit be held in abeyance until completion of the COE's *Ongoing Commencement Bay Study/ Environmental Impact Assessment*. Other areas of conflict and misunderstanding centered on the proponents' failure to comply with the terms of two previously issued COE permits. The U.S. Fish and Wildlife Service (FWS) felt that Paxport should be required to satisfy the conditions of both previous permits before being issued other permits.

Finally, fourteen months after making application, a meeting was held at which the applicant submitted sufficient information to establish that the mill was water-dependent and that there were no alternatives other than filling the tidelands that would accomplish their purpose. It was also at this meeting that mitigation was first proposed. Up until this time the concept of mitigation had not been applied in Puget Sound. The proposal came from a FWS biologist who was new to the area and to the project. The biologist in question had only recently been transferred by the FWS from the Chesapeake Bay area where he had gained considerable experience in wetlands mitigation. This in part explains why the concept of mitigation had not been introduced earlier. In addition the FWS had just recently adopted an official mitigation policy (U.S. Fish and Wildlife Service 1981), and none of the other agencies involved had a formal policy at that time.

At this stage, the focus moved to the development of a mitigation plan, which took about three months. The permit was finally issued three months after approval of the plan, fully two years and seven months after the application was first submitted to the COE.

Unfortunately the Paxport Mills controversy did not end with permit approval. Nearly two years after the permit was granted, none of the resource agencies had evaluated the success of Paxport's mitigation project. According to the approved plan, Paxport was to create approximately 0.5 acres of new intertidal habitat at a site just north of the proposed fill project. The new intertidal area was to be created by recutting the bank and by placing concrete rubble rip-rap, clean new fill, and clean dredge materials.

In preparation for another project that was to involve a similar mitigation scheme, the FWS required that the Port of Tacoma hire a consultant to sample the Paxport mitigation site. The results of the consultant's field study were not expected by any of the parties involved. The report stated that a large percentage of the enhancement area consisted of anaerobic (black) mudflats with a number of surface seeps high in hydrogen sulfide running into the area. The report concluded that the site would be slow to recover and would probably never become a highly productive mudflat.

There are now a lot of unanswered questions surrounding the project. Whereas the consultant's report may have been conducted too early to make accurate predictions, it did serve to alert resource agencies to a potential problem. Resource agencies have asserted that Paxport violated the terms of the permit by using material from the adjacent log yard instead of clean fill. This generated questions about what should be done with the site. Should the fill be removed and replaced with clean material or capped over to prevent further leaching? Is the site's productivity sufficient to benefit juvenile salmon? As a means of resolving the matter, additional benthic monitoring was conducted by a consultant hired by Paxport Mills. This study was completed in July of 1986 and is awaiting agency review. At the time of this writing, the final determination of the fate of the enhancement site has yet to be made. An additional problem for Paxport is that the fill project authorized under the permit has not been constructed. The original permit has now expired and a new permit must be approved in order to do the work.

### The Port of Tacoma

Although the Port of Tacoma was not directly involved in the Paxport project, it is possible that the Paxport situation set the stage for the Port's response in the next case. This case begins with the decision of Sea-Land Services Investments Inc., a major ocean steamship line, to give serious consideration to moving their facilities from the Port of Seattle to the Port of Tacoma, if Tacoma could meet a number of criteria set by Sea-Land. In order to meet Sea-Land's requests, the Port would need to create a site for an administration, operation, and maintenance center. Creation of this site involved the filling of approximately 10 acres of wetlands adjacent to the Puyallup River. This wetland (known as the 11th St. wetland) is behind a dike that runs along the northeastern shore of the river. It originally existed due to groundwater seepage; however, a broken tidegate allowed a portion of the site to experience tidal inundation from the low salinity waters of the river.

The sequence of events connected with this project begins with a complaint filed with the COE by the Puyallup Indian Tribe. The complaint reported the placement of fill material into a wetland, that the work had been ongoing for the past year, and alleged that the Port of Tacoma was the violator. The COE confirmed the violation with an on-site inspection on December 18, 1981. A formal letter was sent to the Port on January 25, 1982 stating that the Port was in violation of federal law through their placement of fill materials on wetlands adjacent to navigable waters without obtaining a Section 404 permit from the Department of the Army. In the letter the COE also provided the definition of the term *wetland* to reduce any confusion.

The Port responded to the letter through their attorney in February 1982. The letter stated that the area in question (known to the Port as Parcel 5), was not an adjacent wetland to navigable waters of the United States since it was separated from the Puyallup River by an intervening dike. In addition, the letter pointed out that the faulty tidegate (which the Port said was the COE's responsibility to maintain) had been repaired by the Port and the area would soon return to its normal dry condition.

This was the beginning of a lengthy dispute between the COE and the Port over the status of the "wetland." Gradually all the environmental resource agencies of the region were drawn into the dispute.[2] Throughout this phase the Port position was that the parcel of land in question was not a wetland under COE or Washington State jurisdiction. However, the COE and the Washington State Department of Ecology disagreed with the Port and claimed that the area was an adjacent wetland under their respective jurisdictions, and provided the Port with the statutes and regulations establishing their authority to do so. Nevertheless the Port continued to disagree with these agencies and openly challenged their authority by conducting additional work in the wetland area on two separate occasions in March and April 1982.

By this time in the debate, the Port was under pressure to settle the dispute because of the status of negotiations with Sea-Land. The Port could no longer afford to invest further time in debating the wetlands issue. At a meeting on June 18, 1982 with the COE, the Port indicated that it would consider an out-of-court settlement of the violation with the U.S. Attorney's office and formally mentioned to the COE the possibility of mitigating for the loss of the 11th St. wetland.

Over the next seventeen months the process dealt with finding an adequate mitigation site and the development of a satisfactory proposal. In addition it was necessary to resolve the issue of how much mitigation was necessary for the violation itself, and how much more would be required in conjunction with an after-the-fact permit for continuation of the filling (if a permit were to be granted). Eventually a 30-acre site was agreed upon, approximately one-half mile upstream from the 11th St. wetland, and enough of a mitigation framework was developed by the Port's consultant to allow processing of an application for continued filling. Under COE procedures a violation must be resolved prior to permit application for continuation of a project.

On November 4, 1983, nearly two years after the initial complaint, the project left the violation phase and entered the permit application phase. With the planning of the mitigation project well underway, processing of the permit took only four months; it was issued on March 26, 1984. The final mitigation plan required under the permit called for creation of a 12-acre tidal marsh in the northern portion of the mitigation site. This would require breaching the existing levee along the Puyallup River to provide a permanent river connection to the area. The Port was required to complete the mitigation project by July 1985.

Although the Port received its permit and finished the Sea-Land terminal on time, the same could not be said of the wetlands mitigation project. Problems were encountered with breaching the Puyallup River levee, a federal flood control project. Other questions arose on the ability of the Port to fill a small freshwater marsh on the mitigation site, the necessity of moving an aviation fuel pipeline, and the potential for river sediments to choke off any created wetlands. Further delays were encountered during excavation when toxic substances were uncovered on the site, which was a former landfill area. It was then necessary to perform additional tests and remove the materials.

Eventually these problems were resolved and the wetland was planted in the spring of 1986 (see Fig. 12.1). Resource agencies appear to be satisfied with the project thus far. The plan is to monitor the wetland for four years.

This case study demonstrates the problems that arise from a lack of agency coordination and applicant cooperation. These factors contributed to an unnecessarily long permit processing time for both projects, which might have been even longer had mitigation not been suggested. The potential failure of the Paxport mitigation project is another problem and will probably lead to much greater caution on the part of resource agencies in the future. In addition, the Port of Tacoma project pro-

**Figure 12.1.** The Port of Tacoma's 11th St. wetland mitigation project. Note the five-finger design allowing for continual water circulation in the area. (Courtesy: Port of Tacoma)

vides examples of the type of unforeseen difficulties that can arise when mitigation is proposed in response to agency demands and not given adequate advance planning.

## *The Columbia River Estuary and the Port of Astoria*

The Port of Astoria is located near the mouth of the Columbia River in northwest Oregon. Astoria is the largest port in the estuary and is heavily dependent on log and lumber exports to support its marine terminals. The Port also administers the Clatsop County Airport, which is the focus of this case study.

The Columbia River is the largest river on the West Coast of the United States. It drains from seven states and two nations. Within its estuary large areas of tidally influenced wetlands have been eliminated by diking to create agricultural land, used primarily for dairy pasturing. The diversity of land and water uses and the proliferation of jurisdictions (two states and ten local government entities) have led to the formation of a regional planning body for the estuary. This body is called the Columbia River Estuary Study Taskforce (CREST).

Throughout the time period spanned by this case study, there were three processes going on almost simultaneously. The first was the development by CREST of a regional plan for the estuary. This process took place from January 1975, the origin of CREST, until June 1979 when the plan was published (McGolgin 1979). Subsequent to publication of the regional plan, CREST refined the mitigation element of the plan and published a separate *Mitigation Plan for the Columbia River Estuary* in June 1983 (Smith 1983).

At the state level another process was going on. In 1976 the state's Land Conservation and Development Commission (LCDC) adopted the Estuarine Resources Goal, making mitigation a formal state requirement. Goal number 16 is one of nineteen statewide planning goals and guidelines that form the basis of both Oregon's land use and coastal zone management programs. In 1979 further legislation was passed and mitigation was incorporated into the State Removal/Fill Statute (Oregon Revised Statutes 541.626) (Quarterman 1985). Mitigation then became a requirement under the State Removal/Fill permit that is administered by the Oregon Division of State Lands (DSL). The late 1970s and early 1980s were then a time when both agencies struggled to formulate administrative rules and criteria for incorporating mitigation into state land use planning and permit procedures. In addition, in 1982 the Oregon Department of Land Conservation and Development (DLCD) received a grant from the federal office of Coastal Zone Management to establish the Oregon Mitigation Fund. This fund made available government seed money to aid in the formation of mitigation banks.

The third process taking place was the review of permit applications for a proposed expansion of the Clatsop County Airport. This proposal first came to light in 1979 with the publication of the *Clatsop County Airport Master Plan—Technical*

*Report*. It is the events connected with the review and approval of this project that form the basis for this case study.

The airport expansion involved the relocation of a dike riverward for the purpose of allowing a sophisticated Instrument Landing System (ILS) to be installed, and to create a sufficient clear zone at the end of the runway to allow jets and larger aircraft to use the airport. Relocation of the dike would result in a 3-acre loss of intertidal wetland. Construction of the project would enhance general and commercial aviation and allow the U.S. Coast Guard to base surveillance jets at the airport. It was projected that considerable community benefits would ensue from the expansion, as well as national benefits associated with airborne fishery enforcement by the Coast Guard in the 200-mile zone.

The first stage in the review process was initiated in June 1979. It involved incorporating parts of the airport master plan, including the expansion, into CREST's regional plan. It was through this process that the airport project became caught up in the official review of the CREST regional plan. In order for CREST to designate as "development" the area slated for the airport expansion, an "exception" was required from the estuarine resources goal discussed above. This exception was necessary since the airport was not a water-dependent use and the expansion would fill in a wetland area.

Although the plan was submitted for review in June of 1979, it was not until January 1981 that the DLCD responded with its conclusion that the exception did not meet one of the required goals. It was then necessary to prepare a revised goal exception, which was completed on November 4, 1981. It is the revised goal exception that first mentions mitigation for the project. It did not, however, include discussion of any specific measures.

Prior to completion of the goal exception, the permitting process for the project was initiated with a Joint Application for Permit filed with the COE and Oregon DSL on August 25, 1981. When the COE public notice was released on November 17, it included a mitigation proposal from the Port, which consisted of lowering the elevation of three acres of high forested swamp riverward of the berm to provide new shallow water habitat. This proposal was rejected by resource agencies because the existing swamp provided productive habitat that could not be improved by lowering its elevation.

Eventually resource agencies advocated a proposal that involved relocating an existing dike landward to restore 15 to 25 acres to tidal influence. The site selected was under Port ownership and was marginally used for agriculture. All the parties involved seemed to support use of the site.

It was then that CREST staff began to suggest a proposal for a mitigation bank, since it was apparent that the opportunity existed for a restoration action larger than that necessary for the fill proposal. "Mitigation banking is the process whereby habitat created, restored, or enhanced in advance of a specific mitigation requirement is used to fulfill that requirement" (Maddux 1986).

CREST staff put considerable effort into promoting the concept of a mitigation bank for the airport site. They produced a detailed feasibility report, which included a cost analysis, an assessment of the biological potential of the banked area, and a carefully planned implementation scheme that defined administrative responsibilities. At the same time CREST was also working with state agencies to formulate banking policies. CREST was the driving force for both the airport project and state-level mitigation bank policies.

Finally, three years after the project was first proposed, the Port received approval of its permit from the Oregon DSL in March of 1982. This permit required only a 3-acre mitigation, a 1:1 compensation for lost wetlands. However, it suggested that it would be advantageous to the Port to expand the mitigation, forming a bank. At the same time the DLCD provided further incentive for the bank by stating that seed money of up to $40,000 could be provided from the Oregon Mitigation Fund.

Even after permit approval, the project was further delayed because it was found that implementation of the mitigation project would entail destruction of catwalks along the dike tops, which were used for maintenance access to landing lights. Redesign and reconstruction of new access to these lights would cost about $100,000, the same as the estimated cost of the entire mitigation bank.

Eventually the Port constructed its expansion project. However, it used an onsite mitigation project, separate from the bank proposal. Although the mitigation bank could have greatly facilitated the review of future port development projects, the Port refused to become involved in implementing the bank, largely because of the considerable delays it had already experienced.

The idea of creating a mitigation bank at the site selected during this process has not died. CREST and federal and state agencies still support the concept. The Oregon DSL is doing some additional site evaluation work requested by the FWS. Although there are now other sites competing for Oregon Mitigation Fund money, there is still a possibility that a bank will be implemented in the forseeable future.

## *Summary of Project-Specific Case Studies*

The first group of case studies have in common the development of a mitigation proposal in response to project-specific impacts. To a large extent two problems dominate throughout: project delays and high, often unexpected costs. Whereas the Port of Miami project was implemented quickly, mitigation costs were much higher than planned. Both Commencement Bay projects encountered permit delays, although admittedly the causes were not always directly related to mitigation. The Astoria project experienced both delays and unforeseen costs. These types of problems are fairly typical of project-specific mitigation.

The second group of case studies describes efforts on behalf of Ports and other agencies to go beyond this project-response type of scenario. The beginnings of this process are seen in the Port of Astoria case study. Mitigation banking is seen by many to be a viable means of decreasing permit processing time since it would remove mitigation planning and construction from the process. However, as suggested above, mitigation banks may be effective only when established prior to review of the project at hand.

The common focus of the following group of case studies is the idea of removing the design and, if possible, the construction of mitigation projects from the permit timeline by means of banking or other measures.

## MITIGATION WITHIN THE PLANNING PROCESS: THREE CASE STUDIES—BALTIMORE, TAMPA, LOS ANGELES AND LONG BEACH

### *The Baltimore Harbor Environmental Enhancement Plan*

The 14-mile-long estuary of the Patapsco River, flowing into Chesapeake Bay, makes up Baltimore Harbor. The Port of Baltimore is located in the upper reaches of the harbor. Historically, the marshy land surrounding the harbor was filled to provide suitable docking for merchant ships, and the waters and sediments were contaminated by industrial wastes, sewage, and agricultural runoff. In the last two decades this situation has started to change. Water quality regulations have greatly reduced the flow of nutrients and sediments into the harbor. There are now state and federal permit requirements for filling of wetlands that have dramatically reduced the rate of filling in the harbor. Unfortunately the permit procedures for filling and dredging are very complex, and the different requirements of federal, state, and local governments have delayed a number of vital projects (Dolan 1985).

This case study discusses the development of an Environmental Enhancement Plan for Baltimore Harbor. The objective of the plan is to identify resource enhancement projects that could serve as mitigation for development impacts and potentially rectify past destruction. The plan was developed by the Regional Planning Council in an effort to reduce the confusion and delay associated with the permit process (Regional Planning Council 1982). The Regional Planning Council is an areawide planning agency of the Maryland State government. Its job is to advise the state and the six-member jurisdictions of the council on planning for the Baltimore metropolitan area.

The process began in 1975 when the COE and the federal environmental resource agencies requested that the state prepare a comprehensive plan to guide permit decisions in the harbor. This led to the *Baltimore Harbor Plan,* which specified recommended development patterns for the harbor. This land-use planning effort was expanded to the entire coastal area of the Baltimore region, and in 1978 the *Baltimore Metropolitan Coastal Area Study* was published. These documents formed the basis for local and state policies to guide dredge and fill activities in the harbor. At the federal level, mitigation guidelines were established, requiring the development of substitute resources for impacts related to federally approved fill projects, if the fill is used for purposes other than erosion control (Dolan 1985).

At both levels of government these policies were developed to address the loss of resources. However, there was sufficient difference between the two sets of requirements to create confusion, causing delay of projects in the public interest. The Regional Planning Council set out to do something about this problem through the formation of the Environmental Enhancement Task Force. The task force was made up of representatives of environmental and economic agencies from federal, state, and local levels. This is the major objective agreed to by the task force: "To facilitate appropriate economic development by speeding up the permit process and improving the aquatic habitat of the harbor" (Dolan 1985).

After preliminary work documenting state and federal policies, and activities in other states, the task force put together the *Baltimore Harbor Environmental Enhancement Plan*. There are four major elements included in the plan: an environmental assessment of existing conditions, an assessment of the potential for environmental enhancement, conceptual plans for five specific enhancement sites, and an implementation plan.

The first three parts were prepared by consulting firms under the direction of the task force. The assessment found that whereas the harbor has serious water quality and sedimentation problems, the development of new wetlands, fish reefs, and the rehabilitation of the existing shoreline could do much to improve the aquatic habitat. Maps were prepared to show the existing conditions and enhancement potential of the entire harbor. Areas of intensive industrial use, or designated for such use, were excluded from consideration as enhancement areas. From this information, five specific sites were chosen by the task force to illustrate the range of enhancement types, sizes, costs, and locations. Concept plans were prepared for each of the five sites and are included in the plan.

When the enhancement plan was published in 1982, it included five recommendations that were directed toward facilitating the permit process and implementing the plan. The idea behind the recommendations was to remove the habitat mitigation design, construction, and management from the permit process timeline. If implemented, the recommendations would have set up a system for funding the enhancement projects described in the plan through the accumulation of money from developers of dredge and fill projects.

The implementation scheme failed for several reasons (Dolan 1985). The greatest obstacle came from the federal permitting and permit review agencies, who did not want to give the appearance that applicants could purchase permits. Therefore, they would not consider a fee in lieu of replacing lost resources (Ashe 1982, 30–

32). The other problems had to do with the mechanics of accumulating money. The existing funds that were identified by the plan could not be accumulated, and there were legal restrictions on their use. For example, the Maryland Board of Public Works collects a fee as monetary compensation for use of state wetlands. However, this fee is considered to be for the transfer of property interests, and the board would not necessarily recognize mitigation as such a transfer. In addition, the board opposed substituting a lease system that would generate ongoing revenues, as was proposed in the plan, for a one-time fee that is much easier to administer.

As a result of these problems, a different, less comprehensive implementation system has been developed since the publication of the plan. Permitting and mitigation are still planned and conducted on a project-by-project basis. However, all the agencies involved have signed a "Letter of Understanding Regarding the Environmental Enhancement of the Baltimore Harbor." In this agreement all local, state, and federal agencies responsible for permit review agree to give consideration to the enhancement sites identified in the plan, should mitigation be warranted for a specific development proposal. The letter also outlines the coordination between agencies. In addition to the Letter of Understanding, the COE has agreed to allow as fulfillment of mitigation requirements, completed projects that are constructed by parties other than the developer. This makes it possible for state, local government, or other groups to cooperate in constructing enhancement projects that could be used as mitigation for development proposals.

The enhancement plan did not succeed in eliminating the design and construction of mitigation from the permit process timeline. Although very few projects have been reviewed since adoption of the plan, it has resulted in faster permit processing and a firm (if unwritten) policy of requiring in-harbor compensation for resource losses. In particular, an interstate highway tunnel project took advantage of some of the identified enhancement sites.

The story of the Baltimore Harbor Environmental Enhancement Plan did not directly involve the Maryland Port Administration. Nevertheless, the Port stands to benefit from the efforts of the Regional Planning Council. The Port was represented throughout the preparation of the enhancement plan, but did not actively participate in the process. Since the adoption of the plan, however, the Port has changed its position and now plans on taking an active role in environmental resource considerations. The Port is currently developing a 20-year master plan for dredging and dredge disposal that will incorporate mitigation into the planning process.

This case study shows that concern about permit processing time extends beyond the port/industrial sector of the harbor. The impetus for improving permit processing came from the Regional Planning Council. Evidently the council's six-member jurisdictions felt that there would be sufficient benefit to all harbor users to warrant tackling the problem. Whereas the original scheme for implementing the plan was not viable, the end result is a fairly successful process that could serve as a model for other entities with similar interests.

## *Tampa Bay and the Port of Tampa, Florida*

Tampa Bay is an estuary located on the west coast of the Florida peninsula. It has an area of 400 square miles and an average depth of 11 feet. The Port of Tampa is located in Hillsborough Bay, in the northeastern portion of Tampa Bay. Because the bay is fairly shallow, channel dredging has long been necessary to facilitate ship access to the port. This activity, together with dredging and filling for residential development, has resulted in the destruction of 100 percent of the historical seagrass beds and 80 percent of the intertidal wetlands in Hillsborough Bay. For Tampa Bay as a whole, the figures are only slightly less dramatic; 80 percent of the seagrass and 44 percent of the intertidal wetlands have been destroyed (Fehring 1982). The overall effect of these impacts has been severe declines in the local fishing industry. This case study documents the extent of the Port of Tampa's involvement in local efforts to rectify environmental destruction in Tampa Bay, at the same time continuing port growth and development (see Fig. 12.2).

When environmental legislation was first enacted in the early 1970s, the Port of Tampa quickly realized the importance of accepting and adjusting to the new legal framework. An initial step taken by the Port was to create the position of director of environmental affairs within the Port administration. This was done instead of enlarging the role of either the port engineer or planning director. In addition, the director of environmental affairs was given both the political and financial support to effectively negotiate the Port's position and to conduct necessary environmental research. The individual hired for this position was a highly trained professional ecologist who played an instrumental part in formulating and implementing the Port's environmental policies.

In 1976 the port authority established an Environmental Advisory Committee (Fehring 1980). The committee was comprised of representatives of local environmental groups, universities, and government agencies, and engineers. It provided technical assistance to the port authority and, in turn, created a mechanism for local environmental interests to have input into the port planning process.

The Environmental Advisory Committee also developed ideas for a number of studies that were funded by the Port. Some of the studies have dealt with habitat creation techniques and have contributed to the design of harbor development and mitigation projects (see Fig. 12.3).

In 1982 the port authority dissolved the Environmental Advisory Committee in response to the creation of a new advisory body called the Agency on Bay Management at the Regional Planning Council. Most of the original members of the defunct advisory committee are represented on the new agency, as is the Port itself. The agency's area of concern has expanded to include all of Tampa Bay, not just the Port area. The role of the agency is strictly advisory and tends to focus on the technical aspects of project proposals.

In addition to incorporating environmental considerations into port planning, the

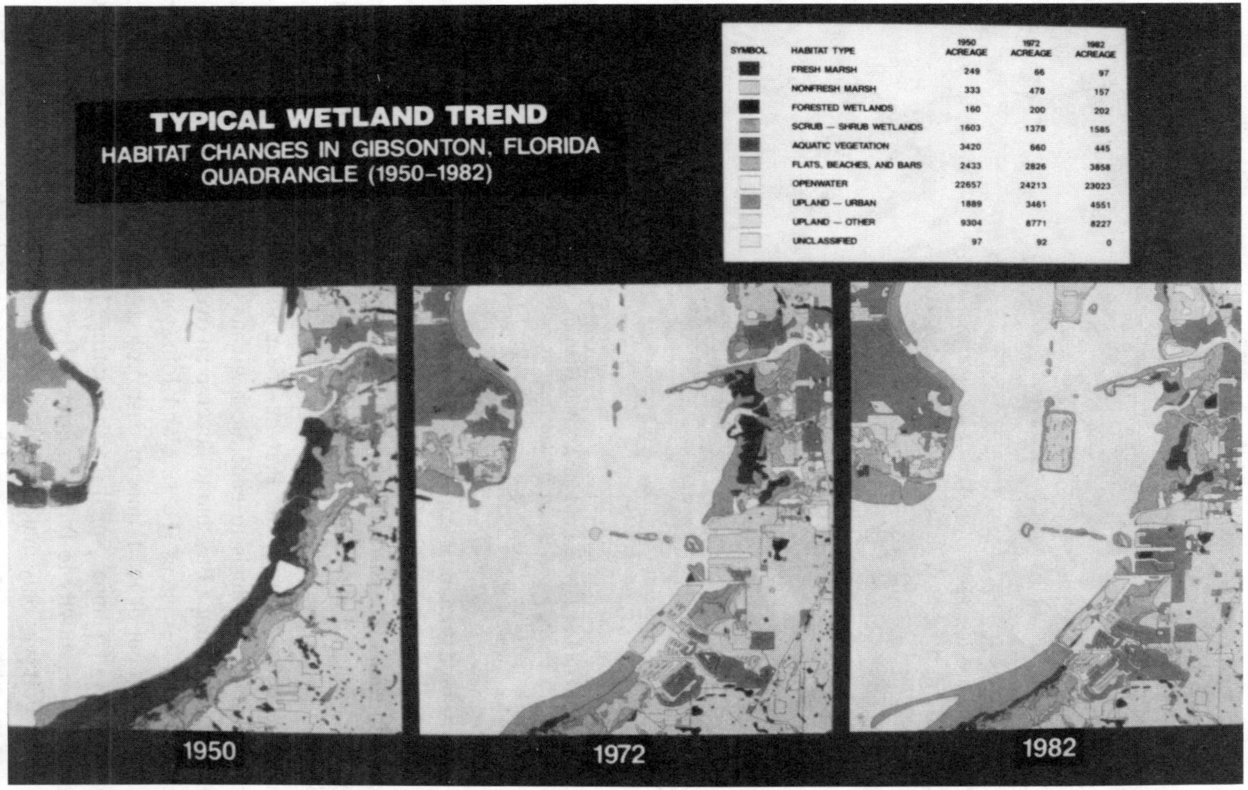


**Figure 12.2.** Wetland vegetation in Tampa Bay has changed dramatically in 30 years as illustrated in Gibsonton, Florida, a segment of Tampa Bay. (Courtesy: William K. Fehring)

**Figure 12.3.** Senator (former governor of Florida) Bob Graham, with students from the Tampa Marine Institute planting cordgrass on a port island created from dredged material. (Courtesy: Port of Tampa)

Port has cosponsored an annual conference on coastal vegetation. Since its initiation in 1975, the program has grown so significantly that the conference is now self-supporting, and financial assistance from the Port is no longer necessary. The wealth of information gathered at the annual conferences has greatly contributed to local mitigation programs.

A more recent activity of the Port has been involvement in an effort among local agencies and individuals to generate a greater understanding of the environmental problems of Tampa Bay and to develop an overall strategy for improving and protecting the bay's resources (Fehring and Johnston 1984). To some extent these efforts have grown out of the failure of past project-specific mitigation efforts conducted by the Port of Tampa. Small marsh plantings have either washed out or were too close to human activity to function effectively. An effort to improve water circulation in the Port was only marginally successful since little was known about the dynamics of the system. One bare-sand bird-nesting area has been successful, but it requires constant maintenance by the Audubon Society to retain its utility. In addition, nearby dredge spoil disposal areas are attracting birds away from the sanctuary, and its future is in doubt. If dredge spoil areas are not kept clear of vegetation, this could result in a loss of bird nesting habitat in the bay.

The failure of mitigation projects generated serious concerns about the utility of the practice in Tampa Bay. Each project involved separate planning, engineering, and mobilization costs. This increased the total cost and also increased the resistance by port and industry interests to do anything more than the minimum. This does not serve to compensate for the hundreds of acres of habitat lost to previous development. A more serious problem was the loss of significant mitigation opportunities as a result of poorly understood resource needs, available options, and the risks involved. For example, a plan to gradually create two large marsh islands out of dredge spoils was rejected because it was thought that maintenance dredging materials would be too polluted and that the sites would be prohibited as open water disposal. As it turned out, the inner harbor dredge materials contained no more contaminants than natural fine sediments in many other estuaries. In addition, it was subsequent to this proposal that the full extent of past resource losses was recognized. According to one observer, a project of this magnitude could have been of great benefit to the bay (Fehring 1982).

A major problem contributing to the failure of project-specific mitigation was the lack of real understanding of the ecology of the estuary as a whole. The recognition of this situation was one factor that led to a one-week symposium in May of 1982. The symposium was sponsored by the port authority, the COE, the Sea Grant Program, and numerous local scientists. Its purpose was to consolidate existing knowledge about the bay. An offspring of the symposium was the Regional Bay Study Committee, which identified and prioritized perceived problems in the bay. The committee identified two priority needs that relate directly to the Port's long-term development potential:

1. Development of a long-term, baywide dredged material disposal strategy.
2. Development of an overall plan for mitigating the environmental impacts of future dredging projects.

The FWS took an interest in the second item and approached the port authority with the concept of a cooperative effort to develop baywide habitat data and potential mitigation options.

In the summer of 1983, the port authority and the FWS entered into an agreement that contained three major objectives:

1. To identify management and mitigation options that will allow development and maintenance of the Port of Tampa to proceed in an environmentally acceptable fashion.
2. To develop an information base in map and text/tabular form that is aimed at analysis and evaluation of mitigation and management options.
3. To develop a management plan to guide the authority in the development and maintenance of the Port (Fehring and Johnston 1984).

In order to accomplish these objectives, five major study tasks were listed. The first two are the preparation of two sets of 1:24,000 scale maps. The first set would depict wetlands, seagrasses, and land use for the periods 1956, 1972, and 1982. The second set is to be an environmental atlas of biological and physical characteristics of the bay. The third task is the development of a geographic information system. This is a computerized management system for the mapped data that can be used to indicate habitat trends and to evaluate mitigation sites. Fourth is the preparation of a synthesis document on the ecology of the bay.

The final task is the preparation of a mitigation options document for Tampa Bay. The document is to be written after all the above data is assimilated and will include an analysis of past mitigation actions. Elements to be included in the document are a feasibility analysis of mitigation options, the identification of sites on maps with the type of mitigation feasible for each site, and site-specific environmental management and restoration recommendations for the bay.

Concurrent with these efforts, the port authority is seeking approval of a twenty-five-year permit for maintenance dredging throughout the port. If approved, it will be the first of its kind in the country and will include a long-term plan for dredge spoil disposal areas. It is hoped that the disposal plan will be tied to the mitigation options to produce a broader dredging, disposal, and mitigation plan that is sensitive to economic, engineering, and environmental considerations.

It is recognized that successful implementation of the findings of the mitigation options study will be contingent upon the ability and willingness of federal and state agencies and the Port to commit to a long-term mitigation agreement. In addition, it will be necessary to develop financing mechanisms for long-term mitigation actions constructed in advance of specific development projects.

This case study provides a good indication of the ways in which a Port can become actively involved in local environmental resource management. The benefits to the Port that have resulted from this level of involvement have been rapid permit approvals and good rapport with environmental interest groups and resource agencies. The Port has also found it to be to their advantage to take the initiative when negotiating environmental requirements. This means being prepared with an accurate analysis of the environmental costs and benefits of a project and its mitigation prior to application for permits. The Port also indicated that showing a willingness to solve problems was helpful in gaining community support for their projects.

## *San Pedro Bay and The Ports of Los Angeles and Long Beach, California*

The Ports of Los Angeles and Long Beach are located adjacent to each other in San Pedro Bay on the Southern California coast. During the last century, Southern California's highly productive coastal embayments have been modified and greatly diminished in extent. In San Pedro Bay fish populations are reported to be diverse and abundant, and a number of migratory birds, including the California least tern, an endangered species, frequent the bay. The Ports of Los Angeles and Long Beach are also very large and rapidly expanding marine cargo terminals serving the enormous metropolitan and regional markets of Southern California.

In accordance with the requirements of the California Coastal Act of 1976, the Ports have jointly prepared a master plan for dredge and fill development called the 2020 Plan, after the date of its planning horizon. In anticipation of a tripling of cargo throughput by the year 2020, the plan projects a need for 2,600 acres of new fill and for dredging 82-foot-deep channels in San Pedro Bay.

In connection with the master development plan, environmental review documents were prepared, and the need for a mitigation framework was quickly recognized as an essential element of the study (Rice 1984a).

In the past, mitigation projects in the two Ports have been conducted in response to specific impacts and were planned and implemented in the context of the COE permit procedures. Mitigation projects of this kind have included creation of a kelp bed along the inside of the inner San Pedro breakwater (Rice 1984b) and provision of a protected nesting site for the California least tern (see Fig. 12.4).

Both Ports also have experience with the creation of mitigation banks. The Port of Los Angeles established a banking system whereby credits are given for creation of additional water area in the harbor, and reductions of water area are drawn against the bank on an acre-for-acre basis. The Port of Long Beach went outside of the immediate harbor area and agreed to create a wetland habitat in the Upper Newport Bay Ecological Reserve (UNBER) located in the city of Newport Beach. The rate of trade-off was established to be 1 acre in the UNBER equals 1.5 acres of deep-water substrate to be covered by fill or pile-supported wharf. Both mitigation banks were established through Memoranda of Understanding (MOU) signed by the Port and appropriate state and federal resource agencies. Each MOU carefully details the extent of habitat to be created, imminent development projects, the responsibilities of the parties involved, and the terms under which the bank may be used. The banks serve the purpose of fulfilling the permit requirements for projects planned for the immediate future and of preventing permit delays for other small projects in the more distant future.

The existing mitigation banks, however, are limited in size and scope and do not address the impacts that will occur as a result of implementation of the 2020 Plan. Instead the COE organized an effort to develop a mitigation procedure that could be applied to projects conducted under the plan. Approximately one year was spent in the effort, which involved both of the Ports and state and federal resource agencies. The objective was to develop a standard method for evaluating biological impacts resulting from development plans. This would enable a particular project under the plan to be reviewed with respect to habitat and a net habitat value change determined. Similar habitat value changes could then be calculated for various types of mitigation options.

The FWS Habitat Evaluation Procedure (HEP)[3] method of assessing impacts was not selected for two reasons. First, HEP is not applicable until both the project site and an appropriate mitigation site are identified and analyzed for their respective resource values. Second, habitat suitability index models are not available for important Southern California marine species.

The method chosen was based on a habitat-area approach that bears some resemblance to HEP.[4] Five basic habitat types were selected by a group of agency representatives. These were then assigned relative "weights" per unit of area by each of the participants. The assigned weights reflected the perceived ecological value of the habitat type in relation to a common denominator habitat (soft bottom habitat, the majority of the habitat to be impacted). A number of potential mitigation options were also identified and were subjected to the same weighting procedure. This process of weighting various habitats amounted to establishing trade-off values for each of the different habitats and mitigation options.

COE biologists then evaluated each of the agencies presentations and attempted to prepare a set of weighted values that represented a compromise of the participants' positions. Unfortunately, the compromise arrived at by the COE was not acceptable to the concerned resource agencies. At that point the project faded and the ports resorted to a modified HEP for resource evaluation in connection with 2020 Plan development. The COE is currently requesting funding from Congress for the completion of the feasibility report for the entire 2020 Plan development. Additional research is to be conducted in connection with the environmental impact statement that will accompany the feasibility report. This will help to clarify the trade-off values established by the COE and may lead to acceptance and utilization of the mitigation procedure.

As mentioned above, both Ports have turned to a modified HEP in response to the initial rejection of the COE habitat evaluation procedure. The modifications to HEP were necessary because of the lack of species models for Southern California marine organisms. Instead, a system of ranking the suitability of the habitat for selected species was used. Species were chosen because they were either common to both the harbor site and the compensation site or were considered ecologically equivalent. Both ports have also identified specific compensation sites that are analyzed in conjunction with the project site. Habitat suitability for each species is ranked by professional biologists representing resource agencies and the port. The assigned values range from 0 (no habitat suitability) to 1.0 (complete suitability). A mean "habitat suitability index" is calculated for each species and multiplied by the number of acres affected to produce the number of habitat units for each species at each site both before and after development. Then the net change in total habitat units at each of the sites is compared. These values must add up to zero or greater in order for the mitigation proposal to be considered adequate compensation. As an example, for the Port of Long Beach's Pier J landfill, a 102.5-acre wetland creation project was agreed to as compensation for a 135-acre fill, based on no net loss of habitat units.

In addition to the difficulties encountered in establishing an evaluation proce-

**Figure 12.4.** Male least tern, with anchovy, courting female. The anchovy "gift" is used to attract a mate (left). Ensuring habitat for this endangered species was a challenge for the Port of Los Angeles. To preserve habitat for the California least tern, the Port of Los Angeles designated a 15-acre protected nesting site shown as the darker vegetated area in the aerial photo. The site is flanked by a 190-acre landfill project. (Courtesy: Port of Los Angeles)

dure, locating sites for mitigation posed significant problems for the Ports of Los Angeles and Long Beach. There are virtually no sites available in the harbor area because it is already heavily developed and because the California Coastal Act requires that future port development be kept within existing port districts. Further, there are not 2,600 acres of property available outside the harbor in the Los Angeles/ Orange County area. Most sites are privately owned and would be available only at exorbitant costs. As a result, resource agencies have allowed the Ports to expand the search area to encompass the area between the U.S.-Mexico border and Ventura County. The Port of Los Angeles has found a single site, at Batiquitos Lagoon near Carlsbad, California, that will potentially provide sufficient restored habitat to allow mitigation for their entire 2020 plan development. The Port of Long Beach is looking at incremental mitigation because it has not been able to find a single site large enough to satisfy its mitigation needs.

This case study demonstrates that resource evaluation is crucial to the mitigation process, and at the same time that it is difficult to arrive at consensus on an appropriate evaluation methodology. This problem is one of the fundamental issues facing environmental resource managers and mitigation proponents.

The study also indicates some of the steps necessary to go beyond project-specific mitigation. The first step is to incorporate mitigation planning into long-range development planning. This allows project designers to anticipate mitigation requirements and avoid the need for additional research and planning during permit review. The second step is to decide on a standardized resource evaluation technique that will allow consistent assessment of project impacts and calculation of adequate compensation measures. A third step, where it is possible, is the creation of a mitigation bank. In this example the Batiquitos site, if approved, will serve as a mitigation bank for the Port of Los Angeles' ongoing development under the 2020 Plan.

## *Summary of Planning Process Case Studies*

The second group of case studies illustrates the effort to avoid designing and implementing mitigation measures on a project-by-project basis. All three studies make use of an ecosystemwide approach to develop a comprehensive resource management scheme within which mitigation plays a central role. In both the Los Angeles/Long Beach and Baltimore case studies, the starting point for the assessment of resource management needs was a port development plan, which formed the basis for projecting future environmental impacts. The Port of Tampa is doing something similar with its 25-year maintenance dredging permit. This is an important first step in the development of a mitigation scheme that adequately addresses ecosystem functions and values.

It should also be observed from these case studies that it is impossible to completely eliminate the project-specific aspects of mitigation. The Baltimore example demonstrates some of the procedural obstacles to this goal. On a more fundamental level, however, is the need to assess project-specific impacts in order to determine the type and nature of mitigation that is appropriate for a particular project. For example, even when a mitigation bank is available from which to draw mitigation credit, it is still necessary to calculate the extent of a project's impacts to be able to determine how much credit to draw from the bank. The direction being taken by Los Angeles/Long Beach acknowledges this reality. The difficult problem of quantifying impacts and establishing trade-offs can be time-consuming, and having a methodology that all parties agree to can speed up the process.

## THE RECURRING ISSUES: TRADE-OFF VALUES, TECHNICAL UNCERTAINTIES, PROCEDURAL PITFALLS

The foregoing case studies provide ample evidence that Ports are getting involved in the planning and implementation of mitigation projects. The level of involvement varies greatly, however. Tampa, Los Angeles, and Long Beach have taken an active role, and mitigation planning has been incorporated into the Ports' management responsibilities. The Port of Baltimore, on the other hand, assumed an advisory role in a mitigation planning effort orchestrated by the Regional Planning Council. The Ports of Miami, Tacoma, and Astoria were all drawn into the development of mitigation projects through the permit process. Although their initial involvement may not have been voluntary, they all acted in good faith towards fulfillment of their responsibilities.

### *Mitigation Policy Issues*

There are three groups of policy issues that affect the successful application of mitigation as a resource management tool. First, and perhaps most significant, is the calculation of trade-off values used to establish mitigation requirements. The second group of issues relates to the technical uncertainty inherent in the creation of artificial habitats. Third, there are a number of resource agency shortcomings that have hampered the success of mitigation projects.

#### **Calculating Trade-off Values**

Trade-off values are used to decide how much compensation and what kind of compensation is necessary to offset the losses caused by project impacts. Getting biologists to agree on trade-off values is extremely difficult. Environmental assessment is not a purely objective science, and there are no accepted, standardized methods for quantifying environmental resource values. Compounding the problem is the multitude of diverse and complex coastal ecosystems, leading to enormous information needs. As a result, most agencies insist on making determinations about mitigation requirements on a case-by-case basis. This can run contrary to a Port's interest in gaining approval for long-term maintenance dredging and/or facilities expansion projects. Of the case studies discussed, the Ports of Los Angeles and Long Beach are making serious efforts to deal with this problem. The modified HEP procedure currently being used by the Ports and the mitigation procedure proposed by the Los Angeles District COE are both mechanisms for incorporating compensatory mitigation into the framework of a long-range development project. The Ports' modified HEP procedure provides for the comparison of habitat values for project-specific impacts and mitigation proposals. The COE mitigation procedure, if approved, would allow the circulation of net habitat value changes for proposed projects and possible mitigation options.

## Technical Uncertainty

Another issue associated with the application of mitigation is the technical uncertainty inherent in habitat creation, restoration, and enhancement. There are two aspects to this problem. The first has to do with inadequate understanding of the habitat needs of certain species and ecosystem functions as a whole. The second involves the actual construction techniques of habitat creation.

In some instances mitigation cannot be implemented because specific species habitat requirements are not known. For example, it is not realistic to consider creating artificial herring spawning beds until it is understood why herring choose their natural spawning areas. On a broader level, sufficient knowledge of ecosystem functions is necessary to be able to construct functioning habitats. This is demonstrated in the Port of Tampa case study where the failure of a number of mitigation projects was caused, in part, by a lack of knowledge about the ecology of the Tampa estuary as a whole.

Although a considerable body of knowledge has been gathered on the construction of certain types of habitats, most notably marshes, the technology of habitat creation is still largely experimental. In recent years projects that have been conducted over the past decade have started to undergo evaluation (Race and Christie 1982; Eliot 1985; Baker 1984). A frequent finding of postconstruction evaluations is that replacement habitats seldom perform as anticipated. These findings are leading to growing concern over the utility of compensatory mitigation as a means of preventing further habitat losses. For example, concern has been expressed by some authors that artificially constructed salt marshes may not function like natural marshes or provide the same habitat value as the marshes they are intended to replace (Race and Christie 1982). In an assessment of San Francisco Bay marsh creation projects, Margaret Race found that "projects have been plagued by multiple problems such as high soil salinities, incorrect slope, improper tidal elevations, incomplete vegetation establishment, channel erosion, sedimentation or poor tidal circulation" (Race 1985). These problems stem largely from inadequate construction techniques. The Port of Miami's seagrass planting efforts are a further example of poor construction techniques contributing to project failure.

Both aspects of the technical uncertainty problem indicate a need for further research into habitat functions and values and habitat construction methods. One question that comes to mind is: Who should be paying for this research? Should it be the resource agencies and the governments they represent, since they are requiring the mitigation and are responsible for resource management? Or should it be the development community since they are responsible for causing impacts to natural habitat? These are difficult questions that are not often given serious consideration until a project proposal is in hand. In the case studies, the Port of Tampa's recognition of this problem led to sponsorship of several research projects and cosponsorship of an annual wetland vegetation conference.

**Resource Agency Shortcomings**

Some authors have pointed out a number of administrative shortcomings that have led to project failure (Eliot 1985; Baker 1984). The first is the lack of clearly established goals for the mitigation. Without specific goals a permittee may complete the required mitigation and still fail to provide any of the resource values of the lost habitat. Another problem has been poor enforcement of permit conditions. In one report several of the projects examined had not been implemented even though four to five years had elapsed since the requirements were established (Baker 1984, 81). A notable deficiency in many mitigation projects is the lack of monitoring and remedial action in the event of project failure. The Paxport Mills project described in the Commencement Bay case study provides an excellent example of the problems that can arise when proper enforcement and postconstruction monitoring do not take place. At the Port of Miami the lack of planned remedial actions in the event of project failure has led to increased costs and many years of trying to complete the permit requirements.

As a result of the types of administrative problems identified above, some environmental organizations have expressed an outright opposition to mitigation (Ortman 1984). Agency representatives, on the other hand, still express optimism about the concept. It is admitted that there have been significant problems with past projects and that the application of mitigation warrants careful scrutiny. Nevertheless, some agency staff are of the opinion that considerable information has been gained from past experiences and that if properly applied, mitigation techniques can successfully compensate for certain types of habitat losses.

In the Pacific Northwest, past experience has led one agency biologist to identify what he considers to be the seven essential elements of an ideal mitigation plan (Cooper 1986). The seven elements are: 1) baseline information for both the project impact zone and mitigation site, 2) environmental goals/objectives that describe the mitigation project purpose(s), 3) a detailed work plan that includes written specifications and description of mitigation techniques, construction sequencing, and site diagrams, 4) performance standards, the specific criteria to either verify fulfillment of environmental goals or to trigger initiation of remedial action or contingency measures, 5) a monitoring program with postproject assessment requirements, survey or sampling methods, and provisions for interagency review, 6) a contingency plan for courses of action or corrective measures to be implemented in the event performance standards are not met, and 7) a performance bond to ensure fullfillment of mitigation and/or contingency measures.

These elements may currently represent a "wishlist" from an agency perspective, but they do present a means of overcoming the agency shortcomings identified above. With the growing awareness of past problems, agencies are likely to enforce elements of the list in an effort to improve success rates.

## *Mitigation Banking*

One mitigation strategy that has been proposed for long-term projects or numerous small-scale projects is mitigation banking. In this strategy, habitats are created, restored, or enhanced prior to their use to fulfill a permit application requirement. Habitat "credits" are accumulated by the creation of appropriate fish and wildlife values, which may then be charged against the habitat losses incurred by future developments.

Mitigation banking has been recommended as a useful tool in the harbor area where there are frequent small projects being constructed in different areas. To mitigate for each project individually is more expensive than constructing one large compensation project. Also, when a mitigation bank is available, the time spent planning a mitigation project can be eliminated from the permit review process. The same benefits hold true for long-term projects such as ongoing maintenance dredging, or, as in the Los Angeles/Long Beach case study, for implementation of a long-range port development plan.

There are, however, difficulties associated with establishing a mitigation bank, as are clearly demonstrated in the Port of Astoria case study. There is the expense of habitat creation or enhancement, which can be considerable. This must be done upfront, prior to gaining any of the economic benefits of new port facilities. In a public port district it may be possible for a port authority to construct a mitigation bank that could be drawn from by private developers or the Port. Parties that draw from the bank could reimburse the port authority for the cost of construction. Another problem has to do with calculation of the value of the habitat "credits" and impact "debits." This is the same as the problem of calculating trade-off values described above.

## MITIGATION PLANNING AND THE BENEFITS TO THE PORT

Environmental mitigation, in principle, is a management tool capable of creating a "win-win" situation for economic development interests and environmental resource managers. Ports that successfully integrate full consideration of environmental resources, including mitigation of unavoidable adverse impacts, into the planning and construction of port development projects stand to benefit from: 1) reduced permit delays and associated costs, 2) reduced uncertainty with respect to governmental approval of permits, and 3) increased public support for port development projects. In turn, environmental resource managers gain the cooperation of a significant set of developers and achieve "no net loss" of environmental resources. In practice, however, there are still significant hurdles to be overcome. The creation of acceptable mitigation proposals is hindered by the difficulties associated with the assessment of environmental impacts and evaluation of habitat and resource values. The physical outcome of mitigation projects has been less than successful for a variety of reasons: insufficient project monitoring, the lack of clear resource goals, and technical problems with actual habitat construction.

Active Port participation can help to improve certain aspects of the planning and implementation of mitigation projects. Public ports have an interest in the greater harbor area, not just specific development sites as is the case with private developers. A harborwide approach to mitigation provides greater opportunities for siting mitigation projects and enhances permit "certainty" for long-range development plans.[5] Ports can conduct or collaborate on necessary research. An improved information base can aid in both impact assessment and establishing trade-off values. Ports can form liaisons with environmental management units and in turn predict their concerns in advance of project review. Port projects that are environmentally sound and successfully compensate for adverse impacts can lead to greater public support for the port authority.

Whereas there may be numerous reasons for Port involvement with mitigation, it is important to keep in mind that compensatory mitigation is not a panacea. The primary project must stand on its own merits. In accordance with the review criteria established by the Section 404(b)(1) guidelines, the proposed project must be water-dependent or water-related, or part of an essential public service; there must be no practicable alternatives to the project that would be less environmentally damaging; and all steps to avoid or minimize impacts must be taken. Compensatory mitigation is then used to offset unavoidable adverse impacts.[6]

## Notes

1. The sources for much of the research and references for the case studies in this chapter are three Master's theses on file at the Institute for Marine Studies, University of Washington. The theses, also available through the UW library system, are: "Environmental Concerns and Port Development" by Ann Wessel (1986); "Seaport Development and Environmental Mitigation: Coastal Permitting as a Bargaining Process" by Richard Morse (1986); and "Estuarine Mitigation Banking: A Chance for Predictability" by Raymond Maddux (1986). The sources cited in these theses include interviews with permitting officials in the Corps of Engineers and in state and city regulatory agencies, personal communications, Memoranda of Agreement, Memoranda of Understanding, correspondence and reports relating to applications, newspaper articles, House Subcommittee testimony, minutes of public hearing, and conference papers.

2. Federal agencies involved included the Environmental Protection Agency, Fisheries and Wildlife Service, National Marine Fisheries Service, and the Corps of Engineers; state and local agencies included Washington State Departments of Fisheries and Game, the city of Tacoma, and the Puyallup Indian Tribe.

3. HEP uses indicator species to calculate the supportive capacity of a particular habitat in the form of Habitat Units:

$$\text{HUs} = \frac{\textit{Study Area Habitat Conditions}}{\text{Optimal Habitat Conditions}} \times \text{Acres}$$

The ratio of study area conditions to optimal conditions is referred to as the Habitat Suitability Index (HSI).

4. For a complete description of the mitigation process, see Donald W. Rice, "Impact and Mitigation," American Society of Civil Engineers Dredging '84 Specialty Conference, Florida, November 14–16, 1984.

5. From a resource manager's perspective, a comprehensive strategy is more likely to address resource needs than an piecemeal approach. In fact, some authors recommend that mitigation not be developed without a comprehensive resource management scheme (Baker 1984; LaRoe 1979).

6. Further, not all types of impacts are mitigable: Habitats that are designated "resource category one" by the USFWS are not eligible for mitigation because that agency has determined that they are of sufficient national or regional importance to be preserved unaltered.

## References

Ashe, Daniel M. 1982. Fish and wildlife mitigation: Description and analysis of estuarine applications. *Coastal Zone Management Journal* **10:**1, 2.

Baker, Gregory F. 1984. An analysis of wetland losses and compensation under the Clean Water Act Section 404 program: Managing natural resources through mitigation. Master's thesis, University of San Francisco.

Cooper, John W. 1986. An overview of estuarine mitigation projects in Washington State.

U.S. Fish and Wildlife Service, Division of Ecological Services, Olympia, Washington, pp. 4–5. Draft of a paper in progress.

Dolan, Mary G. 1985. A Baltimore Harbor enhancement plan. In *Coastal zone '85*. Symposium Proceedings, Baltimore, July 30–August 2. Baltimore: American Society of Civil Engineers, pp. 447–52.

Eliot, Wendy. 1985. Implementing mitigation policies in San Francisco Bay, a critique. Oakland: California State Coastal Conservancy.

Fehring, William K. 1980. Mitigation of port development at Tampa, Florida. In *Coastal zone '80*. Symposium proceedings, Hollywood, Florida, November 17–20, American Society of Civil Engineers, pp. 1608–20.

———. 1982. Ports, industry, and fisheries—can they coexist? In *Improving multiple use of coastal and marine resources,* ed. John J. Reintjes. Proceedings of a symposium jointly sponsored by the International Association of Fish and Wildlife Agencies and the American Fisheries Society, South Carolina, September 22, pp. 30–38.

———, and J. Johnston. 1984. Development of an estuary-wide mitigation strategy for future dredging at the Port of Tampa, Florida. Dredging '84 Specialty Conference, November 14–16, American Society of Civil Engineers.

Gaby, Ronald, and Stephen Langley. 1985. Seagrass mitigation in Biscayne Bay, Florida. In *Coastal zone '85*. Symposium Proceedings, Baltimore, July 30–August 2, American Society of Civil Engineers, pp. 904–19.

Inoue, Satoshi. 1984. Urban coastal zone and port development in Japan. *Coastal Zone Management Journal*. **12**(1):57–81.

Krulitz, Leo M. 1979. Federal legal background for mitigation. In *The mitigation symposium,* ed. G. A. Swanson. General Technical Report RM-65, Rocky Mountain Forest and Range Experiment Station, Fort Collins, Colorado, U.S. Forest Service, pp. 19–26.

LaRoe, Edward T. 1979. The biological basis for estuarine ecosystem mitigation. *The mitigation symposium,* ed. G. A. Swanson. General Technical Report RM-65, Rocky Mountain Forest and Range Experiment Station, Fort Collins, Colorado, U.S. Forest Service, pp. 90–92.

McGolgin, Ian, ed. 1979. Columbia River Estuary regional management plan. CREST Council, Astoria, Oregon.

Maddux, Raymond. 1986. Estuarine mitigation banking: A chance for predictability. Master's thesis, Institute for Marine Studies, University of Washington.

Morse, Richard. 1986. Seaport development and environmental mitigation: Coastal permitting as a bargaining process. Master's thesis, Institute for Marine Studies, University of Washington.

Neupert, Mary. 1982. Who benefits, who pays, who controls? A study of public enterprise accountability. Ph.D. diss., University of Washington.

Ortman, David E. 1984. Conservation representative, N.W. Office, Friends of the Earth. Prepared statement submitted as testimony before the Subcommittee on Fisheries and Wildlife Conservation and the Environment of the Committee on Merchant Marine and Fisheries, House of Representatives, 98th Congress, June 6.

Perez, Armando I., Luis Ajamil, Carmen Lunetta, Marilyn Reed, and Nancy Brown. 1983. Environmental protection program for the expansion of the Port of Miami, Florida. In *Coastal zone '83*. Symposium Proceedings, San Diego, California, June 1–4, American Society of Civil Engineers, pp. 1090–1104.

Perry, Ernest L. 1984. Fish and wildlife mitigation: Costs and implications for port development in the 80's, January. Submitted as testimony before the Subcommittee on Fisheries and Wildlife Conservation and the Environment of the Committee on Merchant Marine and Fisheries, House of Representatives, 98th Congress, June.

Quarterman, Philip J. 1985. Estuarine mitigation policy in Oregon: A history. In *Coastal zone '85*. Symposium Proceedings, Baltimore: July 30–August 2, American Society of Civil Engineers, pp. 875–94.

Race, Margaret Seluk. 1985. Critique of present wetlands mitigation policies in the United States based on an analysis of past restoration policies in San Francisco Bay. *Environmental Management*. **9**(1):71–82.

———, and Donna R. Christie. 1982. Coastal zone development: Mitigation, marsh creation, and decision making. *Environmental Management*, **6**(4):317–28.

Regional Planning Council. 1982. The Baltimore Harbor environmental enhancement plan. Baltimore, September.

Rice, Donald W. 1984a. Impact and mitigation: Los Angeles/Long Beach landfill plan. Dredging '84 Specialty Conference, American Society of Civil Engineers, Florida, November 14–16.

———. 1984b. Los Angeles Harbor kelp transplant project. In *Gambling with the shore*. Proceedings of the 9th annual conference of The Coastal Society, October 14–17, Atlantic City, New Jersey, 303–309.

Smith, Scott E., ed. 1983. A mitigation plan for the Columbia River Estuary. CREST Council, Astoria, Oregon.

U.S. Fish and Wildlife Service. 1981. Mitigation policy. *Federal Register* **46**(15):7644–56.

Wessel, Ann. 1986. Environmental concerns and port development. Master's thesis, Institute for Marine Studies, University of Washington.

## *Additional Reading*

American Association of Port Authorities. 1986. *Mitigation of environmental impacts—New challenges, new opportunities*. Alexandria, VA: AAPA.

———. Mitigation Subcommittee. 1984. Regulatory reform for Section 10 and Section 404 permits. Draft of subcommittee report, Alexandria, VA: AAPA, 4–5.

# Chapter 13

# WATERFRONT REVITALIZATION:

## Ways to Retain Maritime Industries

*Robert F. Goodwin**

---

Since the mid-1970s waterfront cities large and small have invested heavily in the revitalization of their urban waterfronts. Abandoned wharves, docks, and piers have been transformed into attractive destinations for tourists and local residents alike. Gone are the sounds of riveting from shipyards, the smell of fish being baled out of fishing boats, the sight of longshoremen crowding the entrance to the union hall. In their place are the sounds of cash registers, the smells of ethnic food markets, and the sight of tourists lining up for ice cream cones and harbor tours. The old harborfront has disappeared, and an extension of the central business district's (CBD) retail and entertainment functions has begun to occupy its former site. On the long-vacant industrial shorelines of New York's Manhattan Island, Baltimore's Inner Harbor (see Figs. 13.1 and 13.2), and Chelsea's naval yard, new developments that return life to the waterfront are tinged only by nostalgia for what has long been lost. But in some cities pockets of surviving maritime industry are succumbing to the "success of revitalization."

A plea for protection against displacement is being heard, paricularly from fishermen and the industries that serve them: boatyards, seafood plants, fish brokers, and auctioneers. Some maritime community leaders fear that erosion of the "critical mass" of support industries necessary to retain local fishing fleets and other maritime activity will result in loss of business and jobs to other ports (see Chapter 8).

After a decade or more of revitalization, it is appropriate to assess those efforts. What policies have shaped successful waterfront redevelopment? Where might failures be averted through better planning and policy design in the future? Central to this assessment is the need to choose between two views of the urban waterfront as

*Partial financial support from Washington Sea Grant Marine Advisory Program under NOAA Grant #NA86AA-D-SG 044, Project #A/CZ-1 is gratefully acknowledged.

**Figure 13.1.** Baltimore's inner harbor in the 1950s. Note the small-size vessels and the downtown prior to rejuvenation. (Courtesy: M. E. Warren)

**Figure 13.2.** Baltimore's inner harbor in the 1980s. An active people place, including a marina, aquarium, science center, history vessel, and market adjacent to a revitalized downtown. (Courtesy: M. E. Warren)

a resource. The first holds that the waterfront is an extension of the central business district, its activities, and amenities; the second views the waterfront as a unique district where activities and amenities derive from the contiguity to navigable water. Predictably, but not uniformly, the former view has adherents representing private property owners, real estate and development interests, and city development agencies seeking private financing of public amenities, whereas the latter view has proponents among fishermen, maritime business interests, and coastal resource management agencies. Public ports are often split over the issue, depending on local circumstances such as Port land holdings, the economic vitality of local marine enterprises, and the political climate. In some cases, Ports have remained uninvolved in waterfront revitalization entirely. The pivotal role to be played by public ports in resolving which view of the urban waterfront is ascendant—planning from the water inland or from the uplands waterward—is explored in this chapter.

## FROM WORKPLACE TO PEOPLE PLACE IN THREE DECADES

Since the end of World War II, three major trends in society have transformed the old working waterfronts of our nation's port cities: the ascendancy of airline carriers over steamships for transporting people (and some goods) within and between continents, the technological revolution in the scale and efficiency of waterborne cargo shipping and handling, and the migration of industry from the central city to the suburbs (*see* Moss 1976).

With the introduction into service of the Boeing 707 in the late 1950s came the demise of the transoceanic passenger liner. The speed of safe, reliable jet transport reduced transatlantic travel from days to hours. Wide-bodied jets now fly nonstop halfway around the globe, whisking tourists and business folk to destinations quite beyond their reach two decades ago. The slow ceremonious leave-taking at dockside, the paper ribbons trailing down from the ships' railings, the gradual loss of recognizability of familiar faces as the vessel moved out of the harbor, the last smudge of smoke on the horizon—these images belong to an age long eclipsed by the airport, its security checks, and hurried handshakes in anonymous terminal buildings. Is this Denver or Dallas? O'Hare or Omaha? In steamship days at least you knew whether you were in the Port of New York or San Francisco.

A better than tenfold increase in the size of bulk carrier vessels, the advent of containerized cargo carried on increasingly large and faster container ships, and the mechanization of dockside cargo handling all have had a profound effect on the location and scale of port activity. The use of major load center ports coupled to intermodalism has resulted in shipping lines abandoning vintage port terminals adjacent to the central business districts of cities and dropping many ports-of-call entirely (see Chapter 6).

Within port cities, docks and piers built for breakbulk freighters and passenger liners have been abandoned; new marginal wharves, designed for containerships and bulk carriers 600 feet and more in length, have been constructed on filled lands or other flat sites well beyond the congestion of the central business district.

Backup land required for staging containers awaiting transshipment is simply unavailable adjacent to the old working waterfront, since each container berth consumes from 10 to 20 acres of level, paved area and must be served efficiently by rail and highway connections (see Chapter 4). The transit sheds on downtown wharves and piers that once were used to store cargoes handled by gangs of longshoremen were quickly rendered obsolete when the pilfer-resistant, sealed container replaced the wooden pallets, rope slings, and handcarts of earlier times.

Between the old working docks and the downtown core of port cities were industries associated with the port: warehouses, wholesale traders, and manufacturers who capitalized on easy proximity to raw materials carried on ships and to vessels leaving regularly for distant markets. Other industries drawn to the local labor pool and easy access to markets within the central business district shared the same area of the downtown periphery. But the multistory warehouses and factories that housed these enterprises lost their advantage when truck transport began to eat into the railroad cargo business. Freed from fixed rail locations, wholesaling and light manufacturing moved away from the CBD to cheap flat sites in the suburbs. Manufacturing processes could now be organized horizontally in large, open-span, single-story structures, and a ubiquitous road network permitted materials and finished products to be moved by truck.

The consequence of these trends—technological changes in passenger transportation, shipping, cargo handling, and in wholesaling and manufacturing—was that by the 1950s a crescent-shaped "zone of discard" had been left around the periphery of the CBD where it touched the waterfront and the railroad yards.

During the 1950s and early 1960s, public works activities dealt another series of blows to the discarded working waterfronts. Urban renewal schemes razed blocks of dilapidated, but in retrospect historically significant, buildings, and urban highway construction plowed a path of least resistance along the waterfronts of Boston, New York, San Francisco, Seattle, and many other port cities. The waterfront, in addition to being abandoned, was now walled off from the core of the CBD by vacant land, elevated highways, and disused railroad rights-of-way. The old working waterfront had become "a ghost area—a deserted, inaccessible, depressing reminder of better days" (Wrenn 1983).

Not quite. Pockets of maritime industry remained: passenger ferry terminals, fish processing and wholesaling, tugboat moorage, garbage scows, and a few water-dependent manufacturing enterprises continued to call the underutilized waterfront home.

But the major source of revenue—cargo handling—had dried up. Funds to maintain seawalls, pilings, and pier sheds were hard to find. Unchecked damage from marine worms, waves, and weather was eroding away the abandoned structures surrounding the residual maritime industries.

Enter the 1960s: the decade of activism, inner city revolt, and neighborhood organization. Urban renewal was declared a dismal failure and new urban freeway projects were killed by articulate, well-organized neighborhood and civic leaders. Historic conservation, "adaptive reuse," and attention to the needs of the pedestrian and inner city residents overrode the interests of the suburban community. Given national focus by Ladybird Johnson and fueled by federal largesse, trees were planted, sidewalks widened, new mass transit schemes were funded, and grants for historic preservation stemmed the loss of noteworthy urban landmarks. Inevitably, attention was focused on the old working waterfronts. Urban renewal schemes, office, retail, and residential developments, and city beautification programs transformed the old commercial districts adjacent to the waterfront into bustling new extensions of the central business district. The waterfront's low land values and adaptable structures became a magnet for public and private development capital. Boston's historic wharves close to the Faneuil Hall redevelopment area were converted to condominiums; revitalization of San Francisco's Ghirardelli Square and Embarcadero district propelled the conversion of Fisherman's Wharf into a tourist attraction; and planning for Baltimore's Inner Harbor got underway.

Passage of the federal Coastal Zone Management Act (CZMA) in 1972 reinforced the nation's new interest in the waterfront. The CZMA was one of dozens of environmental measures enacted by state and federal legislatures during the early 1970s (see Chapter 5). Certain salient features of the CZMA are worth noting. Unlike the National Environmental Policy Act, the Clean Air Act, and the Water Pollution Control Act, the CZMA is not only a statute designed to conserve environmental resources, it is first and foremost a measure designed to encourage coastal and Great Lakes states to *plan and manage uses* of the land and water in the nation's coastal zone. Ports and maritime industry are singled out for attention; and through designating urban waterfronts as Geographical Areas of Particular Concern, states have focused resources on planning for their redevelopment (Hershman et al. 1978). Other parts of the CZMA identify national policies for the conservation of historic and cultural resources, the restoration of urban waterfronts for public uses, and, importantly, for reserving land for water-dependent industry. These national policies, coupled to the overriding concern for providing public access to the shore, constituted a broad mandate to revitalize deteriorated urban waterfronts in a comprehensive way.

The paucity of public access to old working waterfronts has been remedied effectively in scores of North American cities. And public access has been broadly conceived to encompass more than *physical* access to the water's edge: *visual* access to working port areas has been achieved through viewpoints, observation towers, and, in one case, jumbo-size periscopes. Signage and storylines at these locations provide *interpretive* access to intrinsically fascinating maritime activities where physical public access would be disruptive or dangerous. Maritime historical museums and waterfront interpretive centers add another dimension to interpretive access: an intellectual and emotional link to the port's past, present, and possible futures.

Historically significant structures that for years had languished into disuse now house residences, offices, retail stores, restaurants, and, too rarely, a modest complement of marine-related businesses. Like the corpses of steam locomotives on display in small-town railroad depots, these old wharves and pier sheds have lost that which gave them authenticity: noise, hot smells, and productive messiness. Whereas most would argue that it was better to reuse them than lose them, there is a growing sentiment that something real has been lost that might have been recaptured: not general cargo handling—that has moved away for good because of containerization—but perhaps some contemporary marine activities of a lighter sort, activities tolerant of inquisitive pedestrians. Fish auctions and fish-handling halls, Coast Guard vessel moorage, research and survey vessels, tug dispatch stations, visiting naval vessels, transient pleasure craft moorage, historic vessels, yacht clubs, and harbor tour docks are among the authentic, contemporary maritime uses that might be courted to coexist with the new, chic, gentrified waterfront. But they may require subsidies to remain or relocate there, since many of these uses do not pay the rent necessary for refurbishing the structures that house them, or to which they moor.

Seattle's city council recently approved a policy requiring owners of piers in the "Harborfront" area adjacent to downtown to upgrade their perimeter docks and provide cleats for vessels to moor there. These requirements are triggered when pier owners apply for a redevelopment permit. The city's intent is to have tourist-serving restaurants and retail stores on the ground floor and the offices on the upper floors provide a subsidy for commercial vessel moorage. What may be achieved is the retention of authentic maritime activity on Seattle's historic Harborfront.

In some U.S. ports the booming cruise ship industry has brought the passenger ship back to the downtown waterfront. One-way Caribbean, Hawaiian, and Mexican coastal cruises have benefited the mainland ports that can best serve those routes: Miami, Canaveral, and Everglades on the East Coast, Los Angeles, San Diego, and San Francisco on the West Coast. The Passenger Services Act has inhibited cruise ship terminals being developed in Pacific Northwest ports[1] and has shunted the lucrative Alaskan cruise business to Canada (see Fig. 13.3).

So much for recapturing maritime authenticity in revitalized downtown waterfronts. But a more serious problem occurs where successful redevelopment of abandoned sections of the waterfront begins to encroach on active marine industrial areas.

## THE DISPLACEMENT ISSUE: SAVING WATERFRONTS FOR MARINE INDUSTRIES

Seasoned observers of the changing urban waterfront have reported that the boom in waterfront redevelopment is threatening the survival of traditional maritime en-

**Figure 13.3.** The cruise ship *Noordam* of Holland America Lines, serving Alaska, moored at Canada Place in Vancouver, B.C. (Courtesy: Holland America Line/Westours Inc.)

terprises in many of our coastal cities (Breen and Rigby 1985). For example, on Seattle's Lake Union, the last accessible pier to which fishing boats could moor while their owners performed repairs on them has become a recreational marina; and on the upland portion of the site, several businesses doing marine as well as non-marine repair work have been replaced by a restaurant/office/retail complex (see Fig. 13.4).

Small-boat building and repair yards in the Marinship district of Sausalito, California, are threatened by gentrification of the surrounding area; tugboats in Jersey City and lobster boats in Boston Harbor are being displaced; in Portland, Maine, the city struggles to maintain the fishing fleet and seafood processors on its downtown waterfront.

Portland, Maine, and Seattle, Washington, are examples of the many cities attempting to hang on to their marine industries through restrictive land-use regulations. Yet in both these cases, municipal policymakers are under great pressure to lift restrictions and allow non-water-dependent uses. Waterfront developers and property owners in both cities claim, with some justification, that the demand for land for water-dependent use does not exist, and they are left with no reasonable use of their property. Static or declining fisheries create ripples through the whole seafood and marine support services industry, and foreign yards outcompete U.S. shipyards for new vessels and repair work in the merchant marine fleet. In the Pacific Northwest a slump in the wood products industry has caused lumber and plywood mills to close down: many of these are on the shoreline. In Seattle commercial real estate brokers identified a surfeit of industrial waterfront land for sale just as the city council adopted a *more* restrictive water-dependency policy for certain segments of the city's industrial shorelines. Some observers in Seattle fear that implementing this "urban maritime" rezoning might expose the city to legal action by property owners who contend that an unconstitutional "taking" of property without compensation would occur. Under the current, less stringent "urban stable" shoreline designation, developers in Seattle attest to having made bona fide attempts to develop waterfront parcels in water-dependent uses, but failed to attract new marine businesses. Similar experiences are reported on the Portland, Maine, central waterfront where the W-1 restrictive maritime zone applies.

**Figure 13.4.** The H.C. Henry Pier was the last dockside repair facility available to commercial vessels on Seattle's Lake Union. In 1985 it became the site of a mixed-use restaurant/retail building and recreational boat moorage with public access. (Courtesy: Robert F. Goodwin)

## *The Urban Land Market*

The urban land market, if left to itself, results in the allocation of land to its "highest and best use." Those uses that gain most from being in the core of the central business district (retail stores and offices) are able to outbid other uses (hospitals, apartment houses, warehouses), which occupy sites farther away from the prime locations in the CBD. The scarcity of land at the core forces the most intensive (vertical) development to occur. Farther out, land is in greater supply, is used less intensively (horizontally), and commands decreasing prices per unit area as distance increases from the urban core. Regulations (zoning, building codes, and such) are imposed on land use and development to address issues that the free market in urban land is not capable of addressing. Most notable of these are the "spillover effects" of one activity on an adjacent one: noxious noise and effluents, view blockage, nighttime commercial operations, traffic congestion. Other motives for regulation concern the creation or retention of public benefits that are not captured in the price of land: preserving wooded hillsides or downtown amenities, such as landscaping and street furniture, is possible only through zoning ordinances and building codes. Historic districts have been designated to preclude wholesale alterations to their unique character. But where regulation leaves landowners no "reasonable use" of their property, courts have ruled the regulations an unconstitutional use of the police power resulting in the taking of private property without just compensation—an action expressly forbidden by the Fifth Amendment to the U.S. Constitution.

Waterfront land, by virtue of three special characteristics, becomes allocated by the urban land market in ways different from land away from the shore. First, being the link between upland and navigable water, it is valued by water-dependent uses discussed earlier. Second, because of their amenity as recreational and scenic sites, shoreline parcels tend to be developed earlier and more intensively than adjacent uplands. And, third, because its supply increases *linearly* with distance away from the urban core, as opposed to ordinary land whose supply increases with the *square* of distance, shoreline land commands premium prices. Evidence of these facts is found not only, for example, along Chicago's North Shore district where high-rise apartments line the lakeshore, but also in rural shorelines far from urban areas. Here, perhaps the only developed land is found ringing lakes and ponds and lining river banks, whereas, one block from the water's edge is agricultural and forest land.

The development pressure on urban shorelines, then, is intense. Without regulation protecting navigation and commerce uses, urban development of the kind found surrounding the core of the CBD would tend to encroach on the water's edge: hotels, office buildings, residential condominiums.

For most marine businesses the option of using scarce, expensive waterfront sites more intensively is severely limited. Boatyards, seafood processors, marinas, and marine repair firms cannot expand vertically. Even stacked dry-storage facilities for small boats are limited to three levels by the handling equipment they utilize.

Moreover, as gentrification comes to the waterfront and nearby blocks, tourists, new residents, and their cars congest streets and cram parking lots. Truck access to waterfront businesses becomes more difficult as a new "people place" displaces the old workplace environment. Fisherman's Wharf in San Francisco and Boston's Fish Pier come readily to mind as examples of working waterfronts under pressure from tourism and recreation crowds.

The common thread running through these and other cases of "displacement" is an economic one: multistory development brings higher returns to waterfront landowners than does less intensive traditional marine uses. Speculation that such uses will eventually materialize drives up property values, tax assessments, and rents. Much like farmers on the rural/urban fringe, marine business owners exhibit a kind of "dual personality" about land use restrictions designed to "protect" their enterprises. Geographer Jean Gottman's pithy observations about farmland preservation are worth repeating, particularly his report of a comment made by one farmer somewhere on the megalopolitan Eastern Seaboard:

> We farmers raise three crops. We go onto a place in youth and raise a *family*. We spend our working years producing *milk*. When we get ready to retire, we harvest enough *capital gain* from the land to keep us in our old age. (Gottman 1964, 274. Emphasis added.)

Marine business proprietors, like our farmer friend, welcome protection for their businesses while they are in production, but want a free market in urban land when it comes time to sell out and retire! Again, returning to Seattle's Lake Union shorelines, landowners, many of whom have marine businesses, are found on the membership list of two organizations representing almost diametrically opposite views on shoreline land use policy. The Seattle Marine Business Coalition, organized in 1983 to stem the conversion of waterfront parcels to non-water-dependent uses (notably restaurants), lobbies vigorously for the mayor's proposed "urban maritime" shoreline zone. The Lake Union Association, a much older organization representing private property owners around the lake, opposes the restrictive zoning with equal vigor. Yet their memberships overlap! Having one's cake and eating it, too is an apt metaphor to describe the mutually exclusive and ultimately destructive dual goals of the marine business/landowner in waterfront cities throughout the nation. Policymakers on city councils, on Port or harbor commissions, and in state coastal management agencies are searching for rational and politically acceptable arguments to resolve this shoreline land use conundrum.

## *Regional Economic Development*

A criterion commonly applied in land use policy evaluation is: What are the regional economic development implications of this decision? Will this policy promote or retard economic development? In the case of marine enterprises, do they contribute enough to the region's economy to warrant protection from the forces of the urban land market?

Marine enterprises, like other economic activities, may be placed in one of two categories: a "basic," or "export" category, or a "nonbasic," or "service" category. A basic activity is one that sells its product or service to customers outside the region (or city) under consideration; a nonbasic activity sells to customers residing within the region. In reality many enterprises fall into both categories, simultaneously selling to both home and distant markets. Revenues from basic enterprises are considered "new" money, which generates local employment and produces income available for spending on locally produced and imported goods and services by local consumers. Nonbasic enterprises are responses to markets generated by basic industries. Economic development councils, public port districts, and chambers of commerce understand this distinction when they court manufacturing firms to locate in their region. Local governments, through their zoning ordinances, support the effort to attract new basic industry by ensuring that an adequate supply of industrially zoned land exists.

As noted in Chapter 8, the seafood industry is a basic activity supporting an array of enterprises that serve it, and many of these firms require waterfront sites. There are industries other than seafood harvesting and processing that have a similar dependency on a waterfront site: boat dealers, marinas, yacht brokers, shipbuilders, and marine cargo terminals are obvious examples. But less obviously, a broad group of manufacturing industries are found concentrated along urban shorelines. These firms use the harbor to receive raw materials, ship finished products, or, in former times, discharge industrial effluents into public waters. Sawmills, pulp and paper manufacturers, oil refineries and petrochemical plants, batch concrete manufacturers, flour mills, marine construction yards, and iron and steel mills exhibit varying degrees of "water-dependency" (Goodwin 1987).

Such industries constitute an important component of the economic base of port cities. They consume large blocks of waterfront land and are often noisy, messy, and hazardous industrial environments. Whereas shipyards, active marine terminals, and heavy manufacturing industries are located usually well away from the central urban waterfront, many of the smaller marine repair, storage, and service firms and seafood processors and wholesalers are found in or close to waterfronts undergoing redevelopment. And it is usually these smaller-scale enterprises that are most threatened by displacement.

By and large, the activities that are displacing traditional marine industries are neither part of the basic sector of a city's economy, nor are they water-dependent. And those non-water-dependent activities that do have a basic or export component—certain office and hotel functions, for example—can be accommodated on upland sites without compromising their ability to function as engines of economic development. The offices housing attorneys or architects serving distant clients, the hotel or restaurant catering to out-of-town business folk and tourists, or the condominiums providing housing services to the local population need not compete for waterfront sites against boat repair yards or marinas. A hotel, condominium apartment, commercial office building, or restaurant may capture higher rents from the amenity value of a waterfront site than from an inland one, but each can prosper away from the immediate water's edge while still gaining advantage of water views. A sound regional economic development argument can therefore be made for restricting the use of shoreline areas to those activities that are water-dependent.

## *Land Use Controls that Protect Maritime Use*

Faced with rising land values and property tax assessments, the attraction of reaping capital gains from their land, and the congestions caused by tourists and residents, traditional marine businesses are unlikely to remain long on redeveloping urban waterfronts. Many of these enterprises are caught in cyclical or boom-bust business oscillations to which competing land uses are more immune. Once displaced by CBD activities housed in major new structures, marine uses are unlikely to return to their former sites, even were their businesses to undergo expansion in future years. Changes in land use to more intensive utilization of sites operate like a "ratchet": they do not easily reverse direction. And whereas the loss of one marine firm may not be unduly consequential (a competitor absorbs the unfulfilled demand for the good or service produced by the displaced firm), the long-term, cumulative losses could undermine an important component of the city's economic base and erase a part of its maritime culture.

The remedies available to city and state agency policymakers are similar to those useful in conserving agricultural lands, historic structures and districts, or urban open space: restrictive zoning, tax abatement, public acquisition of critical parcels through fee-simple or less than fee-simple purchases, or transfer of development rights to surrounding developable lands.

Restrictive zoning will work only as long as marine businesses remain viable and some protection from the spillover effects of tourist-serving uses is in place. It will not work on vacant land parcels for which there are no new marine uses "waiting in the wings"—such a regulatory restriction invites legal challenges under these circumstances.

Property tax abatements have been used in attempts to induce farmers to maintain land in agriculture at the rural/urban fringe. Typically, a participating farmer agrees to hold land in agricultural use for a set time period (ten years in Washington State), during which the land is assessed on a "current use" basis. If this condition is violated and the farmer sells the land to a developer, the difference between the property taxes actually paid and what would have been paid is immediately assessed on the farmer together with a penalty tax. Critics like Gottman (1964) point to two flaws in such "open space taxation" schemes: first, farmers facing the chance to accrue large capital gains seldom participate. Second, when they do participate, it is often worth paying retroactive taxes plus penalties to reap the capital gain early; and even where a farmer fulfills the conditions, the agreement eventually expires. Tax abatement measures are not permanent solutions to preserving land in existing uses. One could surmise that they would be no more successful when applied to waterfront parcels in marine uses than when applied to agricultural lands at the edge of the city.

To achieve coastal management objectives, several authors have recommended the use of "transferable development rights" (TDRs). Usually, the objective is to preserve wetlands or other critical habitats from encroaching development. TDRs have been used successfully for preservation of historic districts. However, setting up and implementing TDRs is difficult, both legally and practically (McGilvray et al. 1985). An area under development pressure is "frozen" or down-zoned, whereas a contiguous or nearby area also under development pressure is up-zoned to permit higher densities or building heights. To build in the "receiving" area, the developer must purchase development rights from landowners in the "sending" area. A free market in development rights rewards the landowner whose property is frozen in existing use, while permitting appropriate additional development to occur on receiving area sites. TDRs were designed with equity in mind: to eliminate the "windfalls" and "wipeouts" induced by restrictive zoning populations in developing areas, and thus avoid running afoul of the "taking issue."

A permanent but more costly tool to protect special areas from development pressure is public acquisition of development rights. A public entity—usually a municipality—pays the landowner the difference in the value of land in existing use and the same land at "highest and best" (market) use. Once implemented, this scheme precludes forever development of the land beyond the uses currently in place.

In King County, Washington, voters passed a bond issue to purchase development rights to certain agricultural lands after a voluntary open-space taxation-abatement measure proved to be unsuccessful. Because the landowners are left with some use of their property and, hence, a residual land value, purchase of development rights is cheaper than the fee-simple acquisition of the property. Other less than fee-simple acquisition schemes include purchasing access easements to the shore.

Where waterfront land is vacant and where marine businesses are not expanding in the near future, it is likely that fee-simple purchase of the property by a public entity will be the only effective way to "bank" waterfront land for anticipated, but not immediately evident, marine uses. Ideally, such a public entity would play an active promotional role in attracting future marine-oriented tenants. Furthermore, that entity could be a co-developer of the property, demolishing dilapidated structures, improving access, repairing bulkheads, and preparing the site for future use. Land "in the bank" could be leased for interim short-term uses pending development for marine enterprises: parks, seasonal fairs and festivals, and storage sites are some examples. But a clear understanding would have to be reached with interim users that these uses could be terminated any time a bona fide marine tenant or buyer was found.

## PUBLIC PORTS AS SHORELAND BANKERS

Public port authorities, whether they are arms of general-purpose municipal governments or special-purpose districts, are uniquely equipped to carry out the function of shoreland banker. Reserving land for future use is common practice in many, if not most, port districts. Land is part of the capital assets of a Port, to be used to attract industry to the region. Ports operate industrial parks, secure permits to fill shorelands for future cargo-handling terminals, and purchase land around airports for expansion of runways and terminal facilities when the demand warrants it (see Chapter 1). Acquiring, improving, and marketing waterfront land for future small marine businesses (marine industrial parks?) is a feasible and appropriate task for the port authority. Marine businesses are a natural constituency of the public port. Their linkages to traditional port functions—fishports, recreational smallcraft harbors, and general harbor management—enmesh small marine businesses within the long-standing, traditional port community. Where the urban land market fails to account for the unique suitability of shoreline sites for water-dependent uses, calling on public ports to correct market imperfections is consistent with the original purposes of public enterprise (see Chapter 4).

Two arguments can be made against public port intervention in this area: ports are running short of capital, and the marine businesses for which waterfront land would be "banked" may not expand in the foreseeable future. Were waterfront parcels to remain vacant for long periods of time, critics maintain that scarce Port capital would be tied up in nonproductive use and no revenue would flow from the Port's investment to defray debt service on bonds issued for such purposes. These objections can be answered in two ways. First, the decision to bank waterfront land for future use is a political one, which the district's voters might support even though it would cost them more in the future. Second, waterfront land adjacent to the CBD constitutes an appreciating asset. If the experiment failed (that is, no marine-oriented business development occurred on the land banked by the Port), the lands could be declared surplus and sold or leased for nonmarine development. By reassembling many small parcels into a few larger ones, the port could realize development of a scale difficult to achieve on private, multiple-ownership lands. And if necessary, the authority's power of eminent domain could be used to acquire key parcels from unwilling sellers.

## MIXED-USE DEVELOPMENT AND THE QUESTION OF COMPATIBILITY

Not all marine businesses are incompatible with tourist and visitor-serving uses. Harbor tour boats, recreational moorage, cruise ship terminals, and retail outlets for wholesale fish buyers are uses that can coexist with restaurants, boutiques, public aquariums or maritime museums, and that might well fail without them. Under these circumstances, mixed-use developments occupying either new or adaptively reused old structures are seen as appropriate ways to redevelop decaying waterfronts, while restoring authentic marine activity. Marine uses that cannot afford the construction costs or land rents so close to the CBD receive an implicit subsidy by being part of a mix of more intensive retail and commercial uses. As described in Chapter 8, this principle is being used on Boston's Fish Pier where commercial office space and seafood handling occur in the same structure (see Fig. 13.5). Planners there are reserving all ground floor space for seafood activities while restricting commercial offices to upper floors. The "mix" occurs because they are under the same roof, but

**Figure 13.5.** Boston's Fish Pier, where the upper floors of the buildings are reserved for office space, and the lower zone is an active fish processing and wholesale market area. (Courtesy: Massport)

the uses themselves are segregated since their needs are so different. And on Seattle's Harborfront, city officials now permit more office use on the upper floors of historic pier sheds where ground floors house a mix of harbor tours, restaurants, and retail uses. They found that office use could mix quite well with tourist-related uses on the deck level. Mixed-use developments, then, will succeed if the uses are made as compatible as possible through flexible regulations and careful design.

Perhaps the use most incompatible with traditional marine industry is housing; and it is housing that developers would most like to locate in revitalizing waterfronts. Once a neighborhood is established and organized, residents will resist existing marine uses that generate noise, glare, dust, or other byproducts of the businesses. Enjoying the views of shipping and maritime activity at a distance is one thing; living next to a noisy, messy boatyard, shipyard, or working seafood dock is quite another! Such attitudes toward marine businesses will likely be felt as vehemently from hotel guests as from condo owners. The fear that overwater housing would displace the fishing industry led Portland, Maine, voters to restrict further mixed-use development on their downtown waterfront. The city council had allowed mixed uses, including housing, on four of the twenty-two downtown waterfront piers, but in May 1987 voters overrode the council through a referendum. The measure, which passed by a 2:1 majority, restricted all twenty-two piers to maritime uses for five years.[2] The consequence, city economic development planners fear, will be insufficient revenues to maintain the necessary waterfront infrastructure to house present and future marine activity in downtown Portland.[3] The city council of Gloucester, Massachusetts has taken similar action to implement a nonbinding referendum passed in 1985, banning all new residential developments on its downtown waterfront.[4]

## HARBORWIDE PLANNING FOR REVITALIZATION

In a theme issue of the *Coastal Zone Management Journal* addressing urban coastal management, special editor Robert Warren comments:

> Currently, city involvement starts and stops with a "showcase" redevelopment project in the central business district portion of the waterfront. . . The miles of additional shoreline that contain recreational areas, retirement communities, wetlands, industry, marinas, high-rise residential and commercial complexes, and open space, frequently including municipally owned land, are never viewed as components of a single coastal resource system or managed by local officials within an integrated framework. . . . If the mix of projects affecting the coastal zone is undertaken on an ad hoc basis, the results constitute *a coastal policy without debate or conscious choice*. (Warren 1979. Emphasis added.)

Warren's criticism of the way cities were managing urban waterfront land and water uses in the late 1970s gets to the nub of the "displacement" issue: planning for the revitalization of one part of the urban shoreline must proceed with an understanding of how such decisions affect the remaining shorelines. In other words, *the urban harbor should be managed as a single unit of land and water space that supports a complex, interlocking complement of uses*. Only then will the unintended adverse consequences of well-intentioned but imperfectly understood policies be averted.

Cities should not attempt to achieve every "good" coastal management goal at all sites (*see* Wagner 1985). It may be unwise to force physical public access into every industrial shoreline. Similarly, attempts to adhere to strict water-dependency criteria for uses in restored pier sheds may result in vacancies and poor maintenance; and overdevelopment of public access and viewpoints could create underused public improvements that are expensive to maintain. Each coastal management goal should be linked to specific site conditions in the framework of a harborwide development strategy.

Conversion of obsolete facilities to house tourist and visitor-serving activities is an appropriate way to achieve revitalization, support economic development goals, and improve environmental quality of the waterfronts. But the zone in which intensive redevelopment occurs should be limited to inactive, deteriorated structures and not be allowed to spread, willy-nilly, into nearby working waterfronts. Nor should spillover effects—traffic and on-street parking congestion, for example—be permitted to disrupt efficient truck access to marine businesses.

Above all, before embarking on "showcase" redevelopment projects, city planning agencies should conduct a thorough inventory of their industrial shorelines, interview marine business organizations to understand the site and locational requirements of their member firms, and designate zones for their protection and future expansion.

Public port authorities can be active partners in planning for the future of their harbors, entering the process during the early stages and speaking out for the needs of their "natural constituencies." City and port officials acting jointly should determine their respective roles in implementing the harbor development plan when its form has emerged. Too often Ports simply react to a plan for the city's waterfront, restricting their role to that of "damage control," protesting unfeasible public access demands, or resisting loss of shoreside sites for future marine cargo-handling facilities. Ports can insist that planning for future harbor developments proceed from navigable waters inland, rather than from the CBD waterward. And even where their landholdings include vacant downtown waterfront sites, Ports can still champion and help implement the goal of bringing back authentic maritime activity to the place where many cities were born: the water's edge.

## Notes

1. The Passenger Services Act prohibits foreign vessels from transporting passengers one-way between two U.S. ports. Since the majority of Alaskan cruise business is one-way and

the major Alaskan cruise line vessels are foreign flag, it is more expedient to embark the passengers at Vancouver, British Columbia, and start the cruise from a Canadian port.

2. Around the coasts, *National Fisherman,* July 1987, 16.

3. Personal communication, Clark M. Neily, city of Portland, Maine, economic development director, July 1987.

4. Around the coasts, *National Fisherman,* July 1987, 16.

## References

Around the coasts. 1987. *National Fisherman* **68:**3, 16.

Breen, Ann Cowey, and Dick Rigby. 1985. *Caution: Working waterfront. The impact of change on marine enterprises.* Washington, DC: The Waterfront Press.

Goodwin, Robert F. 1987. Measuring water-dependency: A Puget Sound example. In Proceedings, *Coastal zone '87.* New York: American Society of Civil Engineers.

Gottman, Jean. 1964. *Megalopolis, the urbanized northeastern seaboard of the United States.* Cambridge, MA: The MIT Press.

Hershman et al. 1978. Under new management: Port growth and emerging coastal management programs. Seattle and London: University of Washington Press, Washington Sea Grant, 212.

McGilvray, Laurie J., Glen D. Anderson, and Niels West. 1985. Managing coastal development: An evaluation of the transfer of development rights approach. *Coastal Zone Management Journal* **3:**1.

Moss, Mitchell L. 1976. The urban port: A hidden resource for the city and the coastal zone. *Coastal Zone Management Journal* **2:**3.

Wagner, Mary Jo. 1985. Water-dependency as a shoreline allocation criterion for siting maritime industries in Puget Sound, Washington. Unpublished Master of Marine Affairs thesis, Institute for Marine Studies, University of Washington.

Warren, Robert. 1979. The role of cities in managing the urban coast. *Coastal Zone Management Journal* **6:**2, 3.

Wrenn, Douglas. 1983. *Urban waterfront development.* Washington, DC: The Urban Land Institute.

# Chapter 14

# PUBLIC PORT ACCOUNTABILITY:

## A Framework for Evaluation

*David J. Olson**

---

Seaport accountability may appear straightforward enough at first glance: If seaports are public agencies created by government, they should be required to render an accounting for their actions and their performance according to well-established standards regularly applied to government departments. This simple formulation fails, however, to address several key questions. To whom are Ports accountable? Is it specifically to those adjacent waterfront communities most impacted by terminal development and operations, or is it more generally to the larger public interest? Which governmental jurisdiction should Ports render an account to? Is it the immediate government of the city abutting the harbor, or immediately affected neighborhood councils, or more distant county and state governments? What is it, exactly, that Ports are expected to be accountable for? Is it simply to render an account, after the fact, for decisions they have made and actions they have taken, or are there objective standards for reviewing their performance and guidelines against which their policy directions may be compared? Are seaports just like regular government agencies just because they are public authorities? Or is there something about the line of business they are in or the products and services they produce that makes them different and requires accountability systems specifically designed for seaports? These questions begin to show how seaport accountability is neither a simple subject nor one lacking in controversy within port communities.

*Although this paper is singly authored, and all responsibility for its content is mine alone, the larger research project of which this chapter is a part was conducted with Professor Michael Denning of Indiana University. The research was supported by the National Oceanic and Atmospheric Administration, Washington Sea Grant Program, project r/CA-15. This chapter was written while the author was a Visiting Professor at the University of Bergen, Norway. Grateful acknowledgment is given to the faculty there at the Institute for Administration and Organization Theory, particularly Alf-Inge Jansen, whose assistance and encouragement is valued.

Port authorities appear everywhere as targets for charges of nonaccountability, as agencies responsible and responsive only to themselves, and as hidebound empires beyond the reach of popular control or official guidance and direction. Critics who allege shortcomings in Port accountability point to four levels where the public trust vested in Ports is said to be violated. The first, and most general, level is where critics say Ports ignore their mandate to act in the "public interest." The best study to date of the Port Authority of New York/New Jersey summarizes the public interest critique: "It is the thesis of this book that public authorities that are supposed to act in the general interest of a state, region, or city frequently do not" (Walsh 1978, 6). Similarly, in 1986 a California civil grand jury in San Joaquin County issued a report charging that at the Port of Stockton, a major West Coast bulk port, the port commissioners "appear to have abdicated their public responsibility in setting appropriate policy."[1]

The relationship between Ports and their supervisory governments provides a second level where accountability frequently is said to be lacking. The city, state, and national governments that originally launch and formally hold responsibility for supervising port authorities quickly lose control over the creatures they have created and over time, it is said, prove even less capable in exercising control or influencing policy directions. On the retirement of Peter C. Goldmark, Jr. as executive director of the Port Authority of New York/New Jersey, Mayor Ed Koch observed: "Our differences were that they were accustomed, before I became mayor, to thinking of New York City as the junior partner and the Port Authority as being the general partner."[2] After appointing five of the seven Massport board members, and the Boston Port still refused his wishes, Governor Michael Dukakis mused: "It seems that any person appointed to that board is mesmerized by its power and strength . . . once they get into the clutches of power, they forget about this office" (Cellineri 1976, 108). As "set aside" agencies given extraordinary independence in their organization, Ports have frustrated many political executives and are commonly viewed by them as free-floating entities about which supervisory governments can do little.

A third level where accountability problems surface is internal to the port authority. A near universal fault line identified by critics is the fissure between governing policy boards and management. Although holding formal responsibility for setting policy, port commissions are said to lack the expertise, information, and the time to formulate and decide Port policy issues, let alone to monitor their implementation. Such boards necessarily become, in the critics' view, dependent on management rather than vice versa. The description of Massport by Cellineri (1976, 106) is typical: "The Board of Directors of Massport are part-time and unpaid, and none are professional port administrators. This arrangement has allowed effective power to devolve to the staff, with most of that power concentrated around the executive director." William Erxleben, speaking for Portwatch, an active group of Port of Seattle critics, noted similar patterns of staff domination: "By the very nature of the setup there has evolved a pattern where commissioners have attended these (twice monthly) meetings, ask a few perfunctory questions, and approve the staff's recommendations."[3]

Relations between port authorities and their surrounding communities provide a fourth level where accountability problems often arise. Adjacent neighborhoods receive many of the negative impacts of Port development, including noise from truck and rail traffic, physical impacts from terminal development, and visual impacts from terminal operations. John Vitagliano, manager of the Little City Hall in East Boston, summarized community antagonism toward Massport's executive director: "King made the people feel like they had no voice at all in determining their own future. From the community standpoint, the impact of King was all negative. In all controversies, he showed no flexibility or alternative" (Cellineri 1976, 126) The *Oakland Tribune* (1981) editorialized to a similar effect about community concern at the Port of Oakland: "In cities across the country the Port of Oakland is considered one of the maritime industry's success stories, thanks to its early development of container terminals. But in its own city the port is commonly perceived as arrogant and insensitive to the community. The port commission continues to operate as if its successes of the past—and they have been considerable—exempt it from the accountability commonly expected of public agencies."

These familiar criticisms are challenged by an equally familiar line of defense, which insists that unusual degrees of independence, flexibility, and managerial freedom are essential for port authorities. The ideal Port is not organized as a typical government agency, defenders claim, because Ports are involved in business transactions requiring autonomous corporate status. If recognized as the public businesses they are, and allowed sufficient independence, Ports will be prepared organizationally to achieve the public interest by enhancing the economic well-being of their local communities. By this account, Ports originally were designed as independent agencies to remove them from the inevitable contamination and corruption of politics that accompany direct governmental supervision. Ports are in the business of shipping where, according to their defenders, economic and not political rules apply. If there is direct political oversight and control of port authorities, businesslike management principles and practices deteriorate and businesslike managers are not attracted to Port executive positions. In their relations with general purpose governments, Ports require unusual degrees of autonomy if they are to succeed in the business of attracting cargo. Port commissions accordingly need freedom from political constraint in order to arrive at timely business decisions. The commercially competitive marketplace within which Ports operate requires quick, and sometimes closed, decisions as dictated by their markets and customers. So, too, with Port management, the argument concludes, where Port executives operate within general policy guidelines set out by port authority boards, but otherwise must have the freedom and flexibility to respond to changing market and business conditions.

As this conflict over Port accountability is usually posed, there is deadlock in the argumentation between Port critics and Port defenders. On one hand, the critics view Ports as public agencies that should be held directly accountable by a supervisory government for their activities as they impact affected communities and for their general performance as it fulfills the public interest. Anything less than direct accountability violates norms of popular control in a democratic society. On the other hand, defenders of Port autonomy rest their case on the view that mixing politics and markets in the Port industry contaminates both, the prevention of which requires independence from direct political supervision for port authorities. Anything less than Port autonomy inevitably leads to political interference and discouragement of cargo movements through the harbor.

In order to move beyond the stalemate between Port critics and defenders, a better understanding of Port organizations is required and the accountability concept requires clarification. This chapter examines the essential characteristics of seaports and how they are uniformly organized as public enterprise in the United States, the variety of their jurisdictional locations and institutional relations with other governments, and then takes up the accountability concept. Traditional usages of accountability are reviewed in order to offer a reconceptualization tailored to the basic characteristics of Ports organized as public enterprise, which, in turn, is applied to the U.S. Port industry.

## PUBLIC PORTS AS PUBLIC ENTERPRISE

To understand the complexities involved in the Port accountability issue, let us turn to the general case of public enterprise, of which port authorities are a particular kind.[4] Public enterprise is a hybrid of the government agency and private enterprise. It is in hybridizing public and private institutional characteristics that public enterprise displays its basic properties. Its *public* character involves three primary features: creation by the government, statutory assignment of powers, and ownership by the public. Each of these features recommends popular control and accountability measures. The *enterprise* character of the hybrid provides its market orientation and includes four qualities: expectations of market efficiencies in operations, commercially defined performance goals, reliance on user fees for operating revenue and capital markets for construction funds rather than general government appropriations, and the absence of partisan intervention in its operations. Each of these features recommends structural independence and autonomy from government. The hybridization thus combines in one institution what are otherwise thought to be quite unlike and even incompatible structural arrangements.

The implications for accountability systems contained in the two basic properties defining public enterprise are conflictual. To the extent that one emphasizes its public character, government control mechanisms are recommended to promote accountability; to the extent that emphasis is given to its enterprise side, independent and autonomous status is recommended to promote economic efficiency. Thus, in combining attributes of the public agency and private firm, public enterprise appears organizationally to contain the paradox between accountability and efficiency. Control state enterprise too directly and rigidly, in order to promote public accountability, and you necessarily compromise its efficiency. Loosen public controls too much, in order for autonomy to promote efficiency, and you inevitably compromise public accountability. This is the classic Gordian knot confronting Ports organized as public enterprise: between accountability and efficiency there is a zero sum calculus, the more of the one, the less of the other.

All but a handful of general cargo Ports in the United States are organized as public enterprise.[5] As such they share in common, for all their many differences otherwise, the twin qualities of public ownership and market orientation that define public enterprise more generally.

The public nature of Ports stems from several attributes. For at least ten centuries seaports have been viewed as entities responsible to the public interest (Polanyi 1957). Their public nature is rooted in the venerable concept of the harbor as a public highway, to which any and all should have equal access (Fair 1954, 6). Ports are publicly owned by the governments that create them, through statutory enabling acts. Enabling statutes also assign public purposes and policy objectives for Ports to fulfill. The policy boards of Ports are publicly selected, either by appointment or by direct election. All Ports, everywhere and for all time, have received substantial public subsidy.[6] More generally, Ports display important public utility attributes, due in part to their natural monopoly character and in part to the definition of their cargo transfer function as essential to local and national economic well-being. Finally, any manner of political controls are placed on ports, intended to serve broader public interests.

Ports are also enterprises that engage in market-oriented behavior in their methods of operation. Structurally, seaports are assigned significant, although not invariable, degrees of autonomy in formal relations with other governmental bodies. In providing facilities for cargo transfer, Ports pursue activities of a commercial character, adopt businesslike management practices, rely on payments from facility users for operating revenue, and borrow investment capital against future earnings from municipal bond markets. Although Ports tend to exercise monopoly prerogatives within harbor areas, they at the same time are often in fierce commercial competition with Ports in neighboring harbors, and increasingly so with their counterparts across entire coasts or continents. Market-oriented behavior also underlies Port investment in harbor facilities and pricing policies for facilities use.

Public port authorities embody the hybrid qualities of government ownership and market orientation that define public enterprise more generally. Accountability for port authorities thus imposes requirements for adherence to democratic norms of popular control in recognition of their public attributes, and requirements for fulfilling the commercial purpose of promoting the economic well-being of local communities in recognition of their enterprise qualities. The jurisdictional location of Ports and their relations with other governments begin to indicate how the dual requirements are structured.

## JURISDICTIONAL LOCATIONS AND INSTITUTIONAL RELATIONS

Public port authorities exhibit great diversity in both their jurisdictional location patterns and their institutional relationships with governments. Both are crucial to understanding public port accountability because it is to general purpose governments that port authorities are constitutionally responsible, and their primary activities take place at the jurisdicational level at which they are created. The jurisdictional location question fixes Ports along the vertical dimension of the U.S. system of federalism. The institutional relations questions views Ports, once fixed as to location, across the horizontal plane defined by their interactions with the sponsoring governments that create them and other governments that regulate, subsidize, promote, monitor, or otherwise interact with Ports. Port jurisdictional locations and institutional relations are sufficiently diverse to complicate further the accountability question for Ports.

The location of Ports within the U.S. system of federalism is exclusively among subnational jurisdictions, where Ports are authorized by statute, receive their powers and authority, and are formally supervised (see Chapter 1). This pattern of localism is in marked contrast to the near uniform centralism in other western industrial nations where national governments create, supervise, and give policy direction to Ports (Goss 1979). In the United States the national government is distinct by its absence both in creating and supervising Ports, even lacking a national Port policy (Marcus 1976). Instead, it is at the state, county, city, and special district jurisdictional levels that Ports are authorized, organized, and formally supervised.

At least eleven forms of Port organization can be distinguished by differentiating their jurisdictional locations. The forms are shown at the left in Table 14.1, with illustrative examples to the right. The continuum along which the forms and examples are listed ranges from the most decentralized at the top, to the most centralized at the bottom. Less than a handful of general cargo Ports are organized at the most decentralized level as private enterprise, including New Haven in Connecticut and Alameda County in California. There are many more Ports, including all the largest California Ports and all except Seward in Alaska, that are organized as harbor departments within municipal government structures. A few Ports are non-departmentalized extensions of city councils, such as Richmond in California up to 1982 and Seward in Alaska after that year. Organized beyond city boundaries, but still quite local, the special district prevails for all seventy-three public ports in Washington State, including its thirteen deep-water Ports, the largest being the Ports of

Seattle and Tacoma. But this form is also found in Chicago, at the Port of Coos Bay, and twenty-one other Oregon Ports, and the California Ports at Humboldt Bay and Port Hueneme in Oxnard. As we move down the continuum, we move up the scale of centralism, where Ports organized as agencies of the county, multicounty arrangements, and unified areawide special districts appear, respectively, in Cleveland, Portland, and San Diego. Maryland, Hawaii, Maine, and North Carolina all choose to organize their Ports as state agencies, whereas state authorities govern Massport in Boston and Burns Harbor in Indiana. Still more centralized are the bi-state authorities in Philadelphia, St. Louis, and at the nation's largest seaport, the Port Authority of New York/New Jersey. Absent representation in the U.S. case, but the norm elsewhere, is a national port or harbor board.

Table 14.1 depicts a confusing array of alternative Port locations at ten different jurisdictional levels in the federal system.[7] It shows Ports organized differently within a single state, as in California where five forms are represented. And it shows Chicago with two quite differently organized port authorities. There are, nonetheless, some commonalities. The West Coast Ports tend to cluster more toward the decentralized forms, whereas the East Coast Ports tend to be somewhat more centralized. A frequency distribution of where all Ports in the nation would appear across the continuum, as opposed to the selected examples provided, would show a clustering

**Table 14.1.** Alternative Port Jurisdictional Arrangements

| Forms | Most Decentralized | Examples |
|---|---|---|
| Private | ↑ | New Haven, CT; Alameda County, CA |
| Local city department | | Los Angeles, CA; Long Beach, CA; Oakland, CA; San Francisco, CA; Milwaukee, WI; Chicago (Navy Pier), IL; Alaskan City Ports |
| Extension of city council | | Richmond, CA; Seward, AK |
| Local special districts | | Seattle, WA; Tacoma, WA; Chicago (Regional Port District), IL; Coos Bay, OR; Humboldt Bay, CA; Oxnard, CA |
| County agency | | Cleveland, OH |
| Multicounty agency | | Portland, OR |
| Unified areawide special district | | San Diego, CA |
| State agencies | | Maryland, Hawaii, Maine, North Carolina |
| State authorities | | Massport; Burns Harbor, IN |
| Bi-state authorities | | New York/New Jersey; Delaware River Port Authority, PA/NJ; St. Louis, MO/KS |
| National boards | ↓ | Absent in the United States |
| | Most Centralized | |

toward the decentralized end (Maritime Transportation Research Board 1976, 41). One such survey of Ports revealed that public ports are primarily an extension of local government agencies, as opposed to state or federal, with the local government agencies "owning an estimated 70 percent of public port facilities" (MARAD 1974, 2). There would be relatively few Ports located toward the lower end, and none appearing under national boards.

The diversity of Port locations along the vertical dimension of the U.S. federal system is more than matched by the variety in Port relationships with other governmental bodies. At the national level, Ports are regulated, subsidized, and promoted by the Army Corps of Engineers, Federal Maritime Commission, Interstate Commerce Commission, Customs Service, Environmental Protection Agency, and the U.S. Coast Guard, among other federal agencies. Ports regularly interact with state government agencies having responsibility for coastal zone management, environmental protection, shorelines management, audit, and fisheries, among other areas. County and city governments also affect Ports in the areas of land use, police, fire, and permitting, to mention but a few (see Chapter 5).

This variety in the horizontal relations Ports have with other governments may be simplified by distinguishing between the "parental" government, which creates and has primary responsibility for supervising Ports, and all other governmental agencies. For the Port of Los Angeles, the parent government is the city government of Los Angeles; for the Port of Baltimore it is Maryland State government; for the Port Authority of New York/New Jersey it is state governments in both New York and New Jersey. Port institutional arrangements with their parental governments vary from their status as departments of regular government, to agencies semi-independent from parental governments, to fully autonomous corporations. Because they derive their operating powers and authority from parental governments and receive their mandated goals from them, the most sustained and regularized interactions are between Ports and their primary parental governments.

The distinction between parental and all other governments with which Ports interact begins to give some order in Port accountability systems. Yet even here Ports are structurally linked to their parental governments by different institutional arrangements. In the important case of how Port policy boards are selected, for example, seven alternative methods can be identified: 1) whether they have a policy board (most do) or not (Seward), 2) although all policy boards are publicly selected, whether this is by appointment (most Ports) or election, 3) if selection is by appointment, whether by the mayor (Los Angeles), by the city manager (Long Beach), by the governor (Massport), or by two governors (New York/New Jersey), 4) again if appointed, whether approval of appointees is required by the legislature (San Francisco) or not (Portland), 5) whether board positions are formally reserved for particular groups like shippers, labor, business, political parties, or areas of the city (Milwaukee) or not (Oakland), 6) if policy board members are elected, whether by districts (Port Hueneme) or at-large (Tacoma), and 7) if such elections are by partisan ballot

(Seattle pre-1960) or by nonpartisan ballot (Seattle post-1960). Variability in policy board selection is but one example of how Ports differ in their relations with supervisory governments. Similar listings could be constructed showing equal diversity in whether Ports are subject to employee civil service systems of their parental governments or have self-contained personnel systems, open versus closed meeting requirements, audit reports, financial restrictions, annual reports, borrowing authority limits, taxing authority, or none of these or any degrees of these.

The seemingly endless stream of differing jurisdictional locations and institutional relations among U.S. seaports creates a formidable challenge for attempts at specifying which organizational arrangements are more or less conducive to Port accountability. Dwelling on the nuances of particular structures may prove less useful in this attempt than turning to the concept of accountability itself and how it has been conceptualized in the traditional literature. Two schools of thought about accountability may be identified, which suggest broader patterns in organizational arrangements for Ports that influence their economic performance and public accountability.

## ACCOUNTABILITY: AN ELUSIVE CONCEPT OF ADMINISTRATION

Accountability is one of those ambiguous terms, like democracy and freedom, that evokes powerful meaning systems. It often is used in public discourse and academic writings as if its meaning is self-evident, when it is not. It is variously used to describe a reporting relationship between agents responsible to principals, or steering relationships between principals directing their agents, or it is said to be a matter of control between dependent agents and their supervisory principals. There simply is no agreement in administrative theory on the meaning of accountability or on the forms accountability takes.[8]

Two contrasting approaches to accountability systems for port authorities are suggested in the literature on public enterprise.[9] The first is a narrowly defined *institutionalist* approach with precise objectives and direct control measures designed to produce procedural accountability. The second is a broadly based *functionalist* approach with, at best, indirect control measures intended to secure accountability in substantive performance. The two approaches recommend opposite organizational designs to induce Port accountability and, in so doing, begin to identify the structural influences affecting Port behavior.

The institutionalist approach emphasizes the public side of public enterprise and attempts to ensure accountability by locating seaports within the hierarchical authority relationships of traditional government departments.[10] It recommends formal control mechanisms enabling political leadership to direct Port policies through appointed or elected boards of directors, to monitor policy execution through budgetary control, and to review Port performance by administrative oversight. Port managers are required to render specialized accounts in matters of audit, budget, annual report, and the like. Some measure of flexibility is granted to Ports, allowing them to transact business exchanges, but these are constrained by political oversight. The accountability problem to institutionalists is identifying just those commercial operations where Ports require latitude, where indirect controls apply; otherwise direct controls similar to regular government departments apply. It is the balance between indirect and direct controls that requires mastery to institutionalists, and once achieved, the symmetry between political control structures and specified areas of independence yields necessary Port accountability without curbing flexibility. Institutionalists tend to arrive at a single best way for organizing port authorities, patterned on the departmental model, and they tend to ignore the markets and types of goods that provide the context within which Ports operate. To this extent, technical and formal accountability may be achieved through uniform institutions of control, but the actual performance by Ports may be ignored.

The second approach is the *functionalist.*[11] Reacting against unduly restrictive constraints they find in the institutional model, functionalists emphasize the enterprise side of public enterprise and attempt to enhance Port performance, particularly its efficiency, by not constraining operating, legal, or financial flexibilities. Political constraint is capable of replacement, in this version, by powerful marketplace constraints evident in the commercial markets and types of goods produced by Ports. Clear, statutory-mandated performance goals are recommended in this approach, as are structural independence for Ports and autonomy for Port managers. Under these conditions, control hierarchies can be relaxed or abolished, and replaced by periodic performance evaluations according to standards set in the stated goals. Such evaluation reviews will provide sufficient accountability, as the functional performance of Ports will reveal whether the public interest has been fulfilled or not.

Institutionalists and functionalists recommend contrasting accountability systems for Ports. Institutionalists argue that Ports ought to be directly accountable to elected politicians, that their organizational form should closely resemble the regular government department, and that key policy, financial, and sometimes operational decisions should be under the control of elected politicians. Functionalists argue that economic efficiency in Port performance requires indirect control measures, organizational autonomy to allow managerial flexibility, and relative freedom from continuous operating controls, interfering politicians, shifting electoral demands, and special interest group claims.

Port authorities on the West Coast provide an ideal test on the merits of the institutionalist and functionalist models. If either is correct about the effects of institutional arrangements, there should be regularly patterned differences in the performance of West Coast port authorities on the basis of their differing organizational forms and administrative relations. These are shown in Table 14.2, with organizational forms reported in the second column and three key administrative relationships

**Table 14.2.** Major West Coast Ports' Organizational Forms and Administrative Relations to Supervisory Governments

| Port | Organizational Form | Policy Board Selection | Personnel System | Form of Subsidy |
|---|---|---|---|---|
| Seattle | municipal corporation | popular election | Port's own | property tax |
| Tacoma | municipal corporation | popular election | Port's own | property tax |
| Portland | municipal corporation | governor appoints | Port's own | property tax |
| Oakland | city department | mayor appoints | city's | various forms (city determines) |
| Long Beach | city department | city manager appoints | city's | various forms (city determines) |
| Los Angeles | city department | mayor appoints | city's | various forms (city determines) |
| San Francisco | state*/city department | governor*/mayor appoints | state's*/city's | various forms (state*/city determines) |

*From 1863 until 1969, the Port of San Francisco was organized as an agency of California State government; from 1969 until present it is organized as a department of the City of San Francisco.
*Source:* AAPA, state port associations, individual ports, (Barnhart 1986).

with supervisory governments in the last three columns. The Pacific Northwest Ports in Washington and Oregon conform organizationally and administratively to the functionalist ideal; California Ports closely approximate the institutionalist model.

Organizationally, all three Pacific Northwest Ports exist as separately established municipal corporations, with the greater autonomy from other governments this provides. All the California Ports are organized as departments of their respective city governments, thereby lacking the formal independence of their northern neighbors. The three administrative relationships similarly vary. Popular election of port commissioners in Seattle and Tacoma links their policy boards to the electorate. In California, Port policy boards are tied directly to city government through mayor or city manager appointed port commissioners. Pacific Northwest Ports internally control their own personnel systems; Ports in California must conform to the personnel systems of their supervisory city governments. Finally, all three Pacific Northwest Ports control their own subsidy, through the property tax that is theirs to extract; California Ports receive various forms of subsidy controlled by other governments. In sum, and across all criteria, Ports in the Pacific Northwest are assigned a degree of autonomy that is lacking in the departmentalized California Ports. If organizational status and administrative structure are the determining factors, as both the institutionalist and functionalist arguments insist, then the performance of the Pacific Northwest Ports should regularly differ from their southern neighbors. They do not, of course.

Neither organizational form or administrative structure, nor the two together, are determinative of Port performance. On most performance measures, the Ports of Seattle and Oakland resemble each other more closely than Ports in their own respective states that are similarly organized. The same applies to the Ports of Portland and San Francisco. In adapting to the container revolution in the mid-1960s, for example (elaborated below), Seattle and Oakland followed similar policy directions by investing early and vigorously in the container technology, whereas Portland and San Francisco were reluctant and late entrants into the container era.

The failure of organizational form and administrative structure to explain Port performance is because they provide only a partial explanation. Each is important to Port performance, but they are neither the only, nor frequently the most important, factors explaining how Ports behave. Where both institutionalist and functionalist approaches have erred is, for all their other differences, in what they share in common. Both schools of thought embrace an explanatory framework that conceptualizes performance in Port efficiency and Port accountability as a function exclusively of organizational form and administrative structure. Without rejecting the importance of the latter, we need to continue the search for additional factors influencing Port accountability and efficiency. A starting point is provided in Mansfield's (1975, 335) advice that there is "no single royal road to accountability," but instead "some dozens or scores of paths more or less traveled, more or less tangled."

## PORT ACCOUNTABILITY: A POLITICAL ECONOMY PERSPECTIVE

Drawing upon institutionalist and functionalist insights, a combined political economy perspective may be devised by identifying those properties in the political and economic environment of ports that act most directly to influence Port behavior.[12] Two basic political structures promoting public accountability in the institutionalist account are Port organizational structures and the public policy goals mandated to Ports by their parent governments. Two economic structures influencing the commercial performance of Ports in the functionalist argument are the markets within which Ports operate and the nature of the economic goods produced by Ports. As reconceptualized, organizational relationships remain important, but they are now one of four structures influencing Port behavior and explaining public accountability. The political economy perspective thus provides a four-dimensional explanation for Port accountability by recognizing the separate and combined influences of organizations, markets, goods, and goals.

As structures in the port environment, the influence of organizations, markets, goods, and goals is either as constraints on Port behavior or as opportunities for Port actions. Each structural influence appears with variable form. Organizational forms vary, as discussed above, between relatively dependent and independent administrative structural arrangements. The market structures within which Ports operate vary among monopoly, oligopoly, and competitive markets. The market structure for any Port is defined by two factors: the number of rival Ports it has and the size of its natural hinterland. The fewer rivals and larger hinterland a Port has, the more it operates in monopoly markets; the more rivals and smaller hinterland a Port has, the more it operates in competitive markets. Ports produce economic goods or services that vary between private and public type goods.[13] A good is private or public depending upon whether customers of the Port consume the good in discriminate amounts and thus can be forced to compensate the Port for its production costs through user fees, in which case the good is private; if not, the good is public and requires collective action to produce it. Policy goals assigned to Ports vary in three respects: their scope (narrow or broad), their clarity (precise or ambiguous), and their number (single or multiple). It is these variable structures in the environment of Ports that explain public accountability in Port performance.

This perspective on accountability could be applied to any number or kind of Port decisions or performance indicators to assess how useful it is. In what follows the perspective is applied to the manner in which Ports accommodated themselves to the container revolution. The technological innovation of containerization is chosen because it had a more profound effect on shipping and on Ports than any other development in the twentieth century.[14] Its impact on the shipping industry was as great as the conversion from sail to steam power before the turn of the century. By unitizing general cargo in container boxes and allowing their transport intact across land as well as sea, the container revolution changed shipping routes across major oceans and increased the size and speed of ships. It also altered port facility requirements, vastly expanded Port hinterlands, and concentrated cargo in a few load center Ports. In little over a decade the container revolutionized the waterfront and transformed the port industry. The future fortunes of Ports were set for decades to come by the decisions they made to accommodate container ships and capture container cargo (see Chapter 6). Because of this choice of the shift to containers as the way to apply the Port accountability framework, the remainder of this chapter deals with the experience in larger commercial Ports rather than specialized Ports such as recreational fishing or bulk-products Ports.

## *Organizations*

Differences in Port organizational structures act to facilitate or encumber their adaptation to changes in their environment. The speed, manner, and ease with which Ports accommodated the container technology are partly, but not exclusively,[15] explained by how organizational arrangements provided opportunities for adaptation to containerization or denied them. The structural effects of organizational arrangements become evident in comparing the differential responses to containerization between the Ports of San Francisco and Oakland on the West Coast, and between the Port Authority of New York/New Jersey and Massport in Boston on the East Coast.

The sudden reversal in dominant port status between the Ports of San Francisco and Oakland in the mid-1960s is the stuff about which legends are made.[16] Legends provide entertaining folklore, but the remarkably different organizational structures of the two Bay Area Ports provide a more persuasive explanation for why Oakland quickly embraced the container trade and soon surpassed its more reluctant San Francisco rival, which refused to adapt, and declined as a result.[17] Organized at the time as a municipal department, the Port of Oakland enjoyed significant freedom within which to act. It was governed by a seven-member Board of Harbor Commissioners appointed by the mayor, and it had semi-independent status.[18] More importantly, under Oakland's city charter, the Port exercised autonomy over its operating and capital budgets, and it had independent authority to issue revenue bonds for developing container terminal facilities. Organizationally, the absence of externally imposed constraints provided the opportunity for adapting to the container.

By comparison, the Port of San Francisco suffered a number of organizational disabilities that constrained its ability to accommodate containerized cargo. Primary among these was its organization as an agency of California state government, which gave the San Francisco Port less independence and fewer degrees of freedom than its cross-Bay rival. Originally organized under the State Board of Harbor Commissioners in 1863, it remained a state agency until 1969.[19] For its first seventy-five years, the Port of San Francisco thrived under state agency status. It was the first public port in California and by far the largest West Coast port through the nineteenth and early twentieth centuries. California legislators willingly appropriated subsidies for its facilities and the statewide electorate regularly voted approval for its bond issues under the calculation that concentration of cargo there reduced per unit shipping costs to the entire state.

With the separate organization of the Ports of Long Beach and Los Angeles in the 1920s, and with Southern California's population explosion in the Great Depression and post-World War II years, the prior calculation was reversed. Now legislators from Southern California, reflecting the classic Southern versus Northern California schism, began refusing San Francisco its historic subsidy, and the demographically more numerous Southern California voters rejected Port bond referenda. The state agency that previously favored Port development at San Francisco had become an organizational encumbrance and a barrier to development. San Francisco's cargo tonnage declined precipitously, from a high of 7.8 million tons in 1947, to 4.2 million tons in 1964, to a mere 2.3 million tons in 1977. The Port of San Francisco approached the container era with a harbor naturally deeper and historically larger than its cross-Bay rival, but constrained by archaic organizational structures that inhibited necessary adaptations to the container. The legacy of Oakland's early and massive accommodation to containerized cargo and San Francisco's refusal or inability to do the same is that a quarter-century later over 90 percent of the container cargo in the Bay Area crosses Oakland's port terminals.

Similar organizational effects appear in container adaptation patterns on the East

Coast in New York and Boston harbors. The bi-state Port Authority of New York/New Jersey acted early to capture container trade and continued to prosper as a result,[20] while at Massport the state agency refused to decide early and suffered accordingly by losing the container business.[21] The bi-state organizational form at the Port Authority of New York/New Jersey imposed a peculiar kind of constraint: the necessity for quid pro quo development across the two states in order to arrive at major decisions. Whenever a major facility is proposed whose development will benefit one state, a similar initiative or one of comparable value for the other state is necessary for any development to occur. As the container era approached, Austin Tobin had nearly two decades tenure as Port director at the New York/New Jersey Authority, sufficient time to master the necessities for regional balance between the two states and to move his Port to act on the container. Among Tobin's many landmark balancing acts was launching the Port Authority's first container terminal in the late 1950s, locating it on the New Jersey side at Elizabeth; at the same time the Port Authority bought two miles of Brooklyn's waterfront for marine terminal development on the New York side (Doig 1985, 39–40).

Massport, by contrast, moved in fits and starts. A comprehensive study of port modernization needs was completed in 1964, which recommended a commitment to developing container terminals. At the same time Massport was investing heavily in developing Logan Airfield as its first priority. It did develop a container facility at Castle Island, but its actual operation was delayed time and again by uncertain Port commitment and labor turmoil. The facility opened in 1970, by which time the container business had gone elsewhere, mainly to New York's harbor. Pronouncing the introduction of container terminals in Boston a failure, Cellineri (1976, 93) says that the episode "approximates a fairly accurate microcosm of the Port's total character."

These comparisons of how differently organized Ports on the West and East coasts adapted to containerization suggest that organizational structures can act to facilitate or retard Port accommodations to major changes in their environment. San Francisco and Massport were constrained by organizational arrangements that precluded timely and substantial changes in their policy directions, whereas Oakland and New York/New Jersey embraced the container technology with relative ease and the absence of organizational barriers. The former Ports continue to suffer trade losses to this day as a result, while the latter Ports prosper.

## Markets

A second key structural influence on Port behavior in the political economy perspective is the nature of the markets within which Ports operate. If markets have an effect on Port behavior, the most important decisions Ports make should be regularly patterned according to whether their markets are competitive in nature or more monopolistic. Building container terminals in the 1960s proved to be crucial investment decisions affecting the way Ports pursued maritime cargo, and with what success, over the succeeding two decades. West Coast seaports responded quite differently to the two waves that define the container revolution.[22] The first wave, in the mid-1960s, when the domestic carriers began to containerize shipping, saw some Ports acting to facilitate the container revolution by building container terminals, while other Ports resisted and refused to do so. The Ports of Oakland and Seattle invested earlier and more massively in container terminals than did other West Coast Ports. The Ports of Los Angeles and San Francisco initially refused to invest in the new technology and thereby resisted the container revolution.

The relative market position held by Ports, as shown in Table 14.3, appears to have influenced their decisions on container terminal investment. Measured by the market share of liner cargo trade they controlled, the largest, most successful Ports proved most unwilling to invest in the container terminal; the smaller, more competitively situated Ports proved most enthusiastic. The Ports of Los Angeles and San Francisco in the critical year of 1966 held dominant market positions that tended more toward the monopoly, giving them a sense of security in existing shipping technology and making them less willing to innovate in what might have appeared to be a risky investment. Oakland and Seattle were comparatively much smaller at the time, the former never having attained major Port status, the latter in decline over the postwar years. Their market shares made them the more competitive and created incentives to take risks in order to overtake their larger rivals. They responded early, invested massively in the new container technology, frequently with no contracted tenant in hand, and thereby transformed their status from minor to major Ports, a set of decisions that continues some two decades later to explain their major Port status.

In the second wave in the late 1960s, when foreign carriers adopted containerized shipping, this same relationship repeated itself. It was the Ports with the least

**Table 14.3.** Market Position and Capital Expenditures for Major West Coast Ports, 1966–1972

| Port | Share of Liner Cargo Trade (percent)* 1966 | Expenditures on Specialized General Cargo Facilities (millions of dollars) 1966/1972 |
|---|---|---|
| Los Angeles | 23.4 | 16.0 |
| San Francisco | 18.1 | 18.8+ |
| Long Beach | 14.2 | 16.6 |
| Portland | 11.6 | 12.4+ |
| Seattle | 9.9 | 35.6 |
| Oakland | 4.6 | 38.5 |

*Additional 18.2 percent handled by other Ports.

+Figures include major expenditures for LASH and RORO facilities.

*Sources:* U.S. Department of Commerce (March, 1974, 28); Hayuth (1978, 179).

cargo, and thereby the most competitively situated, that proved most eager to attract the container business. The Ports with the largest market shares, secure in their strong market positions, proved most reluctant and even resistant to invest in container terminals. The investment patterns described above are reported more fully in the second column of Table 14.3, indicating expenditures in millions of dollars in general cargo facilities over the time period 1966–1972, covering both the domestic and foreign waves of containerization. Here we see the lesser Ports of Oakland and Seattle investing at a rate twice as great as the larger, more dominant Ports of Los Angeles and San Francisco.

The critical decisions by Ports to invest in container terminal facilities suggest that such investment actions were strongly influenced by Port market structures. The market-powerful Ports were the most conservative, interested more in protecting their existing market position and less pressured to pursue new sources of cargo to innovate. Because they lacked captive cargo, the more competitive Ports were early and eager enthusiasts in the container innovation. They took risks by investing massively in the new technology, and they were pressured to do so by local communities interested in overtaking larger rival Ports for the economic well-being that would ensue. Market structures thus influence the performance of Ports and the policy directions they take and for which they are publicly accountable.

## *Goods and Services*

A third structural influence on Port behavior is based on the goods and services produced by Ports. Containerization altered the goods and services Ports are required to provide to facilitate cargo transfer. Prior to the mid-1960s, the primary mode for shipping general cargo was in "breakbulk" form. As the term implies, this cargo is transferred in small and various size lots, usually one piece at a time, thus making breakbulk cargo terminals highly labor intensive and costly operations. They are inherently unprofitable operations for Ports, which are unable to recover the costs of providing terminal facilities through their tariff, dockage, and wharfage charges. Ports provide such facilities to attract cargo and realize secondary economic benefits for their communities. Because of these features, breakbulk cargo tends to be more of a public, or semi-collective type, good. And as a public good, breakbulk terminal facilities are priced by Ports through administered price formulae that require public subsidy.

The container revolution changed Port pricing practice. Administered tariff charges were replaced by long-term leases giving preferential use to steamship companies. Compared to harbor tariffs, the use of terminal leases to assess user fees is more remunerative to Ports, market sensitive, and it is potentially profitable, thus easing the requirement for subsidy. These consequences follow from the more private-type good containerization introduced. We may summarize these effects: Port pricing policy is influenced by the nature of the economic good provided, and as this economic good changed from breakbulk terminal facilities to container terminal facilities, the character of the good provided by Ports changed from more public to more private, and so, too, did Port pricing policy change from the harbor tariff to terminal leases.[23]

The nature of the good may also be seen to interact with market characteristics to influence Port pricing decisions. The movement from tariff to lease pricing methods produced different lease strategies among West Coast Ports. Three types of container terminal leases, based on compensation formulas, may be outlined (Dowd 1984). The first is the *flat rate,* which sets a constant user fee for the length of the lease. Because the rate is fixed, this lease type yields the least revenue to the Port, but it induces the steamship tenant to put through greater cargo volumes. It was, and is, the favored leasing arrangement at the Ports of Seattle and Tacoma and was relied on by Oakland and Long Beach early in the container era. Each Port employing the flat rate lease was, or is, in more of a competitive market setting. The *mini-max* is a second lease type. It guarantees a minimum compensation to the Port and increases rent from the minimum on the basis of cargo volume, up to a maximum beyond which cargo shipped is a free good to the steamship tenant. The mini-max lease type reflects a more balanced relationship between Port and tenant, with the latter induced to either keep volumes at the minimum or to greatly exceed the maximum. After the Port of Oakland surpassed San Francisco in the late 1960s, it adopted this lease type and withdrew from the flat rate in recognition of its strengthened market position. The *revenue sharing* arrangement is a third lease type, essentially replicating the mini-max formula without a capping ceiling. Instead, a sliding scale of charges informs the compensation formula with no maximum. It increases revenue to the Port, but it does not encourage cargo. Here the Port is in a stronger bargaining position, extracting monopoly-type rents. The Port of Los Angeles has always favored this type of lease, as has Long Beach after 1980 when it reached parity with Los Angeles. With respect to pricing policies, it can be concluded that Ports are simultaneously subject to structural influences deriving from the goods and services they provide and the market structures within which they are provided. Achieving public accountability for Ports requires recognition of the effects these structures have on Port decisions and policy directions.

## Goals

The fourth and final structural influence in the political economy perspective is the set of goals assigned to Ports. Because Ports are public entities, the stated goals of port authorities derive from the purposes and objectives assigned to Ports in enabling legislation by their sponsoring governments. Besides providing the means for cargo transfer between water and land, Ports are multipurpose organizations with responsibility for a diverse set of activities. Seaports build and operate marine terminals, airports, bridges, tunnels, mass transit systems, pleasure boat marinas, fisheries terminals, commercial and industrial real estate parks, foreign trade zones, hotels, waterfront tourism facilities, parks, greenbelts, golf courses, convention and entertainment centers, trolleys, belt line and trunk railroads, and highways and ferries. The substantial diversity of Port activities creates multiple goal structures where Ports are expected to pursue a number of ends simultaneously. Often these are conflictual[24] (see Chapter 7).

Seaports do not attempt to fulfill all their goals with equal force, nor are they expected to do so by their sponsoring governments. Instead, there is an ordering of goals between primary and secondary by both the sponsoring governments and seaports themselves. The primary goal of seaports remains the maximization of waterborne commerce.[25] The most consistent measure applied both inside and outside port authorities to determine success or failure in goal fulfillment is the market share and/or trade volume controlled by Ports.

The relative success among West Coast Ports in fulfilling their primary goal of cargo maximization can be assessed for the recent period by examining market shares and trade volumes across the major Ports. The trade volume in tonnage and the market share for liner cargo are reported in Table 14.4 as these were distributed across the four leading West Coast Ports during the period 1966–1982, when the containerization of cargo had its most visible effects. The Ports of Oakland, Seattle, and Long Beach all improved dramatically in fulfilling their primary goal of maximizing waterborne cargo. Measured by improvement in tonnage and market share, the performance of Oakland is the most impressive as it registered an eightfold increase in tonnage and more than quadrupled its market share. It continues to be heralded as one of the nation's most successful Ports because of its performance. Nearly as impressive is the Port of Seattle's performance. It quadrupled its tonnage and more than doubled its market share. The Port of Long Beach almost quadrupled its tonnage and almost doubled its market share. Perhaps more important, Long Beach first reached parity with its rival Port across San Pedro Harbor, and eventully sur-

**Table 14.4.** Market Share of Liner Cargoes for Major West Coast Ports, 1966–1982 (millions of metric tons and percent of West Coast total)

| | 1966 | | 1974 | | 1978 | | 1982* | |
|---|---|---|---|---|---|---|---|---|
| Port | mmt | percent | mmt | percent | mmt | percent | mmt | percent |
| Los Angeles | 1.73 | 23.4 | 1.94 | 23.1 | 2.68 | 23.0 | 2.85 | 19.4 |
| Long Beach | 1.05 | 14.2 | 1.49 | 17.7 | 2.33 | 20.0 | 3.71 | 25.2 |
| Oakland | .34 | 4.6 | 1.13 | 13.4 | 2.40 | 20.6 | 2.97 | 20.2 |
| Seattle | .73 | 9.9 | 1.25 | 14.9 | 2.30 | 19.8 | 2.95 | 20.1 |
| Subtotal | 3.85 | 52.0 | 5.81 | 69.1 | 9.71 | 83.5 | 12.48 | 84.8 |
| Others | 3.55 | 48.0 | 2.60 | 30.9 | 1.92 | 16.5 | 2.23 | 15.2 |
| Total | 7.40 | 100.0 | 8.41 | 100.0 | 11.63 | 100.0 | 14.71 | 100.0 |

*An estimate based on actual results for first 9 months.

*Source:* Port of Seattle (April 1983, 607), Hayuth (1978, 179).

passed the Port of Los Angeles in both tonnage and market share. The Port of Los Angeles registered a consistent increase in its tonnage, but an equally consistent decline in market share. On such comparisons are assessments of Port success or failure based. Whether it is internal to the port industry, within the Port's surrounding communities, or in national arenas, the extent to which Ports fulfill their primary goal is the main measure of their performance.

All structural influences act with differing effects on seaports, including the goal structure. The relatively successful Ports included in Table 14.4 maintained the maximization of cargo goal as primary, even while affirming secondary goals in competition for Port resources and in conflict with the primary goal. Goal maintenance was fostered in part by the uniformity with which port managers and policy boards viewed ever-increasing cargo volumes and market shares as critical litmus tests of their stewardship of the enterprises. A more important factor, which itself guided managerial and commissioner policy objectives, was pressure on Ports from their local communities to maximize cargo throughput for the derivative economic benefits this would realize.[26]

Maintenance of the cargo maximization goal as primary frequently required West Coast Ports to pay increased attention to social policies secondary in their goal structure. When it reached record cargo traffic levels in 1971, for example, the Port of Seattle received harsh criticism from community groups adjacent to the harbor.[27] In order to continue rapid cargo increases, a campaign was launched to "peopleize" the Port. The Port built and maintained a number of waterfront parks, constructed a public fishing pier, and provided costly public access and viewpoint facilities. At the Port of Oakland a similar responsiveness to secondary social goals occurred after blacks came to power in Oakland city politics in 1977 and began to question Port development impacts. The Port responded by rebuilding deteriorated Port backlands along the central city waterfront and by contributing excess Port revenues to an impoverished municipal treasury. Fulfilling these more socially oriented goals may be viewed as the price Ports pay to maintain their cargo maximization goals as primary. Confirming this observation is the unrelenting resistance these same Ports show when their communities encroach on their primary goal. Thus the Port of Seattle insistently refuses the wishes of the powerful Queen Anne and Magnolia communities to forego development of an adjacent terminal for container traffic. And the Port of Oakland rebuffed attempts by the black-controlled city hall to terminate trade with South Africa, even though such trade represented less than 1 percent of its cargo traffic.

All the Ports included in Table 14.4 succeeded in greater or lesser degrees in maintaining and fulfilling their primary goal. The two West Coast Ports failing to maximize the throughput of cargo over this time period are the Ports of San Francisco and Portland. As reported in Table 14.2, in 1966 they together attracted nearly 30 percent of the West Coast liner trade. By 1978 their combined market share declined to 16 percent and thereafter it declined even further (Denning 1985). As the Port that lost the greatest share of traffic, the Port of San Francisco represents the most obvious case of failure to maximize the primary goal in the recent period.

Over the past two decades the Port of San Francisco has received mixed signals from its parental government about what exactly its primary goal is. One set of messages signals maximization of cargo by modernizing waterfront terminals to accommodate container traffic. This goal is in competition for primacy with a conflicting set of messages, which recommends abandonment of lost container cargo. Instead, it instructs the Port to redevelop the waterfront for public amenity purposes with priority given to recreational uses for city residents and accommodations for the city's important tourist industry. The second goal set places the waterfront's highest and best use as a function of real estate markets, which recommend Port development of restaurants, tourist shops, hotels, entertainment facilities and commercial office space. Confused over which of the primary goals to pursue, the Port of San Francisco adopted uncertain and sometimes conflictual policy directions, and its performance in maximizing cargo has suffered as a result.[28]

As a structural influence on Ports, goal structures thus tend to be multiple in number, broad in scope, and often conflicting in content. Yet for most Ports, for most of their existence, the goal of maximizing cargo traffic through their harbors remains primary. Because of uniform consent on improvements to local economic well-being this goal promotes, most community groups and political leaders are supportive of cargo maximization as the primary goal for their Ports. When community groups or political leaders do attempt to intervene directly in Port operations related to the primary goal, seaports tend to be sharply unresponsive and insistent on autonomy in the decisions they take and in their execution. With respect to secondary goals, usually of a social and noncommercial nature, Ports tend to be quite responsive to outside influences and pressures. This pattern of responsiveness toward others on secondary goals appears to be a price Ports are willing to pay for continued autonomy in their commercial operations.

## THE SEVERAL PATHS TO PORT ACCOUNTABILITY: A SUMMARY

Accountability for public port authorities always has been and always will be a controversial subject. Ports are forever accused of being nonaccountable, and they are as frequently defended as independent agencies. In an attempt to move beyond the seemingly unresolvable deadlock over Port accountability, the concept of Ports as public enterprise introduced both the public characteristics of Ports and their enterprise features that are at the heart of the controversy. As publicly owned entities, Ports require popular control measures to ensure public accountability. As commercial enterprises, they require measures of flexibility and degrees of autonomy unlike regular government agencies. Here the contradiction between the public and enterprise characteristics of port authorities appeared to be a puzzle, the answer to which first required the pieces to be sorted out.

Ports are uniformly organized as public enterprise in the United States. Beyond that, they are highly diverse and complex. A review of their jurisdictional location showed different levels of government sponsoring Ports and acting as the parental governments to whom Ports are formally and primarily accountable. An equally divergent set of organizational forms and administrative structures appeared in the formal organization of Ports and in their institutional relationships with general purpose governments. In order to comprehend this diversity and complexity, two contrasting approaches to seaport organization and accountability were introduced. Neither school of thought alone provided a compelling resolution for the public port accountability puzzle. Yet each offered insight into basic properties in need of understanding. The institutionalists emphasized Port organizational structures and public policy goals; the functionalists stressed Port market structures and economic goods. These insights allowed Port accountability to be reconceptualized according to the constraints and opportunities represented by the structural arrangement of organizations, markets, goods, and goals. When applied to the actual behavior of seaports, each structure appeared with significant influence on the performance of public ports. There are several implications that flow from this discussion.

A first implication is that there is no single or royal road to public port accountability. Instead, there are several paths. One path is organizational structure. The way in which Ports are organized does influence their economic performance and their public accountability. But it would be, and frequently is, a mistake to view organizational structure as the only, or even the most important, path to accountability. The character of the markets within which Ports operate, the nature of the economic good they produce, and the structure of goals that guide their policy directions all influence the public accountability of Ports. Instead of focusing on any one of these, this discussion recommends a multidimensional approach to the complexities involved in Port accountability.

A second implication is to verify that public accountability and public control are not one and the same for public port authorities or for public enterprise more generally. A Port may be directly and fully controlled by public processes and supervisory governments and still fail to perform its intended or assigned function of enhancing the economic well-being of its local community. Viewing the accountability question exclusively in terms of public control structures and associated Port responsiveness ignores the commercial performance of Ports in varying markets and in producing varying goods. Ports may be fully accountable procedurally to their supervising government and still fail miserably in being accountable for their economic purpose. To stress either the public side to the exclusion of the enterprise side, or vice versa, is to address only one-half of the Port accountability question.

A final implication is that the dilemma between Port accountability and efficiency may not be so straightforward a trade-off as is commonly thought. Conventional wisdom has it that public accountability structures inevitably lead to dependency and the loss of efficiency, whereas Port independence and autonomy lead to a lack of public accountability. The trade-off between accountability is neither a necessary nor inevitable one if the full range of structures influencing Port performance is taken into account. When the varying structural influences of organizations, markets, goods, and goals are correctly understood and applied, efficiency in economic performance and public accountability may be complementary in public port authorities.

## Notes

1. The Port of Stockton, in turn, filed suit against the jury asking that criticism of the Port be struck from the record. *New York Journal of Commerce* (1986).

2. Koch is quoted in the *New York Times* (1984). The *New York Post* (1984) ran a series of articles suggesting that the mayor had not completely reversed the partnership arrangement by characterizing the Port as an "empire" that "runs unchecked" and by quoting New York City's deputy mayor, Kenneth Lipper, as complaining that the Port was "answerable to nobody."

3. Erxleben is quoted in the *Seattle Times* (1983). Four years later, Jim Wright, Port of Seattle Commission president, leaked to the press a letter he had directed to the Port Executive Director James Dwyer, saying in part, "The community is beginning to view the port commissioners as merely rubber stamps. This is intolerable and I must have your cooperation." *Seattle Post Intelligencer* (1987).

4. The literature on public enterprise is substantial and has been reviewed elsewhere (Walsh 1978; Denning and Olson 1981).

5. Isolated, and relatively unimportant, exceptions to public ownership do occur, such as the privately owned port in New Haven, Connecticut, and within harbor areas specialized terminal facilities, usually serving bulk cargo, are frequently provided by privately owned firms.

6. Only the forms vary, not whether there is subsidy. The most direct forms of subsidy are appropriations from the general government, taxing authority, grants, and donations of real property, such as tidelands and submerged lands. The more indirect subsidy forms include property and corporate tax exemption, low-cost financing through access to the municipal bond market, grants of right of way, and gratuitous provision of various services from other governments.

7. This diversity is the result of two primary factors: the historical time period when ports were founded, during which alternative ideas were in currency about the state of the art; and the different characteristics of the reform movement from place to place which gave rise to port locational structures. For a more complete discussion, see Olson (1980b).

8. Three primary usages have been suggested by Orlans (1980, xviii): for financial accounting, disclosure of administrative and operating standards, and the requirements for substantive or programmatic accomplishments. Mansfield (1975, 325) offers four standards for assessing various kinds and degrees of accountability: fidelity, efficiency, policy functions, and loyalty. Walsh (1978, 11–12, 41–44) rejects efficiency as an evaluative standard because of its inappropriateness for public enterprise.

9. Walsh (1978, 38) similarly identifies two general forms of accountability that inform most approaches: one is based on hierarchy, with neat authority lines reporting upward; the other is more decentralist and indirect, along lines of the private firm.

10. The best scholarship in the institutionalist tradition is by Seidman (1952, 1954, 1955, 1959). Also included here are Musolf (1959), Smith (1975), Walsh (1978), Moe (1979), and Sharkansky (1979).

11. Most of the classic writings were functionalist attempts to articulate the "theory of autonomous corporations." See Willoughby (1927), Dimock (1949), and Pritchett (1941, 1946). Also fitting here are Hamilton and Hamilton (1981) and Doig (1983).

12. The development of the political economy perspective on ports is a joint undertaking with my co-researcher, Michael Denning. What follows in the text draws upon, and is an elaboration of, several earlier pieces. Denning and Olson (1981), Denning (1984, 1985), Olson (1980a, b), Olson and Denning (1982).

13. For a more complete discussion of the distinction between public and private goods, see Bish and Ostrom (1973) and Savas (1982).

14. The history, economics, and impact of containerization are more than a story in themselves. For the best accounts of containerization, see Hayuth (1978). Johnson and Garnett (1971), Kendall (1979), Laing (1975), Marcus (1974), National Research Council (1976), Oram and Baker (1971), and Whittaker (1975).

15. The error of the institutionalists and functionalists is to treat organizational structures as a mono-causal variable.

16. Indeed, Ben Nutter, Oakland's Port director at the time, remains a legendary figure in port communities for gambling on the container trade while others balked, and seeing his port reap the subsequent benefits.

17. This account is drawn from a more extensive discussion of these issues in Olson (1980a).

18. Harbor commissioners appointed in the 1960s reflected the interests of an influential central business alliance and Oakland's conservative mass media, both centered on the *Oakland Tribune*.

19. Although formally devolved from the state to the city in 1969, the state continued to control the Port in two key respects: any capital construction over a quarter-million dollars required state finance department approval, and 85 percent of all yearly revenue surpluses over a quarter-million dollars reverted to California's general fund. It was not until 1979 that these two restrictions were finally terminated.

20. This account of the Port Authority of New York/New Jersey is drawn from Doig (1985).

21. The Massport case is found in Cellineri (1976).

22. The structural influence of markets and economic goods on port behavior and performance is developed much more fully in Denning (1985).

23. In this discussion, the effects of the nature of economic goods produced by ports has been confined to breakbulk and container terminals as they vary between public and private goods. For a more expanded discussion of the full range of goods produced by ports, varying between public goods of waterfront tourist centers and public access areas and private goods of a more commercial character, see Olson and Denning (1982).

24. The Port of Seattle's *Statement of Purposes and Objectives*, for example, contains well over two dozen separately listed goals, many of which are in conflict with each other. See Port of Seattle (1981).

25. Out of the over two dozen goals announced in the "Statement of Purposes and Objectives" by the Port of Seattle, for instance, is the clear identification of its primary goal: "The basic purpose of the Port, and the reason for the public involvement in this enterprise, is the enhancement and orderly management of the flow of waterborne commerce." Port of Seattle (1981).

26. The logic underlying such community pressures and promoting port development goals is spelled out by Peterson (1981), who views all cities in competition for economic development as a function of their location within the fragmented federal system and the free market economy.

27. Burke (1976, 126) describes the enraged citizenry: "They chastised the Port for a wide variety of 'abuses,' ranging from an 'unresponsive' attitude toward local community problems to a 'reckless' pursuit of profits and business regardless of the resulting environmental or social havoc."

28. The San Franciso identity confusion is described more fully in Gilliam (1980).

## References

Barnhart, Thomas F. 1986. Port authorities of the West Coast. *The Pacific Maritime Magazine* August **1**:2, 11–23.

Bish, Robert L., and Vincent Ostrom. 1973. *Understanding urban government*. Washington, DC: The American Enterprise Institute.

Burke, Padraic. 1976. *A history of the Port of Seattle*. Washington: Port of Seattle.

Cellineri, Louis, E. 1976. *Seaport dynamics: A regional perspective*. Lexington, MA: Lexington Books.

Denning, Michael. 1984. *The structure and performance of government enterprise*. Ph.D. diss., University of Washington.

———. 1985. Structure, policies and performance: Public ports as public enterprise. Paper presented at the American Political Science Association meeting, New Orleans.

———, and David J. Olson. 1981. The public enterprise and the emerging character of state service provision: Application to public ports. Paper presented at the American Political Science Association meetings, New York.

Doig, James W. 1983. If I see a murderous fellow sharpening a knife cleverly. *Public Administration Review* July/August 292–304.

———. 1985. Public authorities: Tensions between entrepreneurship and accountability. Paper presented at the American Political Science Association meetings, New Orleans.

Dimock, Marshall E. 1949. Government corporations: A focus on policy and administration. *American Political Science Review* October 43:899–921; December, 43:1145–64.

Dowd, Thomas J. 1984. Container terminal leasing/pricing methods and their economic effect: A discussion paper. Institute for Marine Studies, University of Washington.

Fair, Marvin L. 1954. *Port administration in the United States*. Cambridge, MA: Cornell Maritime Press.

Gilliam, Harold. 1980. San Francisco Bay: Mystique versus economics. In National Research Council, *Urban waterfront lands*. Washington, DC: National Academy of Science.

Goss, R. O. 1979. *A comparative study of seaport management and administration*. London: Government Economic Service.

Hamilton, Neil W., and Peter R. Hamilton. 1981. *Government of public enterprise: A case study of urban mass transit*. Lexington, MA: Lexington Books.

Hayuth, Yehuda. 1978. Containerization and the load center concept. Ph.D. diss., University of Washington.

Johnson, K. M., and H. C. Garnett. 1971. *The economics of containerization*. London: George Allen & Unwin.

Kendall, Lane C. 1979. *The business of shipping*. Cambridge. Cornell Maritime Press, Inc.

Laing, E. T. 1975. *Containers and their competitors: The economics of deep sea general cargo shipping in the 1970s*. Liverpool, England: Marine Transportation Center, University of Liverpool.

Mansfield, Harvey C. 1975. Independence and accountability for federal contractors and grantees. In *The new political economy,* ed. Bruce L. R. Smith. New York: John Wiley & Sons.

Marcus, Henry S. 1974. *Planning ship replacement in the containerization era*. Lexington, MA: Lexington Books.

———. 1976. *Federal port policy in the United States*. Cambridge, MA: MIT Press.

Maritime Administration (MARAD). 1974. *Public port financing in the United States*. Washington, DC: U.S. Department of Commerce.

Maritime Transportation Research Board. 1976. *Port development in the United States*. Washington, DC: National Academy of Sciences.

Moe, Ronald C. 1979. Government corporations and the erosion of accountability: The case of the proposed energy security corporation. *Public Administration Review,* November/December, **39,** 6.

Musolf, Lloyd. 1959. *Public ownership and accountability*. Cambridge, MA: Harvard University Press.

National Research Council (NRC). 1976. *Port development in the U.S.* Washington, DC: National Academy of Sciences.

*New York Journal of Commerce,* August 12, 1986.

*New York Post,* April 30, 1984.

*New York Times,* August 27, 1984.

*Oakland Tribune,* February 12, 1981.

Olson, David J. 1980a. California ports. Port of Seattle, Washington.

———. 1980b. Regionalism in port development. In *Port systems study update,* CH2M Hill. Seattle: Washington Public Ports Association, pp. 167–82.

———, and Michael Denning. 1982. Economic development and the role of public enterprise. Paper presented at the American Political Science Association meetings, Denver.

Oram, R. B., and C. C. R. Baker. 1971. *The efficient port*. Oxford: Pergamon Press.

Orlans, Harold. 1980. *Nonprofit organizations: A government managing tool*. New York: Praeger.

Peterson, Paul E. 1981. *City limits*. Chicago: University of Chicago Press.

Polanyi, Karl. 1957. The great transformation. Boston: Beacon Press.

Port of Seattle. 1981. Statement of principles and objectives. Port of Seattle, Washington.

———. 1983. Diversionary trends in the U.S. waterborne trade in containerizable transpacific cargoes 1974–1982. Port of Seattle, Washington.

Pritchett, C. Herman. 1941. The paradox of the government corporation. *Public Administration Review* 1:381–89.

———. 1946. The Government Corporation Control Act of 1945. *American Political Science Review* 40:495–509.

Savas, Emanuel S. 1982. *Privatizing the public sector: How to shrink government*. Chatham, NJ: Chatham House.

*Seattle Post-Intelligencer,* May 3, 1987.

*Seattle Times,* November 29, 1983.

Seidman, Harold. 1952. The theory of the autonomous government corporation: A critical appraisal. *Public Administration Review,* Spring, 89–96.

———. 1954. The government corporation: Organization and controls. *Public Administration Review* Summer, 14:183–92.

———. 1955. The government corporation: Its place in the federal structure. In A. H. Hansen, ed., *Public enterprise: A study of its organization and management in various countries*. Brussels: International Institute of Administrative Sciences.

———. 1959. The government corporation in the United States. *Public Administration Review* Summer, 37:103–14.

Sharkansky, Ira. 1979. *Whither the state: Politics and public enterprise in three countries*. Chatham, NJ: Chatham House.

Smith, Bruce L. R. 1975. *The new political economy*. New York: John Wiley & Sons.

U.S. Department of Commerce. 1974. *North American port development expenditure survey*. Washington, DC: Government Printing Office.

Walsh, Annmarie Hauck. 1978. *The public's business: The politics and practices of government corporations*. Cambridge, MA: MIT Press.

Whittaker, J. R. 1975. *Containerization*. New York: Halstead Press.

Willoughby, W. F. 1927. The national government as a holding corporation: The question of subsidiary budgets. *Political Science Quarterly* 32.

# Glossary

ABSORPTION. The practice of transportation carriers to route cargo to a load center port rather than the nearest port to achieve economies of scale (sometimes called port equalization).

ACCOUNTABILITY. Judging the responsiveness of semiautonomous public ports to the intent of their enabling law or the will of their supervisory government or to the wishes of the people.

ANTITRUST IMMUNITY. The exemption by law of an organization or firm from prosecution under part or all of the U.S. antitrust laws.

ANYPORT. A model of changes in port installations over time and their shift in location within a harbor developed by geographer James H. Bird of the University of Southampton, United Kingdom.

ARTIFICIAL HARBORS. Harbors created in open water bodies by the construction of breakwaters and dredging of basins.

BEACH NOURISHMENT. Depositing sediment on beaches to make up for loss through erosion.

BENTHOS. Plants and animals living on or in the waterbottom. Often at risk during dredging and dredged material disposal in harbors.

BOTTOMFISH. Abundant pelagic fish, such as cod, halibut, hake, and sole, often found in large schools near the seabed of coastal waters.

BREAKBULK CARGO. Heterogeneous general cargo items transported to and from vessels as single parcels, on pallet boards, or by slings.

CAPTIVE CARGO. Cargo whose origin or destination is so close to a port that it flows, of necessity, through that port.

CARGO DIVERSION. See absorption.

CARRIERS. Organizations or individuals providing transportation by any mode.

CARRYING CAPACITY. In ecology, the limit to the amount of life, in numbers or mass, that can be supported by a habitat.

CHASSIS. A steel frame mounted on wheels designed to carry containers on a highway.

COFC (CONTAINER-ON-FLATCAR.) Shipping a container by rail by locking it to a railroad flatcar.

COMMON CARRIERS. Individuals or organizations offering transportation to the general public, normally on a first-come, first-served basis, and subject to a published tariff.

COMPENSATORY MITIGATION. In environmental management, the practice of creating additional fish and wildlife habitat to replace unavoidable habitat losses, so that the legal requirement of no net loss of habitat can be fulfilled.

CONFERENCES. A cartel of steamship lines in particular foreign trades in which rates may be fixed, cargo pooled, or services provided jointly, permitted by law and eligible for a grant of antitrust immunity by the FMC.

CONTAINER. Large, standardized, reusable boxes for quick load/unload of ocean vessels that substantially increases shipping productivity.

CONTAINER LOAD CENTER. A port at which containerized cargo movement is concentrated in order to reduce the number of port calls the containership makes and to take advantage of land transport connections.

CONTAINERSHIP. A class of oceangoing vessel designed to carry standard size containers within cells and stacked on deck, which are loaded/off-loaded by special container cranes. Ships can be full or partial containerships.

COST-SHARING. The requirement that part of the costs of federal water resources projects be borne by nonfederal participants and beneficiaries; the percentages vary widely depending on the size and purpose of the project.

CROSS-SUBSIDIZATION. A practice within an organization where the costs of one activity are partly or wholly covered by the revenue generated by another activity.

CURRENTS. Measurable drift of waters caused by wind, tide, river discharge, or other physical force.

CUSTOMS BROKERS. Individuals or organizations that assist shippers in complying with custom laws and regulations and other requirements affecting the import and export of goods.

CZMA (COASTAL ZONE MANAGEMENT ACT.) A federal law, originally passed in 1972, giving state and local governments incentives to upgrade the regulation of land and water resources in the coastal zone.

DEEP-DRAFT PROJECT. Navigation improvement projects of the federal government where the controlling depth of channels would be 45 feet or more.

DEMURRAGE. A fee charged to the owner of cargo that remains in the terminal area beyond a stipulated period of time after off-loading from the vessel.

DISCHARGE OF DREDGED OR FILL MATERIAL. Each discharge into U.S. waters must be reviewed by the federal Environmental Protection Agency and others to ensure that water quality is protected.

DISCONTINUOUS HINTERLAND. The area to which a port's exports are destined or where its imports originate.

DISECONOMY. An activity within a business organization where revenues do not cover costs, requiring internal, or cross-subsidization.

DISPLACEMENT. In harbor management, the removal of working maritime industry by an urban use (residence, office, retail) that can pay higher rent.

DISTANT WATER FLEET. A group of vessels that harvest fish on fishing grounds distant from their homeport.

DOCKAGE. A fee charged against the vessel for the privilege of mooring at a dock.

DOUBLE-STACK INTERMODAL CAR. Specially designed railroad car that holds containers stacked two-high on each of the five segments of the articulated car.

DRAYAGE. A fee charged by a trucking firm for transporting cargo on or off the terminal, or to or from a warehouse or land transport terminal.

ECONOMIC IMPACT STUDIES. Studies done by Ports that describe the positive benefits of port business on the regional economy through the generation of employment, income, and demand for additional goods and services produced in the region.

ENHANCEMENT. In environmental management, providing additional environmental resources as part of a public works project beyond those that would compensate for unavoidable environmental losses.

ESTUARIES. A semi-isolated portion of the ocean, which is diluted by freshwater drainage from land.

FACTORY TRAWLER. A vessel that harvests fish by dragging a trawl net through the water column and that processes the fish at sea into a saleable product.

FEDERAL NAVIGABLE WATERWAY. Navigable U.S. waterbodies developed and maintained in large part by the U.S. Army Corps of Engineers, designed to serve interstate and foreign waterborne commerce and national defense needs.

FEEDER PORTS. Ports from which locally produced or destined international cargo is transshipped to or from a load center.

FISHERIES CONSERVATION ZONE. A band of ocean area adjacent to U.S. territory that extends from the seaward limit of the 3-mile territorial sea to a distance 200 miles from the coastline.

FISHPORT. A homeport for fishing fleets that provides a wide range of services, such as fish processing and handling, and boat repair and provisioning.

FLAT-RATE LEASE. One type of container terminal lease in which the operator (tenant) pays a stipulated yearly rate regardless of the amount of cargo passing through the terminal.

FORELAND. See discontinuous hinterland.

FREIGHT CONSOLIDATION. Putting cargoes from different sources into one shipping unit, usually a container or truck trailer, to reduce transportation costs.

FREIGHT FORWARDING. A function performed on behalf of a shipper or cargo owner where goods completing one transport mode are reshipped by another mode to their destination.

GANTRY CRANES. A steel framework installed on a terminal that includes a superstructure that spans over a vessel's hold and carries a traveling crane for moving cargo or containers.

GENERAL CARGO. A wide range of commodities individually packaged for shipment by container, neobulk or breakbulk mode; distinguished from bulk cargo which is uniform and fungible such as oil or grains.

GENERAL OBLIGATION FUNDS. A debt instrument that pledges the full faith and credit of the issuing government entities, including tax revenue.

GENERAL PERMIT. Authorization in advance by the U.S. Corps of Engineers for a class of activities in navigable waters that are minor and that have minimal environmental impact. General permits can be either "nationwide" or "regional."

HABITAT CREATION. A form of environmental mitigation in which natural resources such as soils, plants, and waterflow are installed for the purpose of increasing the amount of wildlife and environmental values produced.

HABITAT SUITABILITY INDEX. The ratio of existing habitat conditions in a study area to the optimal habitat conditions of the area where "habitat conditions" is the capacity of the habitat environment to support indicator species.

HABITAT UNITS. A value derived from multiplying the Habitat Suitability Index for an evaluation species by the size of the study area. The habitat unit provides a standardized basis for comparing habitat changes over time and space.

HARBOR. A natural or artificial body of water where vessels can get shelter from the open water and that often contains port facilities and anchorages.

HARBOR COMMISSIONS. A form of harbor management in which municipal-level commissions oversee day-to-day harbor issues including anchorage, moorage, public access, water use, environmental protection, and safety.

HARBOR DUES. Fees which nonfederal port interests are permitted by law to collect from commercial vessels to help recover the non-federal share of the cost of navigation improvement projects.

HARBORFRONT. That part of a city's downtown waterfront that faces the main commercial harbor; often synonymous with "central waterfront."

HARBOR MANAGEMENT. A scheme for achieving multiple use objectives and resolving multiple use problems through coordinated land and water use policies and public and private investment decisions within the harbor area.

HEP (HABITAT EVALUATION PROCEDURE). A standardized procedure for conducting environmental impact assessments that was developed by the U.S. Fish and Wildlife Service. Using this procedure, the project site and the mitigation site can be compared by determining respective habitat value.

HINTERLAND. The area served by or tributary to a port where a port's exports originate and its imports are distributed.

HOMEPORT. The primary base for a fleet of vessels (military, commercial, fisheries, or recreational), which provides repair and supply services to the fleet.

INDEPENDENT ACTION. The right of a steamship company that is a member of a conference that has agreed jointly on rates and service to alter rates and services unilaterally without need of conference approval.

INFRASTRUCTURE. In port development, the channels, seawalls, decks, aprons, electricity, water, sewerage, road access and other installations essential for vessel and cargo operations.

INTERMODAL RATES. A single financial charge for transportation of goods where two or more modes of transport are used.

INTERMODALISM. The ability of cargo packed in a standard-size container to be moved rapidly between ship, truck, and rail in any order.

JOINT SERVICES. A single type of transportation service that two or more carriers agree to provide, in which the equipment of both is used, and marketing, scheduling, and pricing are done jointly.

JOINT VENTURE. In fisheries, an arrangement whereby fish harvested by U.S. fishermen are sold and delivered to foreign processing vessels operating within the Fisheries Conservation Zone of the United States.

LANDBRIDGE. That part of an ocean transportation shipment that crosses a continent by land transport; e.g., Japan to Europe using rail to cross the North American continent.

LANDLORD PORT. A public port agency that leases the majority of its facilities to others for operations.

LASH (LIGHTER-ABOARD-SHIP). A type of oceangoing vessel that carries barges or sealed floatable boxes in which the cargo is loaded; the barges or boxes are moved on and off the vessel by ship-mounted cranes.

LINER AND LINER SERVICE. A transportation service between two points that adheres to a published schedule and tariff.

LITTORAL CURRENT. A measurable drift of water running parallel and adjacent to the shore; synonymous with "longshore current."

LOAD CENTER. A transportation node and set of terminals where people or cargo throughput is concentrated in order to achieve economies of scale and facilitate connection with feeder transport systems and other transport modes.

LONGSHORE LABOR. Workers hired to move cargo between the terminal and the vessel, and to manage vessel stowage and cargo marshaling.

MARAD. See U.S. Maritime Administration.

MICROBRIDGE. That part of an ocean transportation shipment that crosses a portion of a continent using land transport; for example, New Orleans to Europe via the Port of Charleston, South Carolina, using rail for the leg between New Orleans and Charleston.

MINIBRIDGE OR MINI-LANDBRIDGE. That part of an ocean transportation shipment that crosses a continent using land transport and begins or ends on the land transport leg; e.g., Japan to New York via the Port of Los Angeles, using rail for the leg from Los Angeles to New York.

MINI-MAX LEASE. In containerport management, a lease arrangement whereby the landlord Port is guaranteed a minimum compensation from the steamship tenant, but earns more as cargo volumes increase up to a maximum. After the maximum is reached, the steamship tenant pays no additional compensation.

MISSION STATEMENT. In port management, a statement adopted by the governing body of the port organization that outlines its "reasons-for-being."

MITIGATION. In environmental management, avoiding, minimizing, rectifying, reducing, and compensating for project-induced resource losses.

MITIGATION BANK. In environmental management, the process whereby habitat is created, restored or enhanced in advance of a specific mitigation requirement and is used to fulfill that requirement.

MULTIPLIER EFFECT. A measure of the indirect impact of port-related wage payments and purchases on the local economy; the regeneration of production as the original wage payments and purchases spread through the economy.

NATIONWIDE PERMITS. A form of general permit issued by the Corps of Engineers authorizing in advance certain classes of minor work in U.S. navigable waters: e.g., maintenance activities, installation of aids to navigation.

NATURALLY TRIBUTARY. A term used in reference to cargo when a Port claims that certain cargo is inherently dependent on their port because of historic transportation routing proximity, economics, or other factors; a technical term used by the FMC in some of its decisions.

NEPA (NATIONAL ENVIRONMENTAL POLICY ACT). The 1970 federal law that required all federal agencies to consider environmental impacts in their decision-making, to avoid them whenever possible, and to prepare environmental impact statements in certain circumstances.

NONCARGO FUNCTIONS. Those activities carried out by the public port that are intended to generate revenues and/or employment, but are not directly related to the cargo-handling mandate of the port authority.

NONFEDERAL SPONSOR. A party to a federal navigation improvements project who initiates the review and approval process, pays part of the costs and who may undertake planning, design, and construction subject to federal approval and subsequent reimbursement.

OPERATING PORT. A public port that manages the day-to-day activities on its terminal by scheduling vessel calls, arranging stevedoring services, employing longshore labor, and other similar functions.

OVERCAPACITY. The ability of a marine terminal, or a set of terminals in a port range, to handle considerably more cargo throughput than is available.

PARENTAL GOVERNMENTS. In port governance, the governmental unit that creates and empowers the port agency and to whom it is in part accountable.

PIGGYBACK SERVICE. Movement of truck trailers on wheels via rail flatbed cars.

POLITICAL ECONOMY. An explanation of Port behavior that relies on an understanding of the political and economic environment influencing the Port.

POOLING OF CARGO. The practice of a group of cargo shippers to combine cargo into larger lots to achieve economies in transportation costs.

POOLING OF TRANSPORTATION SERVICES. The practice of a group of carriers to jointly market their services in order to achieve fuller capacity.

PORT. A place where there is an interchange of cargo and/or passenger traffic between ships/boats and land.

PUBLIC ACCESS. In coastal management, pathways, parks, viewpoints, view corridors that allow the general public to reach the water's edge for recreation or to see the water from different vantage points.

PUBLIC ENTERPRISE. A form of governmental unit that combines features of a traditional governmental agency (e.g., public ownership and statutorily assigned powers) with features of private enterprise (for example, reliance on private capital and autonomy in decision-making).

RATE AGREEMENT/SERVICE AGREEMENT. A contract between a shipper and carrier that guarantees a certain volume of cargo within a stipulated time in exchange for discounted rates. "Rate agreement" is used primarily in regard to railroads; "service agreements" with ocean carriers.

REDUNDANCY. In regard to port facilities, the existence of similar cargo-handling terminals at competing ports with insufficient cargo business to allow either to run profitably.

REGIONAL PORT STUDIES. A collaborative study by a number of independent port agencies within a region indentifying projected cargo flows, existing terminal and berth capacity of the region, and the shortfall in capacity (if any), thus suggesting what new port facilities are needed within the region.

REGIONALISM. In port management, the concept of greater cooperation among competing ports, including such actions as joint promotion and marketing, collaborative studies, regional planning, and merger of port entities.

REVENUE BONDS. A debt instrument issued by a public port agency, which is secured by a pledge of the revenue due to the Port from its ongoing business activity.

REVENUE SHARING LEASE. A lease between the landlord Port and marine terminal operator in which the Port receives a minimum payment up to a set level of cargo throughput, after which the Port receives a percentage of the tariff revenue received for additional cargo with no maximum and often on a sliding scale.

RIGHT TO CARGO. Claims by some Ports that U.S. law guarantees them a right to cargo-handling business when that cargo is "naturally tributary" to the Port.

RIVERS AND CANAL HARBORS. A type of harbor that physically is located on rivers or canals.

ROADSTEADS. An area of water where the shoreline configuration offers only minimal protection from wind and wave.

RORO (ROLL-ON/ROLL-OFF). A cargo vessel that loads/unloads containerized or unitized cargo by wheeled trailers driven on or off the vessel by trucks/tractors using ramps that connect the vessel and berth.

SAMP (SPECIAL AREA MANAGEMENT PLANNING). A form of harbor management in which public agency representatives work collaboratively, through consensus decision-making, to develop a long-term growth and conservation plan for the harbor, where implementation rests with existing agencies.

SCEND. The upward heaving of the bow of a ship that is underway which is a factor in calculating the depth needed in a channel.

SECTION 10. That part of the U.S. Rivers and Harbors Act of 1899 that gives the U.S. Army Corps of Engineers its primary authority to regulate the use of the navigable waters of the United States.

SECTION 404. That part of the U.S. Clean Water Act of 1977 that gives the U.S. Army Corps of Engineers and the U.S. Environmental Protection Agency authority to regulate the discharge of dredged or fill material into the waters of the United States.

SHIP CHANDLERS. A person or organization that sells provisions, tools, and equipment to vessels.

SHIPPERS. A person or organization responsible for arranging the movement of goods between two points.

SHIPPERS ASSOCIATIONS. Authorization by law for shippers to form nonprofit associations for the purpose of consolidating freight in order to secure time/volume discounts and service contracts from carriers.

SHORELAND BANKING. In harbor management, the purchase, use, or resale of a parcel of waterfront property by a public entity such as a Port in order to reserve it for a certain class of use; for example, water-dependent industry.

SINGLE BILL OF LADING. One contract issued to a shipper by a transportation company promising the transport of goods between two points even though more than one mode of transport is involved.

SITE. In harbor management, those physical characteristics of the local area that affect port development, for example, topography, weather, water depth.

SITUATION. In harbor management, the relationship a harbor or port has with respect to existing or potential cargo/people traffic, for example, ease of transportation with hinterland, proximity to sea.

SPOIL. In harbor management, the material dredged to make a channel, basin or dock that must be disposed of in an environmentally acceptable location.

SQUAT. The tendency of the stern of a ship underway to press lower into the water, which is a factor in calculating the depth needed in a channel.

STEAMSHIP AGENTS. A representative of a vessel's owner or charterer that arranges port services for the vessel while it is in port, and books cargo onto the vessel.

STEVEDORING. The function of loading and unloading a ship, often done by stevedoring companies who then employ longshore labor for specific tasks.

STRANDING. Running a ship aground; sometimes done intentionally to aid in loading/unloading or repair, where there are no port facilities.

STRATEGIC PLANNING. Matching an institution's internal strengths and weaknesses with the full range of its external threats and opportunities, then devising goals and objectives and a plan of action for the institution.

TARIFF RATES. In transportation, a list or scale of prices and charges for various transportation services. In a Port, a list or scale of prices and charges for various port services (for example, wharfage and dockage).

TELEPORT. An array of large-scale satellite communication dishes located in an area with little or no electronic interference designed to send and receive large volumes of commercial data via satellite; some port agencies are promoting and developing these facilities.

TEU (20-FOOT EQUIVALENT UNIT). A standard measure of container capacity aboard a containership or in a containerport in terms of a container.

THROUGHPUT. The amount of cargo that is processed or handled through a port terminal or facility, as measured in tons/year, or TEUs per year.

TIDES. The daily rise and fall of water height due to the gravitational attraction of the moon.

TIME/VOLUME CONTRACT. An agreement between a shipper and a carrier that provides discounted rates for carrying cargo in exchange for a guaranteed volume of cargo over a stipulated time.

TOFC (TRAILER-ON-FLATCAR). Container and chassis shipped on railroad flatcars.

TRAMP. A freight vessel that operates on no particular schedule but picks up cargo or passengers wherever they may be.

TRANSSHIPMENT. The transfer of cargo from one carrier to another for reshipment.

UNFAIR BUSINESS PRACTICES. A general standard used in U.S. transportation law to connote an activity considered detrimental to U.S. commerce and that can be prevented by action of regulatory agencies.

U.S. MARITIME ADMINISTRATION. An agency within the U.S. Department of Transportation with primary responsibility for promoting the U.S. Merchant Marine, subsidizing vessel operation, operating academies, and other maritime functions.

USER FEES. In harbor management, the authority by law to charge a fee against vessels that benefit from the use of newly developed channels, proceeds to be used to offset the cost of the nonfederal share of the project.

VESSEL PRODUCTIVITY. The amount of cargo-ton miles traveled by a cargo ship in one year; as port time decreased with the growth of containerization, vessel productivity increased.

VLCC (VERY LARGE CRUDE CARRIER). Crude oil tanker ships ranging from 80,000 to 250,000 DWT capacity.

WATER-DEPENDENCY. In harbor management, the policy expressed in federal and state laws that scarce shoreland space should be reserved for uses that are dependent on a shoreline location.

WATERFRONT REVITALIZATION. In harbor management, the renovation, reuse, or redevelopment of obsolete shoreland industrial areas for urban-oriented residential, retail, or recreational uses.

WETDOCKS. A basin for mooring vessels that is separated from the main water body by a tidal gate, thus allowing the basin water level to remain stationary and protecting vessels from waves.

WETLAND. An area having one or more of the following three attributes: 1) at least periodically the substrate is dominated by hydrophytes, or plants that grow in water or saturated earth; 2) the substrate is predominantly hydric soils, or those that are frequently wet and sometimes anaerobic, or without oxygen; 3) the substrate is nonsoil and is either saturated with or covered by shallow water at some time during the growing season. Typical vegetated wetlands include marshes, swamps, bogs, bottomlands, and tundra.

WHARFAGE. A fee charged against the cargo for moving it across a wharf or terminal.

# Index

# About the Authors

**ROBIN S. BITTNER** received a master of marine affairs degree from the Institute for Marine Studies, University of Washington. Currently, he is employed by American President Lines as Terminal Manager in Dutch Harbor, Alaska. He was an intern at the Port of Everett, Washington, and helped design and construct maritime displays for The Waterlink exhibit in Seattle.

**DANIEL JACK CHASAN** is a journalist and historian whose books include *The Water Link*, an economic and environmental history of Puget Sound; *Klondike 70*, which describes the early political battles over the TransAlaska Pipeline; *Up for Grabs*, which analyzes conflicts over natural resources; and *The Fall of the House of WPPSS*, an account of the multibillion-dollar Washington Public Power Supply System nuclear fiasco. His articles have appeared in many magazines, including *Smithsonian*, *Audubon*, and *The New Yorker*. He has served as business editor of *The Weekly, Seattle's Newsmagazine*, and editor of *Ocean Leader*.

**PAUL W. CHILCOTE** trained as a transportation geographer, currently is the director of strategic planning at the Port of Tacoma, where he coordinates current Port development, performs long-range planning, and manages the Capital Improvement Program. Previously, he consulted on Washington State port issues and did intermodal business planning for Southern Pacific Railroad in San Francisco. For twelve years he worked as long-range transportation planner for the Port of Seattle.

**THOMAS J. DOWD** had a first career in investment banking and bond counseling, which led to his interest in the financial issues of port development. After receiving a master's degree in marine affairs, he became port industry specialist for Washington Sea Grant and affiliate professor of marine studies at the University of Washington. Dowd concentrates his research, writing, and teaching on port and marine transportation finance, management, and planning.

**DOUGLAS K. FLEMING** is professor of geography at the University of Washington. He was a chartering broker and fleet allocator for an American steamship line in an earlier career, which gave him practical experience in the geography of commerce. He serves on the editorial board of *Maritime Policy and Management* and has published articles on such topics as European coastal steel production, and U.S. air travel patterns.

**ROBERT F. GOODWIN** is Washington Sea Grant's coastal resources specialist at the Institute for Marine Studies, University of Washington. He trained as an architect and an urban geographer, and he applies these fields to the study of the urban coastal zone. Currently he conducts research and educational programs on the revitalization of smaller urban waterfronts in the Pacific Northwest, and the industrial structure of Puget Sound's urban harbors. Goodwin is an avid photographer whose images of coastal subjects have won regional and national awards.

**MARC J. HERSHMAN** is professor of marine studies and adjunct professor of law at the Institute for Marine Studies and the Law School, University of Washington. His work deals with such subjects as shoreline industrial siting, port development, environmental protection, fish and wildlife mitigation, urban waterfront revitalization, and national coast and marine law and policy. He directs a Sea Grant research and advisory services program within the Institute and is author of a reader on coastal management. He is editor-in-chief of the journal, *Coastal Management*, and is a past president of The Coastal Society.

**MARIANNE KORY** received a master's degree in law and marine affairs from the University of Washington Law School and a J.D. degree from New York University Law School. Formerly an attorney with the federal government, she has litigated occupational safety and health cases for the U.S. Department of Labor and was an attorney-advisor to administrative law judges in the U.S. Department of Health and Human Services. She presently works on coastal law and planning issues in the Puget Sound region.

**HAROLD M. MAYER** has been a geography professor for thirty-eight years at both the University of Chicago and the University of Wisconsin-Milwaukee. In addition to an extremely active academic career, including many books and 150 articles, he has been a port commissioner, regional planning board member, and participant in many government transportation studies. He is best known for his seminal books on urban growth and land-use management.

**DAVID J. OLSON** currently is chairman and professor in the department of Political Science, University of Washington. He has studied public enterprise generally, focusing on political and economic performance, and has done particular work on ports as public enterprise. He is author of five books and numerous articles and has consulted for Canada's Board of Harbour Authorities, the Alaska State Department of Transportation, the Port of Seattle, and the New England River Basin Commission.

**JAMES E. RANDALL** is assistant professor of geography at Wilfrid Laurier University in Waterloo, Canada. He currently concentrates his research and writing on issues of community economic development. He has written previously on the role of the waterfront in urban industrial decentralization and on the household economy in industrialized societies. Randall's doctoral degree is from the Department of Geography at the University of Washington.

**ANN E. WESSEL** received a master of marine affairs degree from the Institute for Marine Studies, University of Washington. For five years, she was responsible for the administration of the Shoreline Management Program for Whatcom County, Washington. Currently, she works on environmental mitigation issues for the Port of Seattle and other public agencies in the Puget Sound region.